Geography in Place 2

Second Edition

Michael Raw

Cirencester Kingshill School

NAME	FORM	DATE OF ISSUE	CONDITION	TEACHER SIGNATURE
Steven Kingsbury	10DH	4/9/02	New	

Collins

Contents

1 River landscapes and processes
- 1.1 Introduction *4*
- 1.2 Drainage basins *5*
- 1.3 Rivers as land-shaping agents *6*
- 1.4 Rivers as flows in the water cycle *15*
- 1.5 The Derwent Valley floods *17*
- 1.6 Kielder Water: the UK's first regional water grid *21*
- 1.7 Summary *23*

2 Glacial landscapes and processes
- 2.1 Introduction *24*
- 2.2 The long winter *24*
- 2.3 Glaciers as systems *26*
- 2.4 Glaciers as land-shaping agents *27*
- 2.5 Landscapes of glacial erosion: Snowdonia *28*
- 2.6 Landscapes of glacial deposition *32*
- 2.7 The human use of glaciated uplands *35*
- 2.8 Skiing in Scotland *36*
- 2.9 Skiing in the Cairngorms *36*
- 2.10 Summary *42*

3 Coastal processes and landforms
- 3.1 Introduction *43*
- 3.2 The coastal system *44*
- 3.3 Waves *44*
- 3.4 Coastal features: South Devon and Dorset *46*
- 3.5 Sea-level changes *53*
- 3.6 The *Sea Empress* oil disaster, Milford Haven *54*
- 3.7 The crumbling cliffs of Holderness *56*
- 3.8 Coastal sand dune erosion and recreation *60*
- 3.9 Summary *61*

4 Population: distribution and change
- 4.1 Introduction *62*
- 4.2 Global population distribution *63*
- 4.3 World population growth *68*
- 4.4 How does population grow? *69*
- 4.5 Population change and development *70*
- 4.6 Age-sex structure *73*
- 4.7 Nigeria *73*
- 4.8 Sweden *74*
- 4.9 Problems of population change *75*
- 4.10 Family planning in Bangladesh *76*
- 4.11 Summary *80*

5 Population: migration and resources
- 5.1 Introduction *81*
- 5.2 What is migration? *81*
- 5.3 Why do people migrate? *81*
- 5.4 Rural-urban migration in Peru *82*
- 5.5 Migration within the UK *84*
- 5.6 Population change in Teesdale *86*
- 5.7 International migration *88*
- 5.8 Population growth: opportunities and problems *92*
- 5.9 India's agricultural miracle *92*
- 5.10 Environmental impact of population growth *93*
- 5.11 Egypt: expanding the resource base *96*
- 5.12 Land reclamation in the Netherlands *98*
- 5.13 Summary *100*

6 Managing natural resources
- 6.1 Introduction *101*
- 6.2 Renewable and non-renewable resources *101*
- 6.3 Natural resources and levels of development *102*
- 6.4 Fossil fuels and global warming *104*
- 6.5 The disappearing ozone layer *109*
- 6.6 Acid rain *110*
- 6.7 Opencast mining in the UK *112*
- 6.8 Alternative energy resources *113*
- 6.9 Recycling and conserving resources *115*
- 6.10 Summary *117*

7 Tourism

- 7.1 Introduction *118*
- 7.2 Resources for tourism *118*
- 7.3 Tourism in the UK *119*
- 7.4 Blackpool: a Victorian seaside resort *119*
- 7.5 Travelling abroad *122*
- 7.6 National parks in England and Wales *122*
- 7.7 The Lake District *124*
- 7.8 Tourism in the economically developing world *129*
- 7.9 Green tourism in Zimbabwe *129*
- 7.10 Tourism and conservation in the Galapagos *133*
- 7.11 Summary *136*

8 Contrasts in development

- 8.1 Introduction *137*
- 8.2 What is development? *139*
- 8.3 Measuring development *139*
- 8.4 Global contrasts in wealth *141*
- 8.5 Global contrasts: water supplies *142*
- 8.6 Improving water in Moyamba, Sierra Leone *142*
- 8.7 Global contrasts: feeding a hungry world *143*
- 8.8 Increasing food production: Koshi Hills, East Nepal *144*
- 8.9 Explaining global contrasts: economic problems *146*
- 8.10 Population problems *148*
- 8.11 Social problems *149*
- 8.12 Political problems *150*
- 8.13 Environmental problems in the Dominican Republic *151*
- 8.14 Diseases and pests *153*
- 8.15 Regional contrasts in development *156*
- 8.16 Italy: a divided country *156*
- 8.17 Summary *159*

- The more-difficult exercises are indicated by *.
- Glossary words are highlighted in bold letters in the text the first time they appear.

9 Trade and aid

- 9.1 Introduction *160*
- 9.2 What is international trade? *161*
- 9.3 Influences on international trade *161*
- 9.4 The pattern of world trade *165*
- 9.5 Foreign aid *168*
- 9.6 Types of foreign aid *169*
- 9.7 Foreign aid: who benefits? *170*
- 9.8 Long-term aid projects *171*
- 9.9 Large-scale aid: Ilisu dam project *172*
- 9.10 Small-scale aid: Ghana *173*
- 9.11 Interdependence *175*
- 9.12 Summary *177*

Revision section:
chapter summaries from Book 1 *178*

Glossary *188*

Index *190*

Key Skills Opportunities
Opportunities for producing evidence of Key Skills are identified in the chapter summaries. The bold reference relates to the QCA Key Skill (given below in full for level 2). The numbers that follow identify the exercises in the chapter that provide the relevant opportunities.

Communication
- C1.1 Take part in discussions
- C1.2 Read and obtain information
- C1/2.3 Write different types of document
- C2.1 Contribute to discussions and give a talk
- C2.2 Read and summarise information

Application of Number
- N1/2.1 Interpret information from different sources
- N1/2.2 Carry out calculations
- N1/2.3 Interpret results and present findings

Information Technology
- IT1/2.1 Find, explore, select, develop information
- IT1.2 Present information including text, numbers and images
- IT2.2 Explore and develop information including text, numbers and images
- IT2.3 Present combined information including text, numbers and images

1 River landscapes and processes

1.1 Introduction

Rivers are natural channels, that drain the surface of the land (Fig. 1.1). However, rivers do more than simply remove water from the land: they also carry huge amounts of sediment. This sediment comes from the breakdown of rocks by **weathering** and **erosion**. Rivers play an important part in this process of erosion. As a result, they create much of their own **sediment load**.

By eroding the land, **transporting** sediment and **depositing** the sediment elsewhere, rivers create new **landforms**. In many parts of the world (including the British Isles) rivers are the most important natural agents shaping the landscape today. Rivers also have an important influence on human activities. As we shall see later in this chapter, rivers are both valuable resources as well as natural hazards.

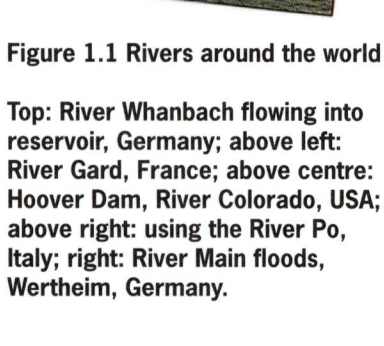

Figure 1.1 Rivers around the world

Top: River Whanbach flowing into reservoir, Germany; above left: River Gard, France; above centre: Hoover Dam, River Colorado, USA; above right: using the River Po, Italy; right: River Main floods, Wertheim, Germany.

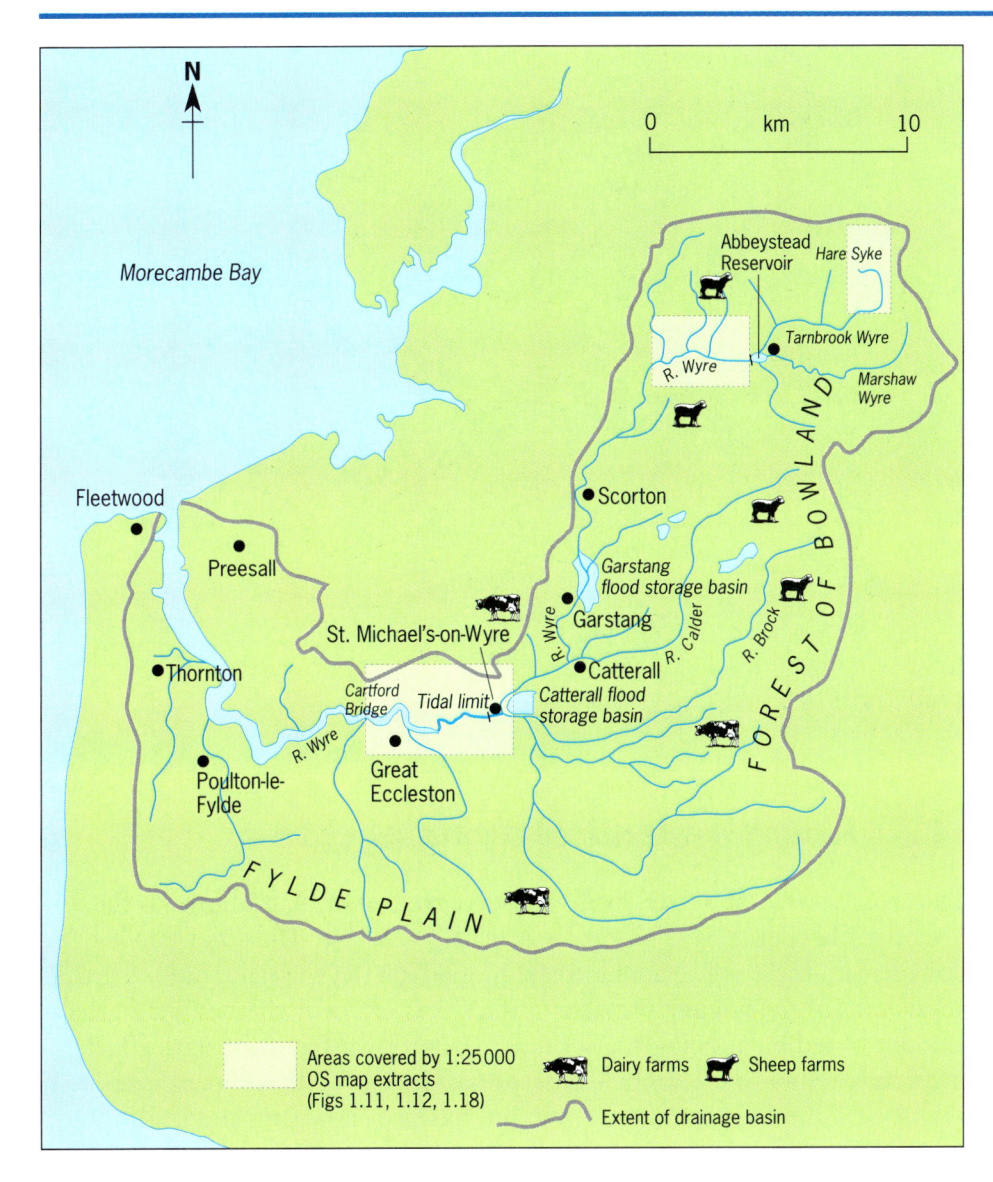

1 River landscapes and processes

Figure 1.2 The Wyre drainage basin

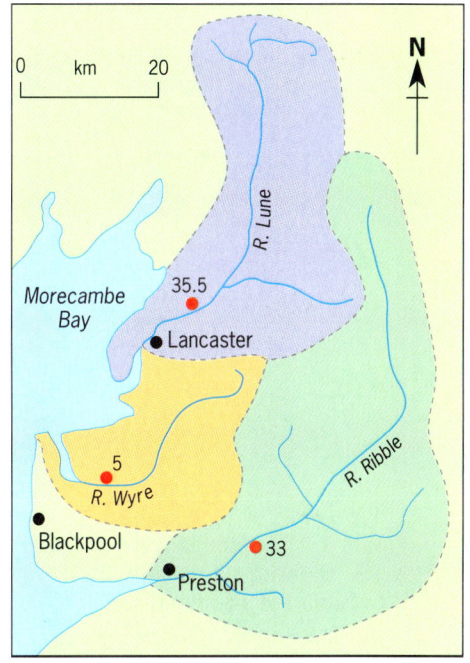

Figure 1.3 (above) Drainage basins in North Lancashire, and average flows (cubic metres per second).

EXERCISES

1 Study Figure 1.1. Explain briefly how rivers can influence human activities.

EXERCISES

2a Using the 1:50 000 OS map covering your school area (you can get a free map at the website www.ordsvy.gov.uk), trace the outline of the drainage basin in which your school is situated. On your outline, show the streams and rivers, their sources, and the main watersheds.
b From the website of the Centre of Ecology and Hydrology (www.nwl.ac.uk), find the discharge of the river that reaches the sea and drains your local area.
3 Study Figure 1.3. Suggest a reason why (1) average flows on the River Wyre are lower than those on the River Ribble and River Lune, (2) average flows are higher for the River Lune than for the River Ribble.

1.2 Drainage basins

A **drainage basin** is the area drained by a river and its tributaries. The boundaries that separate one drainage basin from another are known as **watersheds** (Fig. 1.3). Rivers have their **source** on the most distant watersheds. Most sources are springs or boggy areas on hillsides. A river's **mouth** is the tidal area where a river meets the sea. On lowland coasts in the British Isles, most river mouths form broad, funnel-shaped **estuaries**.

FACTFILE

- The River Wyre is a small river in Lancashire which drains an area of just 450 sq km (Fig. 1.2).
- From its source to its mouth, the River Wyre is approximately 55 km long.
- The source of the River Wyre is the boggy, peat-covered moorlands of the Forest of Bowland. Here finger-tip tributaries such as Hare Syke start.
- From its source, the River Wyre flows south-westwards to the Irish Sea at Fleetwood.
- At its mouth, the River Wyre forms a small estuary.

1 River landscapes and processes

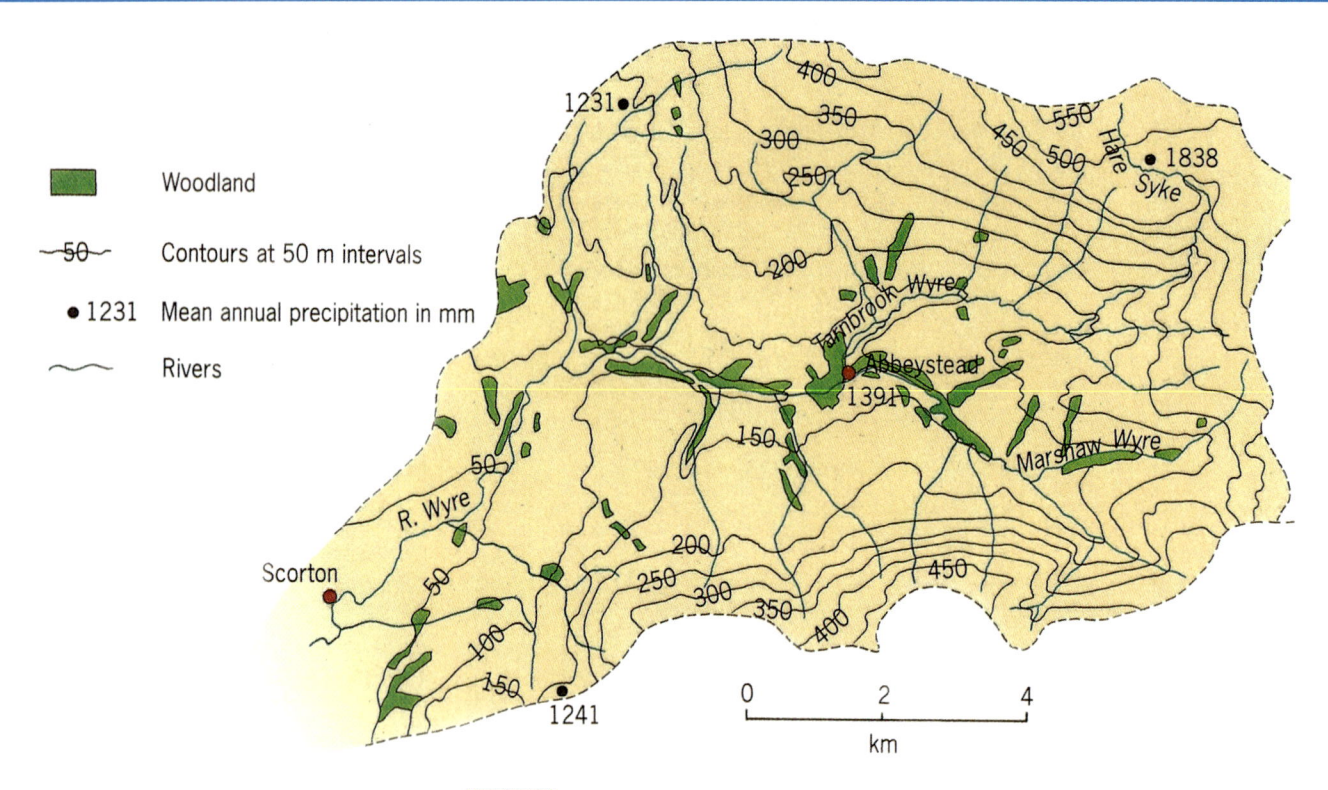

Figure 1.4 The upper Wyre drainage basin, showing slopes and the distribution of woodland

Figure 1.5 Long profile of the River Wyre

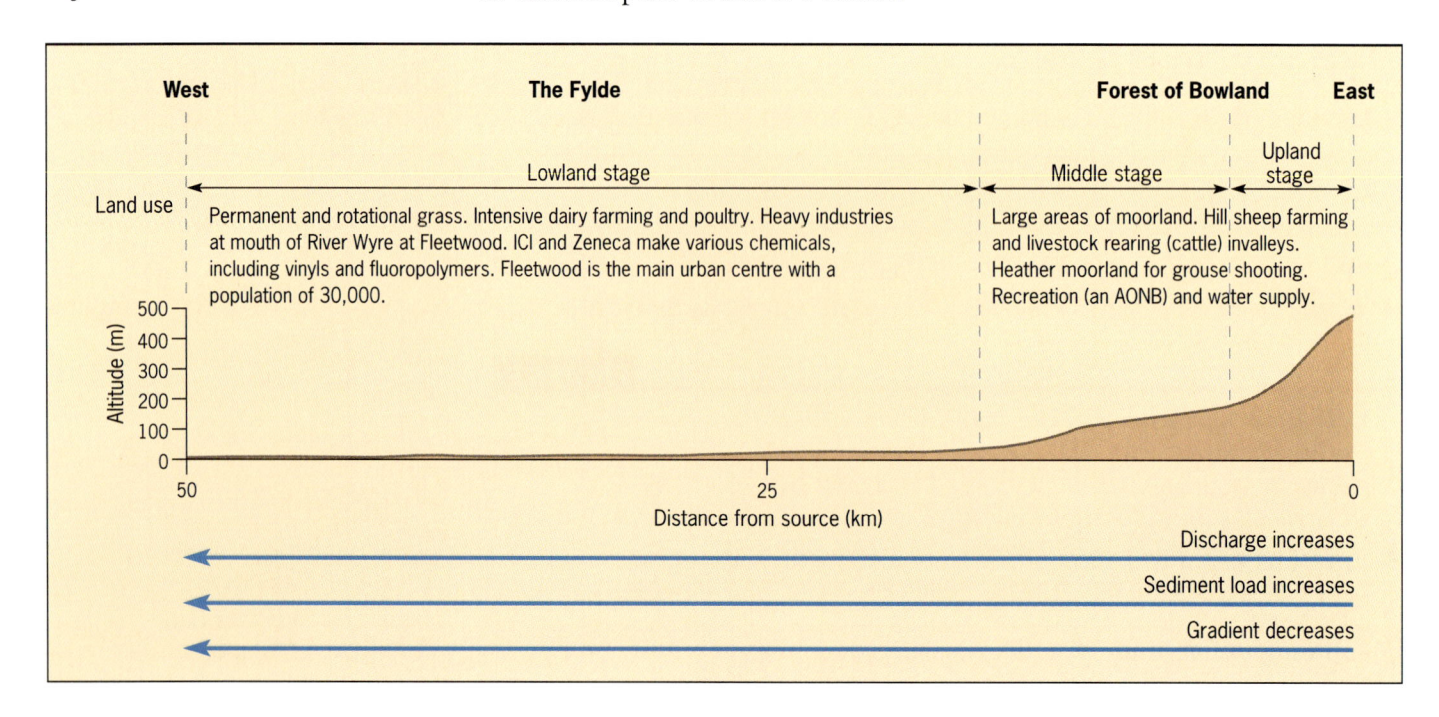

1.3 Rivers as land-shaping agents

The cross-section of a river, from its source to its mouth, is known as the **long profile** (Fig. 1.5). This profile is usually concave. Thus, as we move downstream there is a steady decrease in gradient. At the same time, there is an increase in the volume of water in the river (i.e. its discharge) and in the amount of sediment being transported. These downstream changes affect the river's ability to erode the landscape and transport its sediment load (Tables 1.1 and 1.2). As a result, we find different processes and landforms in different parts of a river's course.

	West	The Fylde	Forest of Bowland	East
		Lowland stage	Middle stage	Upland stage
Land use		Permanent and rotational grass. Intensive dairy farming and poultry. Heavy industries at mouth of River Wyre at Fleetwood. ICI and Zeneca make various chemicals, including vinyls and fluoropolymers. Fleetwood is the main urban centre with a population of 30,000.	Large areas of moorland. Hill sheep farming and livestock rearing (cattle) in valleys. Heather moorland for grouse shooting. Recreation (an AONB) and water supply.	

Discharge increases
Sediment load increases
Gradient decreases

1 River landscapes and processes

Table 1.1 River erosional processes

Abrasion/ corrasion	Coarse sediments (bedload), transported by the river, wear away the bed and banks. This process is most effective during flooding, and if the material is caught in hollows it will erode the river bed to form potholes.
Attrition	As the bedload is transported, it too is gradually worn down (by scraping along the river bed and by collision with other sediments). Sharp edges are removed and sediments become both rounder and smaller.
Hydraulic action	The force of running water alone can erode soft rocks like clay, sands and gravels. The pressure of water moving into cracks and bedding planes can remove slabs of rocks from the channel bed and sides.
Solution	Rocks like chalk and limestone can be dissolved by acid river water and so removed in solution.

Table 1.2 The river's load

Bedload	Coarse sediments (boulders, cobble stones, pebbles) rolled and dragged along the river bed. Smaller, sand-sized particles are bounced along the bed by saltation.
Suspended load	Fine silt and clay particles suspended in water and transported downstream.
Dissolved load	Rocks that are soluble, e.g. limestone, are transported in solution. Unlike bedload and suspended load (only transported at high flow), the dissolved load is transported all the time.

The upland stage

The upland stage of the River Wyre is dominated by steep slopes and hills rising to over 500 m (Fig. 1.4). The hills force moist air from the Irish Sea to rise, drenching the highest ground with more than 1800 mm of precipitation a year. With high altitude, steep slopes and a lot of water, streams in this area have plenty of energy. Most of this energy is used to transport water and sediment and overcome friction caused by the river's bed and banks. Any energy that is left over is then spent on erosion.

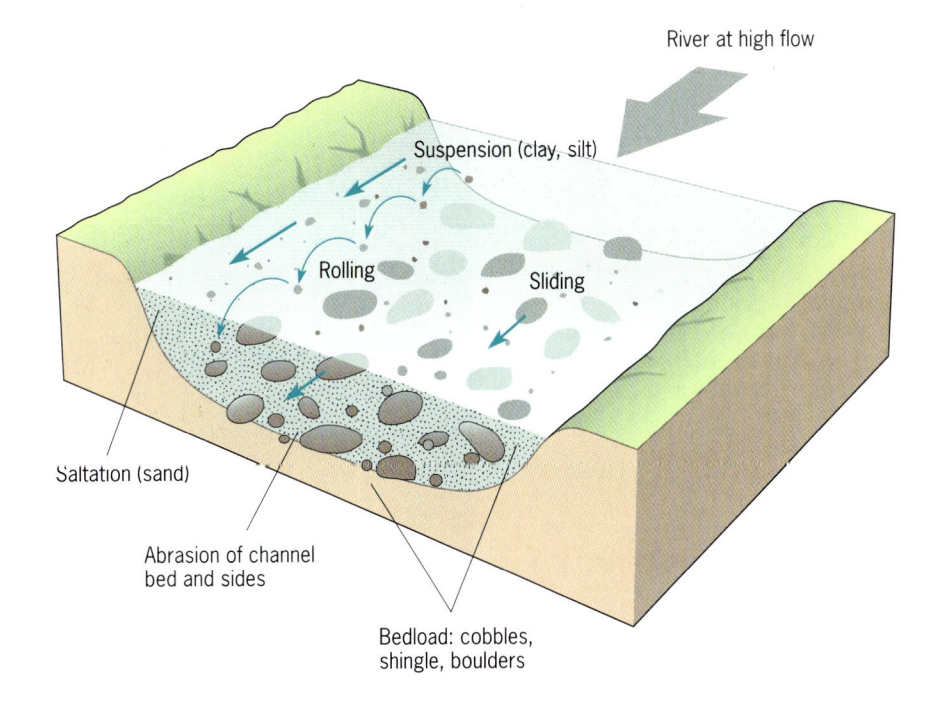

Figure 1.6 How a river transports its load

Erosion

Erosion is the wearing away of the land surface and the removal of rock debris by rivers (and by glaciers, winds and waves). In Britain's uplands, rivers are important agents of erosion. There are two reasons for this:
- Steep slopes mean that rivers have surplus energy;
- Weathering and erosion produce large amounts of coarse rock fragments.

Coarse rock fragments are the river's erosional 'tools'. After heavy rain, when rivers become raging torrents, pebbles, cobbles and even boulders are swept along the river bed (Table 1.2). On these occasions, the river acts like a giant grinding machine, cutting vertically into its bed and eroding the land.

Rapid vertical erosion (Table 1.1) produces the following features:
- Deep, steep-sided valleys with a **V-shaped** cross section (Fig. 1.7);
- **Interlocking spurs**, formed when a river with a winding course cuts down into the land (Fig. 1.8);
- Circular holes in the river bed known as **potholes** – drilled in solid rock by pebbles trapped in swirling eddies in the river's flow.

1 River landscapes and processes

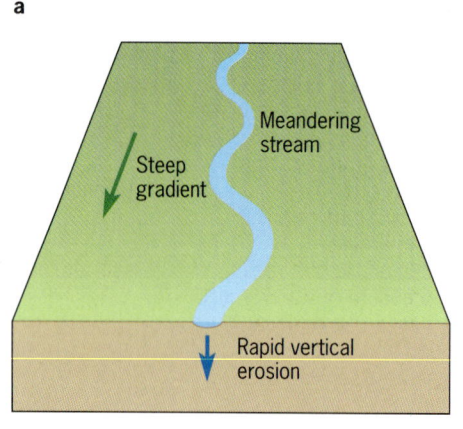

Figure 1.8 (above) Formation of (a) a V-shaped valley and (b) interlocking spurs

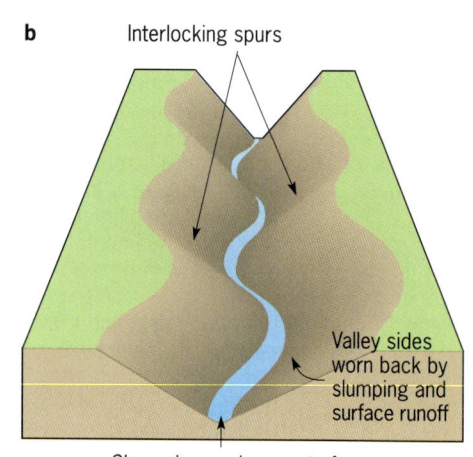

Diagram a labels: Steep gradient; Meandering stream; Rapid vertical erosion

Diagram b labels: Interlocking spurs; Valley sides worn back by slumping and surface runoff; Channel occupies most of the valley floor: no lateral erosion to widen valley

Figure 1.7 (above) V-shaped valley and interlocking spurs

Figure 1.9 (right) High Force, Teesdale, England's highest waterfall, formed where a band of hard igneous rock (The Great Whin Sill) crosses the River Tees.

Figure 1.10 (below) Formation of a waterfall and gorge

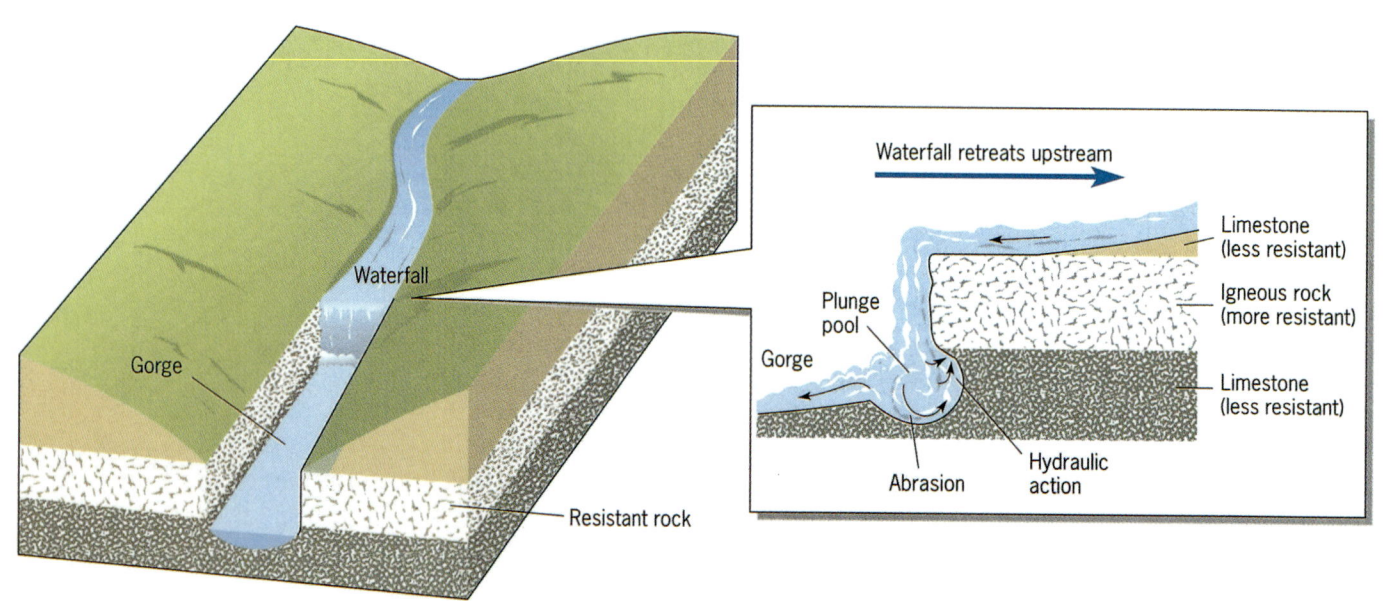

Labels: Waterfall; Gorge; Resistant rock; Waterfall retreats upstream; Plunge pool; Gorge; Abrasion; Hydraulic action; Limestone (less resistant); Igneous rock (more resistant); Limestone (less resistant)

8

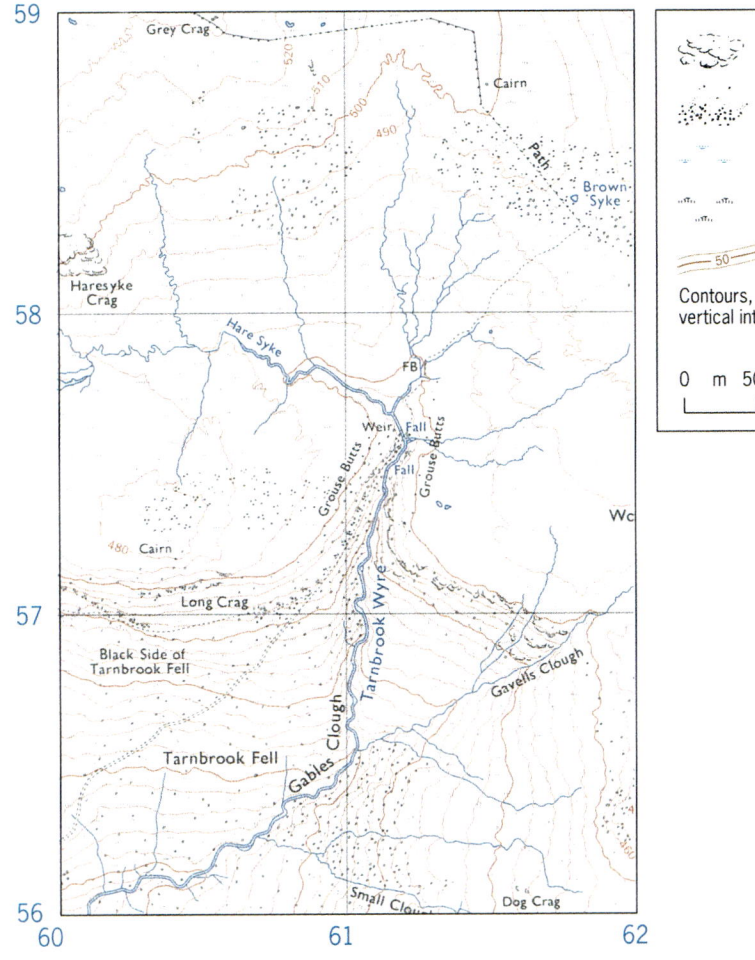

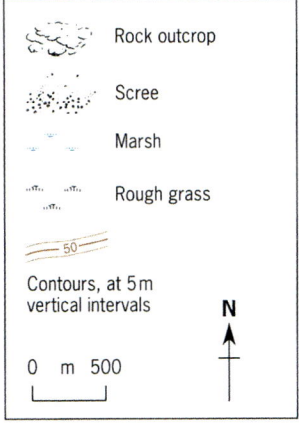

1 River landscapes and processes

Figure 1.11 River Wyre: upper course © Crown copyright

> **EXERCISES**
>
> **4** Use Figure 1.11.
> **a** Draw a cross-section of the valley of Hare Sike between grid references 606574 and 615574.
> **b** Describe the main features of the valley between 616565 and 610577 by answering questions 1 to 9 in Table 1.3.
> **c*** Explain the main features of the valley between 610565 and 610577.
> **d** What is the main type of land use shown in Figure 1.11?
> **e*** Suggest two kinds of economic activity that might be found in the upper course of the River Wyre.

Where bands of harder rock cross the channel, waterfalls and rapids form. At the foot of a waterfall, less resistant (softer) rock is eroded by abrasion and hydraulic action (see Table 1.1). Eventually the harder rock is undercut and collapses. The waterfall therefore slowly retreats upstream. As it does so, it leaves behind a narrow gorge (Figs. 1.9, 1.10).

Between upland and lowland stages

Five or six kilometres from its source, the River Wyre enters a gentler landscape. The river has now been joined by many tributaries (see Fig. 1.2) and has a greater discharge. Other changes have occurred too. The river is at a lower altitude, its gradient is less steep, and its channel has a more regular shape. Although the river transports a heavier load, the sediment is finer and so more is carried in suspension (see Fig. 1.6).

The shape of the river valley in cross-section has also changed. Now the valley is much wider and has a broad, flat floor (Figs. 1.12, 1.13). The valley sides are lower angled and form gentle **bluffs**, while all traces of interlocking spurs have been removed (Fig. 1.14). How have these changes happened? The simple answer is that different processes operate in this part of the river's course.

Table 1.3 OS map skills: how to describe a river and its valley

1. In which direction does the valley run?
2. Is the river and its valley in an upland or lowland region?
3. What is the shape of the valley in cross-section?
4. How steep are the valley sides?
5. How steep is the long profile of the valley?
6. Does the river/valley meander or is it straight?
7. Are there any interlocking spurs?
8. How wide is the valley floor?
9. Are there any waterfalls or rapids?
10. Is there a flood plain?
11. Are there ox-bow lakes and levées?

1 River landscapes and processes

Figure 1.12 River Wyre: middle course © Crown copyright

> **EXERCISES**
> 5 Which part of a river's load is represented by: • point bar deposits • fine silt deposits?
> 6a Using Figure 1.12, draw a cross-section of the Wyre valley between grid references 536537 and 536548. Compare the shape and size of the valley in your cross-section with that of Hare Syke (exercise 4).
> b Describe the River Wyre and its valley in Figure 1.12 by answering questions 1 to 11 in Table 1.3.

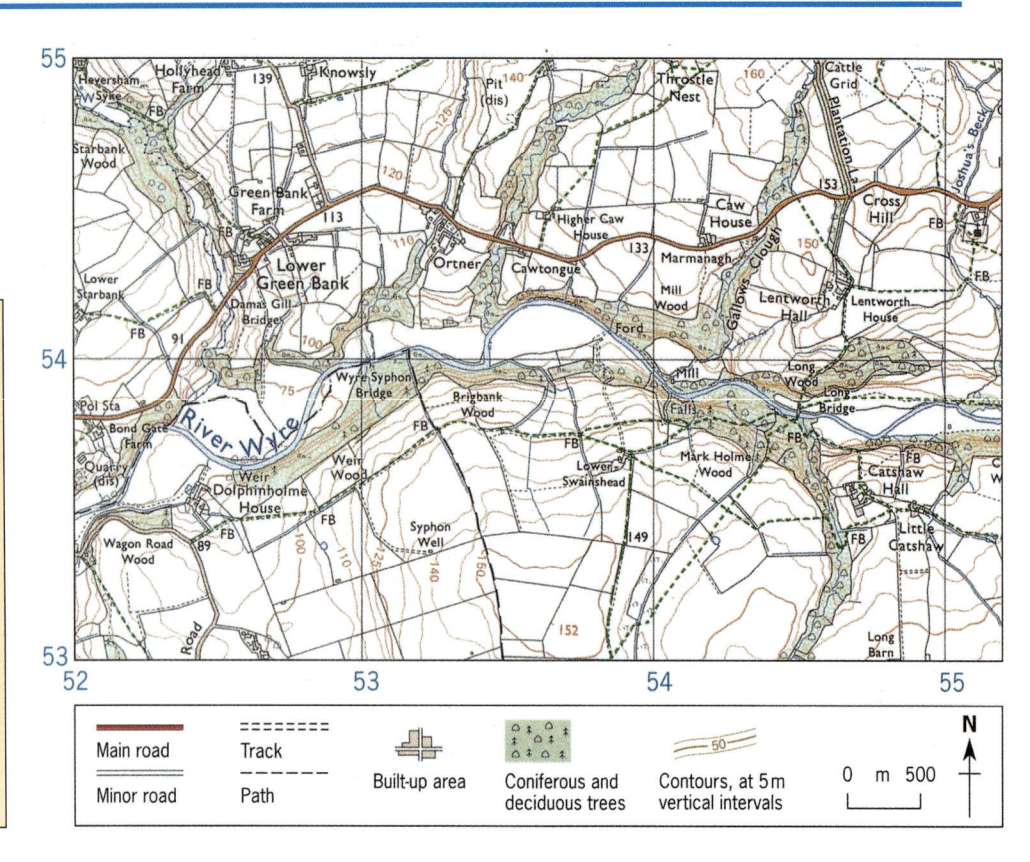

Figure 1.13 Marshaw Wyre valley

Meanders

Vertical erosion, so important in the uplands, has little significance when the river moves to lower ground. This is because the river has a gentler gradient and its tools of erosion (i.e. bedload) are much smaller. With a gentler gradient, the river develops wider meanders, which swing from one side of the valley to the other. Meandering is the key to understanding the valley shape.

♦ Water moving around a meander flows fastest on the outside of the meander bend.
♦ Here, the current undercuts the bank and causes erosion (Fig. 1.14).
♦ If the channel is right up against the edge of the valley, this undercutting causes the valley side to collapse – thus forming a steep river cliff.

In this way, the valley is gradually widened. We call this **lateral erosion**. Meanwhile, the eroded material becomes part of the river's load. Opposite the undercut bank is a lower bank made of coarse shingle and sand. These deposits form a feature called a **point bar**.

Over hundreds of years, the meanders slowly migrate down the valley. This happens because the valley slopes downstream, making erosion more effective on the downstream side of meander bends. As the meanders migrate, they remove any interlocking spurs.

1 River landscapes and processes

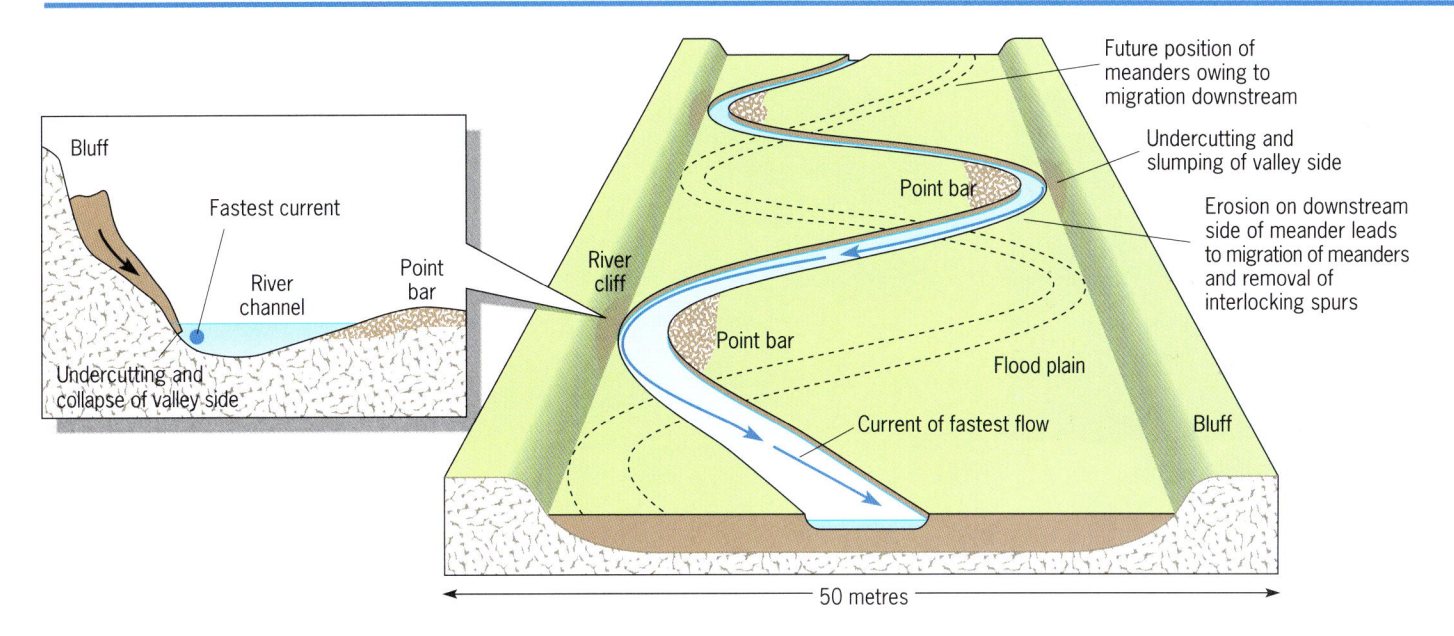

Figure 1.14 The main features of a river's middle stage

Flood plain

We refer to the river's wide, flat valley floor as its **flood plain**. The flood plain consists of river sediment or **alluvium**, which is made up of both coarse and fine sediments. These are deposited over time as meanders criss-cross the valley. Because of this migration, the point bar deposits will cover the entire flood plain. In addition, fine silt deposits are added to the flood plain whenever the river spills out of its channel and floods the valley floor.

Terraces

Terraces are common landforms on flood plains (Fig. 1.15). Matching pairs of terraces on either side of the valley result from rapid vertical erosion by the river (see Table 1.1). This happens when there is a sudden increase in the river's energy. We call this **rejuvenation**. It may be due to a steepening of the river's gradient (owing to a fall in sea level), uplift of the land, or an increase in discharge. The resulting paired terraces are known as **rejuvenation terraces** (Fig. 1.16). Sometimes, where a river has a meandering course within a solid rock channel, rejuvenation leads to vertical erosion and gorge-like valley sides. We call this feature an **incised meander** (Fig. 1.17).

> *EXERCISES*
>
> 7 Study Figure 1.13.
> a Draw a sketch of the area then name and label the features A, B and C and active areas of erosion and deposition.
> b* Explain briefly how the features A–C have formed.
> c Suggest reasons for the areas of active erosion and deposition.

REMEMBER
Rivers erode the land surface and transport and deposit sediment in both upland and lowland environments.

Figure 1.15 River terraces in the Lune Valley

11

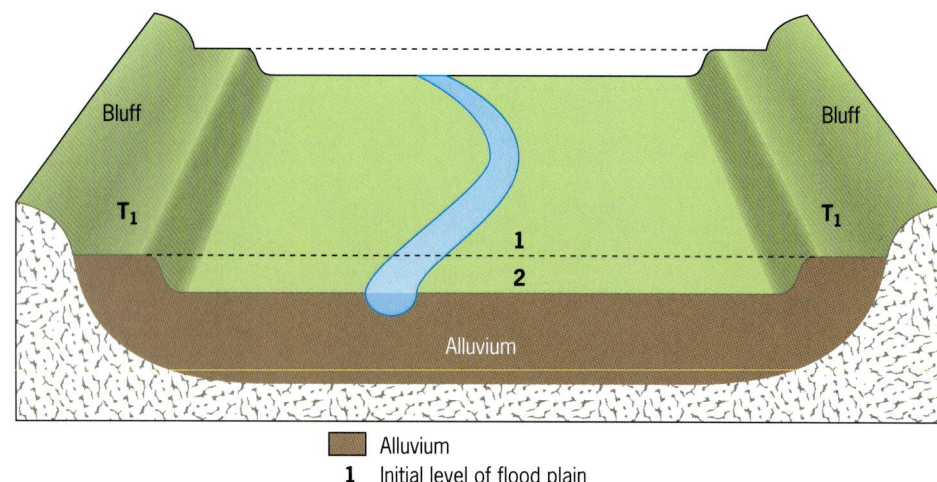

1 River landscapes and processes

Figure 1.16 (above right) Rejuvenation terraces

Figure 1.17 (above left) Incised, or entrenched, meanders on the San Juan River, Utah, USA. They have formed where the river has incised into the rising Colorado Plateau

Alluvium
1 Initial level of flood plain
2 Modern flood plain level
T_1 Matching terraces at flood plain level **1**

The lowland stage

In the last 25 km of its journey to the Irish Sea, the River Wyre falls only 20 m. On leaving the uplands it crosses the gently sloping Fylde Plain (see Fig. 1.2). Here, the river occupies a broad, shallow valley: so broad that the meanders no longer reach from one side of the valley to the other (Fig. 1.18).

The river now has its largest flow, having received water from most of the tributaries in its **catchment**. As a result, the river's load has increased. However, even though its gradient is very gentle, the river is able to transport this increased load because most of it is very fine suspended sediment.

◆ In the lower course, erosion is confined to the river's channel. Where meanders form an almost complete circle, the river may cut across the narrow meander neck, straighten the channel and so follow a more direct

Figure 1.18 (below) River Wyre: lower course © Crown copyright

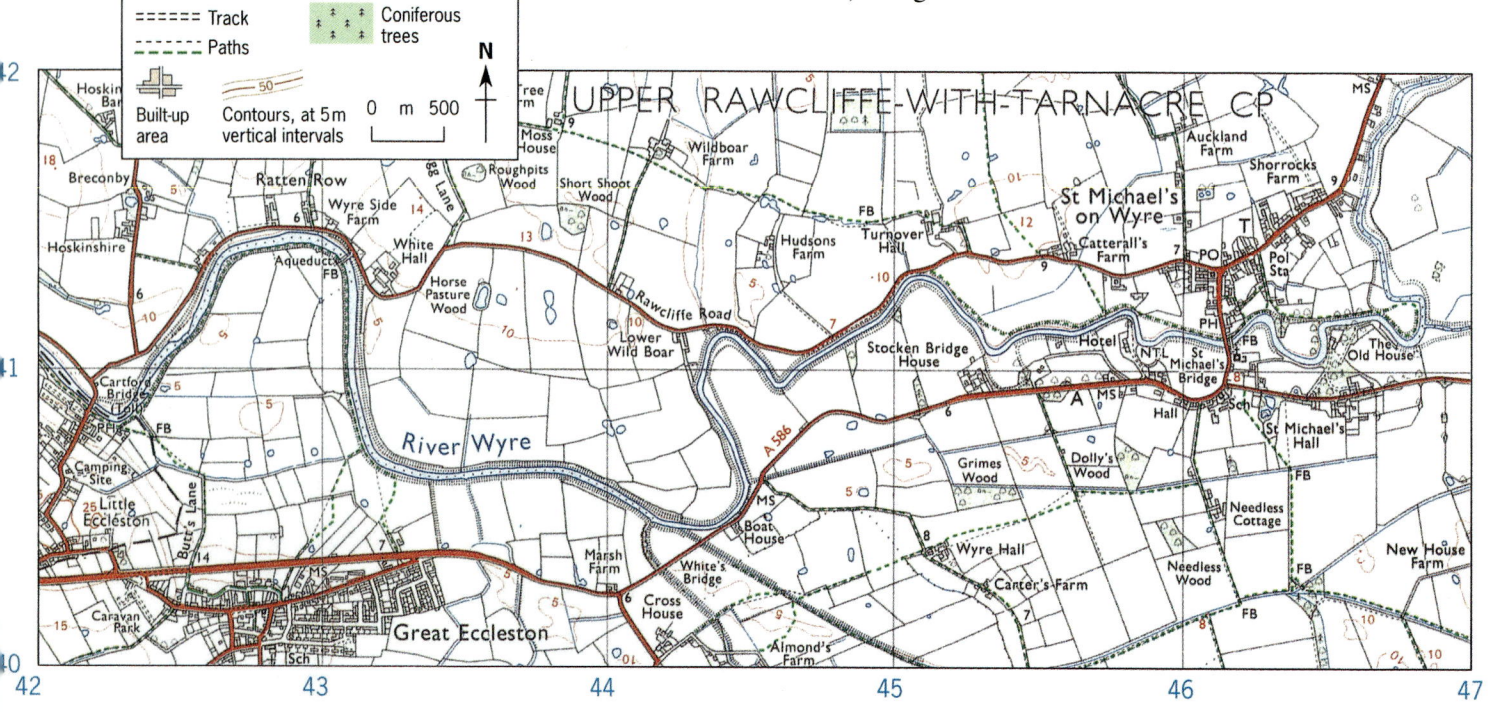

1 River landscapes and processes

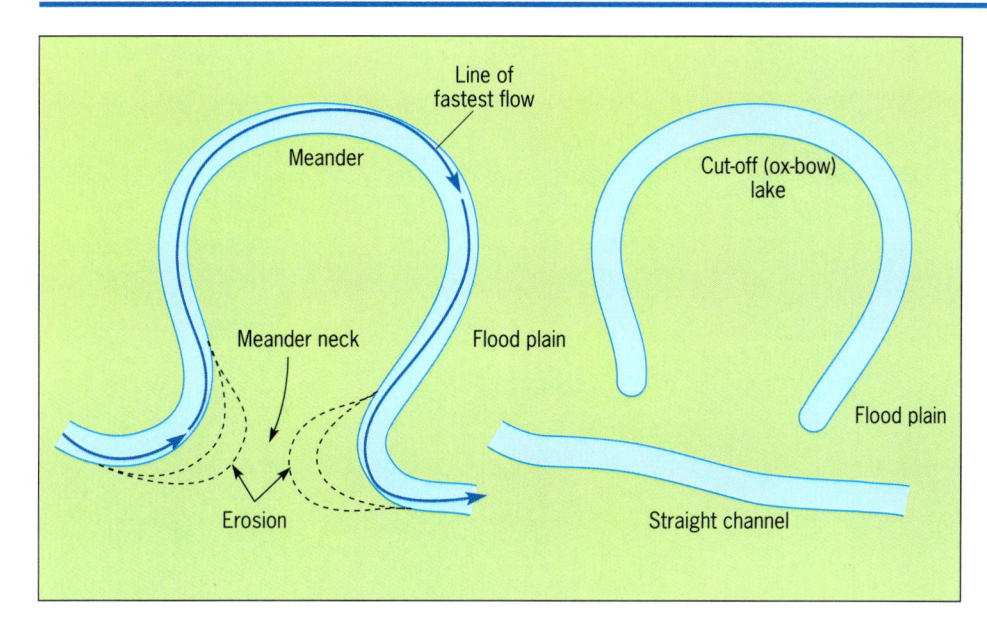

Figure 1.19 Formation of ox-bow lakes

course (Fig. 1.19). The abandoned meander, left isolated on the flood plain, is known as an **ox-bow lake** (cut-off lake).
- Levées form alongside the channel. These are natural embankments built up by deposition of the river's suspended load. This deposition occurs during times of flood when there is a sudden loss of energy as the river spills out of its channel.

Estuaries
The River Wyre finally meets the Irish Sea at Fleetwood (see Fig. 1.2). Here it forms a distinctive, funnel-shaped river mouth or estuary. Ten thousand years ago, the estuary would have looked very different. At that time, sea level was 100 m lower than it is today; most of the Irish Sea was dry land; and the Wyre estuary was a broad, shallow river valley.

As the ice age came to a close, glaciers and ice sheets melted. Sea level began to rise, flooding lowland valleys like the Wyre. Eventually, about 6000 years ago, the sea reached its present level. Since then, the drowned valley at the mouth of the Wyre has been partly filled in by mud and silt brought by the tide. These deposits, together with the scouring action of the tide, have given the Wyre estuary its smooth outline of today (Fig. 1.20).

Deltas
Not all rivers have estuaries. Instead, some reach the sea in **deltas**. These are, in fact, river mouths choked with sediment. This causes the main river channel to split up into hundreds of smaller channels, or **distributaries**.

Figure 1.20 Formation of the Wyre estuary

EXERCISES
8a Use an atlas to find into which major estuaries the following British rivers flow:
- Trent • Great Ouse • Eden
- Medway • Wye • Ouse
- Orwell • Test.

b Rivers meander in all sections of their courses. Using Figures 1.11, 1.12 and 1.18, describe how the meanders change from the upper to the lower course of the River Wyre.

c What evidence in Figure 1.18 suggests that the area around St Michael's and Great Eccleston might be liable to flooding?

d How has the risk of flooding influenced the siting of settlements in the area around Great Eccleston?

e* Suggest a possible crop that farmers might grow in the fields close to the River Wyre between St Michael's and Great Eccleston. Explain your answer.

1 River landscapes and processes

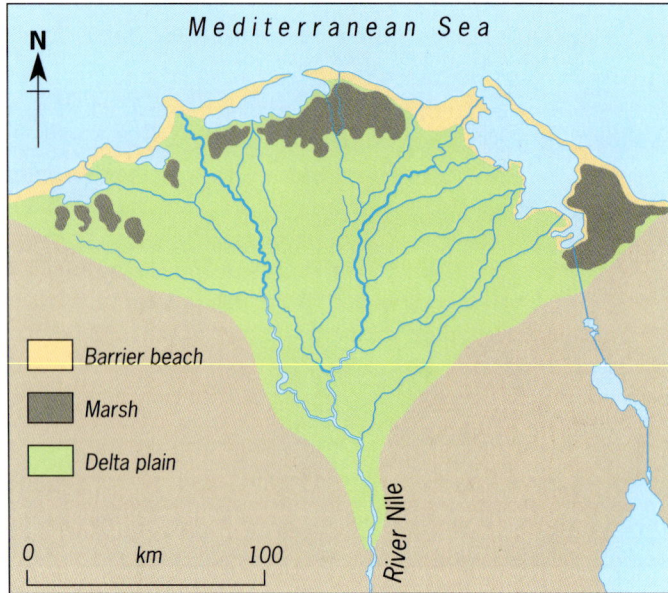

Figure 1.21 (above left) Satellite image of the Nile delta

Figure 1.22 (above right) The Nile (arcuate) delta

EXERCISES

9a Draw a sketch diagram of the Nile Delta (Fig.1.21).
b Use Figure 1.22 to help you label features on your sketch.
10a Use an atlas to find out which of the following rivers have estuaries and which have deltas: Seine, Danube, Po, Severn, Niger, Mekong, Ganges-Brahmaputra, Loire, Zaire, Plate.
b Of those rivers that have deltas, which ones: • flow into tideless seas (i.e. the Mediterranean and Black Seas),
• have major mountain ranges in their basins and possibly very heavy sediment loads?

Deltas form because rivers deposit their silt faster than waves and tides can take it away. There are two circumstances where this is likely to happen:
◆ In tideless seas like the Mediterranean, where the Nile delta in Egypt (Fig. 1.21) and the Rhône delta in France have formed.
◆ Where the sediment load is very large (e.g. Ganges and Brahmaputra rivers in Bangladesh), and waves and tidal action are unable to remove all the silt deposited by the river.

There are two common types of delta: arcuate and bird's-foot. The River Nile is the classic triangular-shaped, or arcuate delta (Fig. 1.22) with a smooth, rounded coastline. It owes its shape to tidal currents, which shift sediment along the coast to form parallel beach bars. Lagoons, which form on the landward side of the bars, are gradually filled with river sediment.

In contrast, the Mississippi has a bird's-foot delta with a distinctive branch-like appearance. This develops as levées form along the distributaries, allowing them to extend seawards. When the river floods, it breaks through the levées, depositing sediment, which eventually fills in the shallow water areas between the distributaries. In this way, the whole delta gradually pushes out beyond the line of the coast (Fig. 1.23).

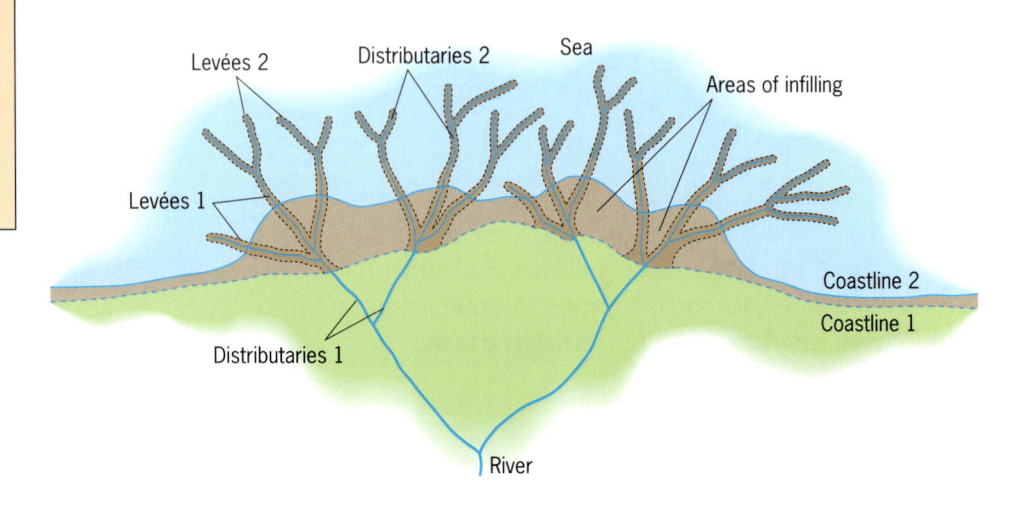

Figure 1.23 Formation of a bird's-foot delta

1 River landscapes and processes

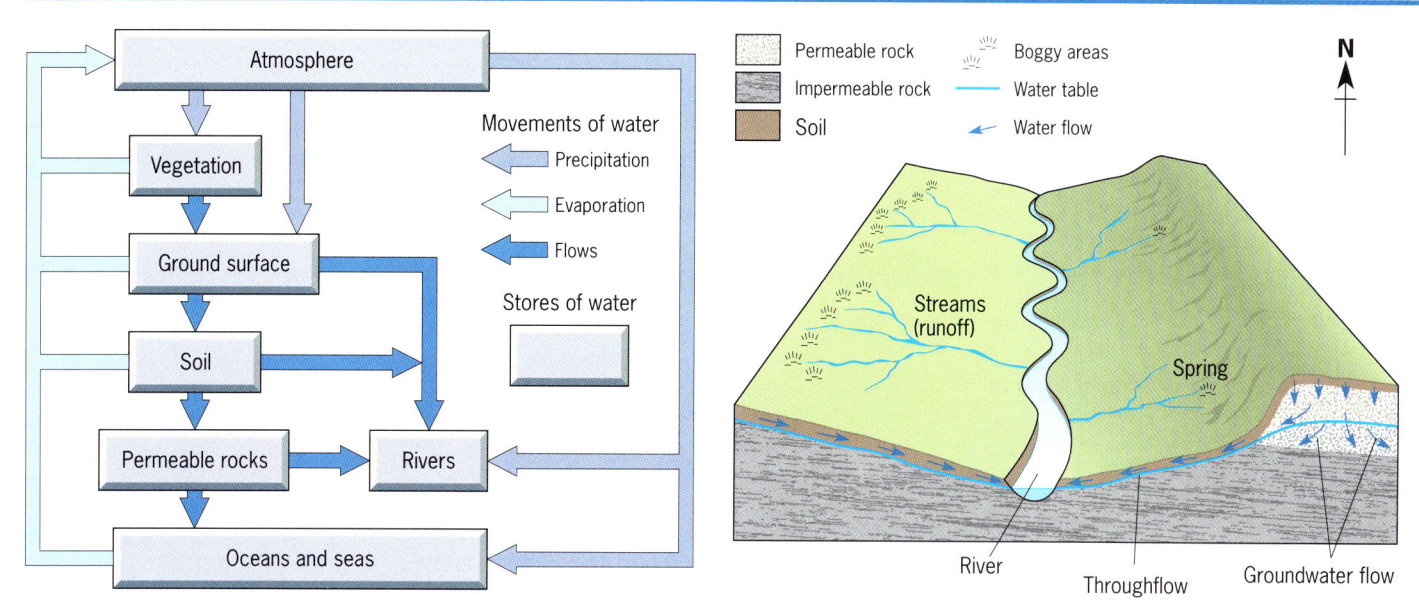

Figure 1.24 The water cycle

Figure 1.25 Sources of water for streams and rivers

1.4 Rivers as flows in the water cycle

Water moves in an endless cycle between the atmosphere and the Earth's surface (Fig. 1.24). Rivers play an important part in this **water cycle**. Once precipitation reaches the ground, it gets into rivers by one of three routes:
- Over the ground surface;
- Through the soil;
- Through porous rocks such as chalk and sandstone.

The speed with which water then reaches rivers depends on the route it takes. For instance, water running off the surface will get into rivers in just a few hours, whereas water flowing through the soil may take several days. This is rapid compared to water moving through porous rocks. It may take months, or even years, before this **groundwater** adds to river flow.

River regimes

The volume of water flowing down a river in a given time is known as the **discharge**. We usually measure discharge in cubic metres per second (cumecs). Over a year, the pattern of a river's discharge varies with the seasons. These seasonal changes of flow are called the **river's regime**. In tropical climates, which have extreme wet and dry seasons, river regimes are dominated by high flows in the wet season. In contrast, in high latitudes and mountainous areas, rivers often have their maximum discharge in the spring when winter snows melt.

Rivers in the British Isles also show marked seasonal variations in flow. The highest flows normally occur in winter, with the lowest in summer (Fig. 1.26). However, the climate of the British Isles does not have wet or dry seasons and snowmelt makes only a small contribution to the flow of most British rivers. Other factors explain the regime of British rivers. The main ones are **evaporation**, **transpiration**, the **interception** of precipitation by vegetation, and the amount of moisture in the soil (Table 1.4).

EXERCISES

11 Study Figure 1.25.
a Name and explain the two sources of streams and rivers.
b From which side of the river (east or west) will rain get into the river faster? Explain.
c* Suggest two possible reasons for the absence of streams and rivers in the upland area shown on Figure 1.25.

Figure 1.26 River Derwent: mean monthly discharge as a percentage of mean monthly rainfall

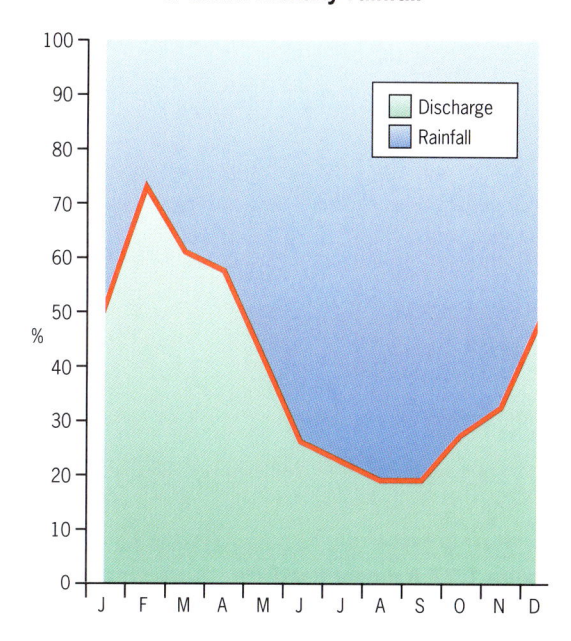

15

1 River landscapes and processes

> **REMEMBER**
> Flooding may occur following a single, heavy storm, or after several days of continuous rain. Intense storms produce flash floods. Prolonged rain gives slow floods, which may last for several days.

Table 1.4 Factors influencing river regimes in the British Isles

Precipitation	Precipitation is fairly evenly distributed all year, though in the north and west, late autumn and winter are often the wettest times. In the south and east, thunderstorms make summer the wettest season in some areas.
Evaporation	Evaporation is highest in summer, so less water is available for river flow.
Transpiration	Plants transpire moisture throughout the growing season (between April and September).
Interception	Plants intercept most precipitation when in full leaf (summer). Moisture trapped on leaves and stems evaporates before reaching the ground.
Soil moisture	By late autumn, soils are usually saturated. Precipitation falling on to these soils simply runs off the surface. In summer, soils are dry and soak up a large proportion of precipitation.

Flood hydrographs

River regimes describe average river flows over a year. In contrast, flood **hydrographs** are concerned with short-term precipitation events (e.g. a thunderstorm or a few hours of precipitation) and their effect on river flow.

A flood hydrograph plots precipitation amounts and river discharge at hourly intervals (Fig. 1.27). You will see on the hydrograph that discharge is made up of slow flow and quick flow.

- Slow flow is water that comes from porous rocks and the soil. These underground sources provide a constant flow of water even in the driest summers.
- Quick flow is water from recent precipitation events. It comes mainly from surface runoff and gets into rivers rapidly.

During a period of precipitation, river levels rise as discharge increases. This is shown as the rising limb of the hydrograph (Fig. 1.27). Eventually maximum flow or peak discharge is reached. Thereafter, water levels (and discharge) start to go down. This is the recession limb of the hydrograph.

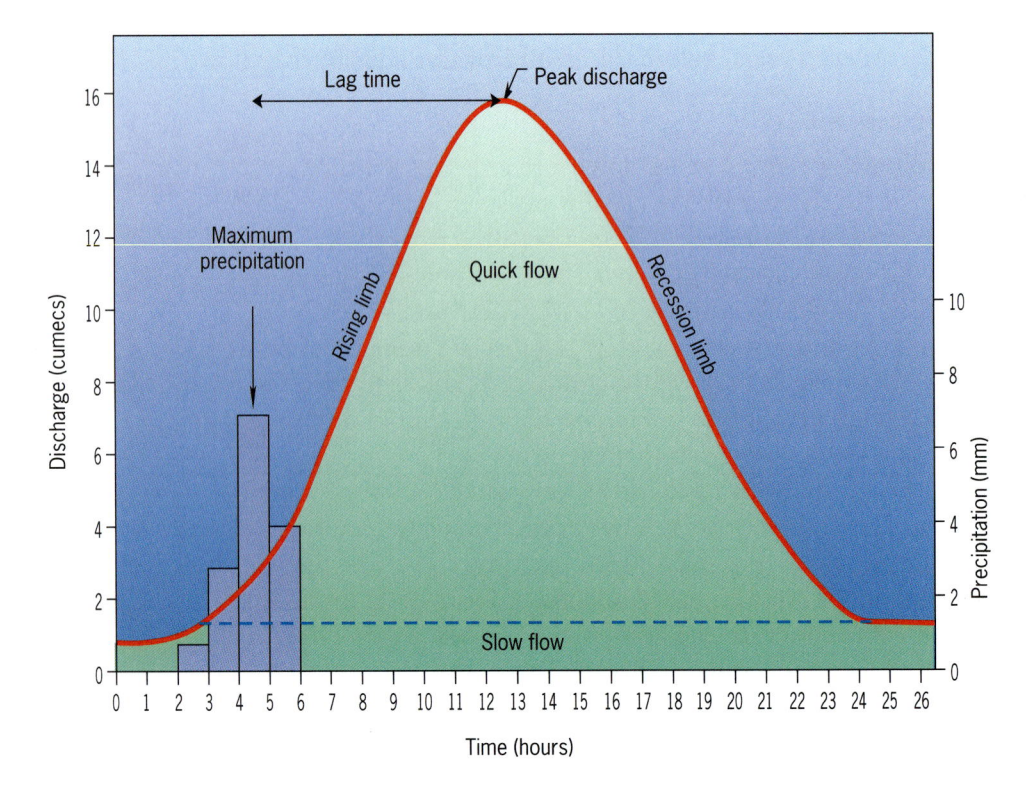

Figure 1.27 A model flood hydrograph

1 River landscapes and processes

Table 1.5 Factors influencing the shape of the flood hydrograph

Precipitation characteristics	• The amount of precipitation per hour (intensity), e.g. high-intensity precipitation is less likely to seep into the soil, and therefore more likely to run off the surface. • The total amount of precipitation. • The type of precipitation, e.g. snow may take days or weeks to melt.
Soil moisture characteristics	• If soil is saturated, precipitation will quickly run off the soil surface into rivers. • Dry soil will absorb precipitation. Water will then move slowly through the soil into rivers.

Drainage basin characteristics	• Rock type:
	  Porous rocks, e.g. chalk and sandstone, will store water and release it slowly. Impermeable rocks, e.g. granite and shale, will allow rapid runoff.
	• Vegetation cover:
	  The denser the vegetation cover, the greater the rate of interception and transpiration. Interception increases evaporation and slows down the movement of water to rivers.
	• Slopes:
	  The steeper the slopes, the more rapid the rate of runoff.
	• Drainage density:
	  The more streams and rivers that drain an area, the faster water will run off.
	• Drainage basin shape:
	  Near-circular drainage basins have higher peak flows than long, narrow drainage basins.

> *EXERCISES*
>
> **12a** Describe the mean monthly discharge of the River Derwent as a percentage of mean monthly rainfall (Fig. 1.26).
>
> **b*** With reference to Table 1.4, suggest possible reasons for seasonal differences in the flow of the River Derwent.
>
> **13** Describe the drainage basin characteristics that are likely to give • shortest lag time and maximum peak discharge
> • longest lag time and minimum peak discharge.

Rivers vary in their speed of response to precipitation (Table 1.5). We measure this as the lag time – the difference in time between maximum rainfall and peak discharge. Rivers with short lag times have steep rising and recession limbs, and high peak discharges. Water therefore reaches these rivers quickly. Such rivers are described as 'flashy' and are particularly liable to flood.

CASE STUDY

1.5 The Derwent Valley floods

In March 1999, a number of towns and villages in the Derwent Valley in North Yorkshire (Fig. 1.28) were hit by severe flooding. Worst affected were the market towns of Malton and Norton, where more than 130 commercial properties and houses were flooded. Damage to surrounding farmland and crops was extensive.

The immediate cause of flooding was above-average rainfall in February, followed by heavy and prolonged rainfall in the first week of March. Snowmelt on the North York Moors made the situation worse. Malton recorded its highest flood levels of the century and the Derwent at Stamford Bridge (Fig. 1.29) had its highest ever flow.

1 River landscapes and processes

FACTFILE

- *Area of the Derwent drainage basin: 1586 sq km.*
- *Geology: Mainly clay, shale and limestone.*
- *Relief: The drainage basin covers a large part of the North York Moors. This upland rises to 454 m above sea level and is dissected by deep, steep-sided valleys. The Vale of Pickering, an extensive clay lowland only a few metres above sea level, also lies within the basin.*
- *Drainage density: Many surface streams and rivers drain the North York Moors. The Vale of Pickering has artificial drainage channels, which remove water quickly from farmland.*
- *Precipitation: Mean annual precipitation for the Derwent basin is 765 mm, ranging from about 1000 mm on the North York Moors to about 600 mm in the Vale of Pickering.*
- *Land use: Moorland, grassland and forestry plantations in the North York Moors; arable land and grassland in the Vale of Pickering.*
- *Average March flow: 22 cumecs.*

EXERCISES

14 Between 15.00 hrs on 4 March and 10.00 hrs on 8 March, 198 mm of rain fell in the Derwent basin. What proportion of the drainage basin's mean annual precipitation fell in this four-day period? (see Factfile)

15 Study Figures 1.29 and 1.30.

a What was (1) the maximum period of continuous rain in the Derwent catchment, (2) the Derwent's peak discharge in cumecs, (3) the approximate date and time of peak discharge, (4) the lag time for the Derwent flood event (i.e. the difference between peak rainfall and peak discharge)?

b For how many days did the River Derwent continue to rise?

c How much greater was Derwent's peak discharge than its average flow for March? (see Factfile)

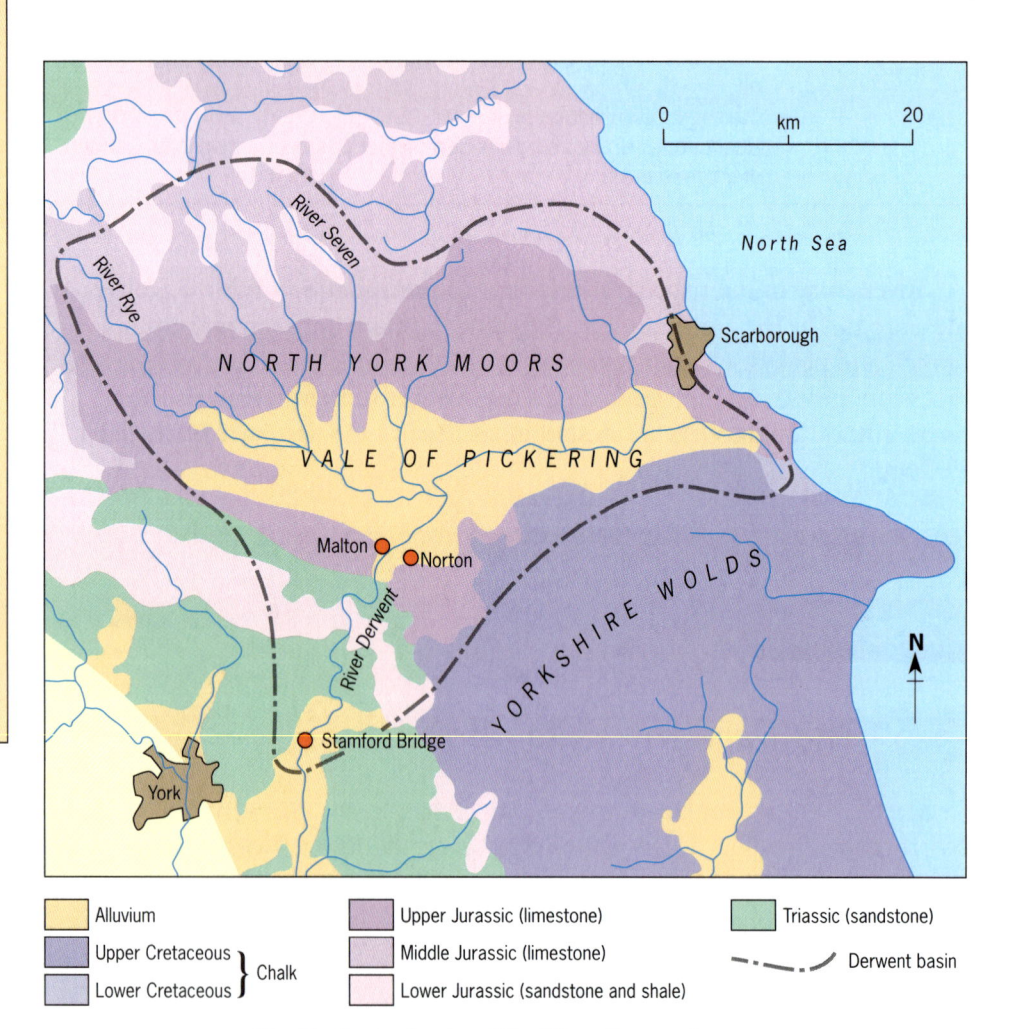

Figure 1.28 Location of the River Derwent valley

Legend: Alluvium | Upper Cretaceous / Lower Cretaceous } Chalk | Upper Jurassic (limestone) | Middle Jurassic (limestone) | Lower Jurassic (sandstone and shale) | Triassic (sandstone) | Derwent basin

The Drainage basin of the River Derwent

1 River landscapes and processes

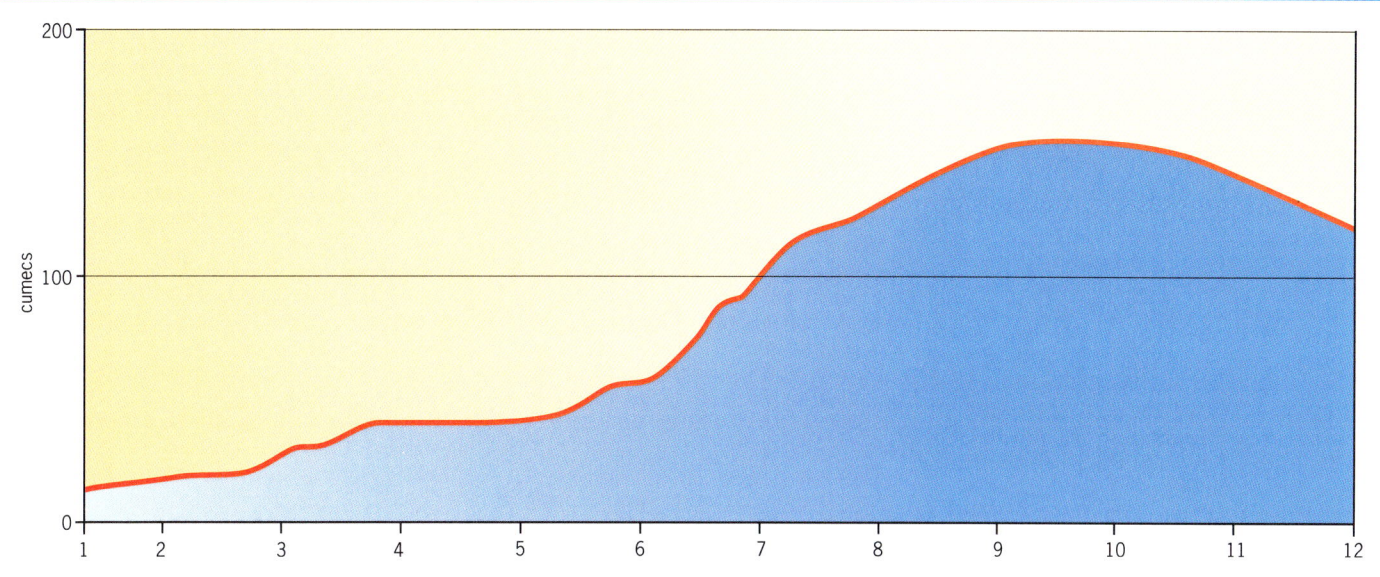

Figure 1.29 (above) River Derwent flow: 1–11 March, 1999

Figure 1.30 (below) Hourly rainfall totals in River Derwent Valley: 1–11 March, 1999

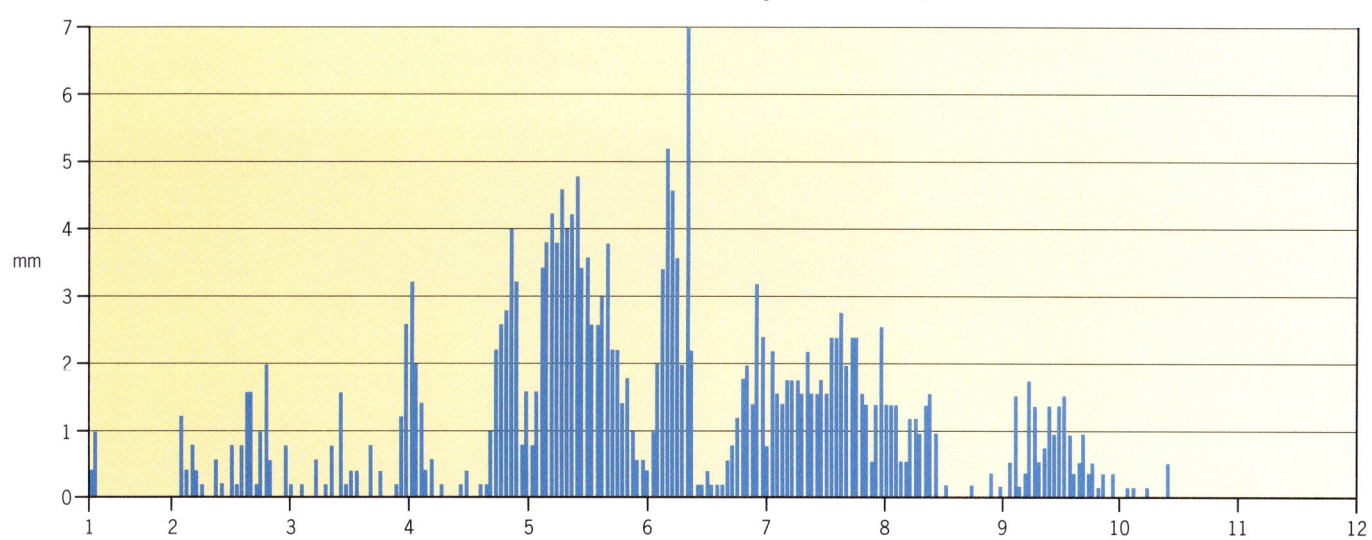

Table 1.6 Two views on the Derwent floods at Malton

Alison Kay, 24-year-old mother whose home was flooded:
  'No one could have predicted that the river would reach our door, but when it kept on rising, surely we had a right to expect more help.'

  'If we'd been alerted sooner, we could have taken steps to help ourselves.'

  'As the river came ever nearer to ours and others' homes, the council failed to react properly. We were offered sandbags but only when it was too late.'

Chief Inspector Dennis Brewer, area police commander:
  'We began issuing warnings as soon as the seriousness of the situation became apparent. Everything possible was done to reduce disruption and hardship.'

  'Our emergency procedures are working well but the reality is we can only do so much when nature chooses to behave in such an extreme way.'

  'I'm not dismissive of those who are suffering hardship, but when people choose to live near a river, they must accept there are certain risks.'

EXERCISES

16 Read the *Yorkshire Post* article (Fig. 1.31) on the Derwent floods and describe the impact of the floods on the area around Malton.

17 Study Figures 1.31 and 1.32 and Table 1.6.

a* What evidence suggests (1) that severe flooding might have been prevented, (2) that the 1999 floods were an exceptional event and could not have been prevented?

b* State your view on the Derwent Valley floods. Give your reasoning.

1 River landscapes and processes

It didn't stop yesterday… again. **John Woodcock** reports on those who have been caught out by the wettest weather for 68 years.

When Alison and Garry Kay bought a bungalow on a main road, how could they have known that three months later they and their baby son would be marooned on an island? No one told them about local history. The estate agent's sales pitch didn't mention the fact that perhaps a couple of times every century one of Britain's loveliest rivers becomes a bloated beast submerging everything in its expanded course.

This time last week a large meadow and a railway line lay between the Kays and the Derwent. Yesterday its muddy waters were more than a foot deep in every room of their £50,000 home.

The couple can just about understand why no one thought to forewarn them about the remote possibility of such severe flooding before they moved in.

But they're furious that as far as they are concerned the local council and Government agencies did not do more to minimise their domestic catastrophe once the scale of the crisis became apparent.

In fact, everyone along the river's route through North and East Yorkshire has been caught out by nature. The Derwent rises on Moorland next to Fylingdales. The tracking station would have warned the UK against incoming missiles but, following six inches of rain since last Friday, combined with melting snow, could not alert towns like Malton and Norton to their worst fate for 68 years.

"No one could have predicted that the river would reach our door but when it kept on rising surely we had a right to expect more help," said Mrs Kay, 24.

She and her 19-year old husband, who moved to the area because he found a job in the local bacon factory have now abandoned their bungalow in Scarborough Road, Norton and with their 11-month old son George moved into a £99 a night hotel with the bills being paid by the insurers.

"Our home is awash, a shambles" added Mrs Kay "We'd fully furnished it after moving in and estimate that £7,500 worth of our possessions have been ruined.

"What's so galling is that as the river came ever nearer to ours and other homes the council failed to react properly. We were offered sandbags but only when it was too late.

If we'd been alerted sooner we could have taken steps to help ourselves. But we were just sitting there watching the water keep on rising. I don't like to whinge but from council staff there was a lot of shrugging of shoulders.

"Some of our neighbours are even more badly hit. One home is under three feet of water and when our friend last saw it his freezer was floating in the dining room."

Chief Inspector Dennis Brewer, the area police commander who was born in Malton is sympathetic but denies that the emergency services were slow to react. "We began issuing warnings as soon as the seriousness of the situation became apparent. Everything possible has been done to reduce disruption and hardship and I think the criticism of the efforts of all those involved is totally inappropriate. Our emergency planning procedures are working well but the reality is we can do only so much when nature chooses to behave in such an extreme way.

"Most people are dealing with problems philosophically. This is a rural area and in a crisis they are used to getting on with life.

"I'm not dismissive of those who are suffering hardship but when people choose to live near a river they must accept there are certain risks."

The floods have divided Malton and Norton and their joint population of 11,000.

A waterway now runs alongside the platform of their railway station that is normally a 400-yard walk from one town to the other. For now, it's a 15 mile detour and a £12 taxi ride. Cabbie Richard Allen said, "I have had three callers put the phone down on me when I told them the price."

For all that, it was normal service as far as possible. Although schools are shut to most pupils, A-level students sat their exams as scheduled at Malton School.

The town's livestock mart was also open for business as usual on a Tuesday. But a sign in the Green Man Hotel told its own sorry tale: "Due to lack of staff caused by the flood this bar is regrettably closed."

The town's horse racing community is also in despair. Trainers are unable to prepare their horses for next week's Cheltenham Festival.

And last night it was still raining…

Figure 1.31 (above) Report by John Woodcock in the *Yorkshire Post*, 10 March 1999

Figure 1.32 (right) Flooding in Malton in 1999.

> **REMEMBER**
> There are two approaches to flood management:
> - prevent floods from happening by afforesting large areas of the upper drainage basin;
> - confine floodwaters to the river channel (e.g. with levées, dams etc.).

Flood defence schemes for the Derwent Valley in future

Following the 1999 floods and criticism from local people, the Environmental Agency is to strengthen flood defences on the River Derwent. A scheme has yet to be devised in detail, but the preferred option is to build flood walls and flood embankments in the towns. Alternatives, such as constructing massive flood storage areas further upstream and increasing maintenance of river banks, were rejected as either too expensive or ineffective.

1 River landscapes and processes

Land-use management
- Afforestation increases interception, evaporation and transpiration. This reduces the amount of water reaching rivers and slows runoff.
- Encouraging grassland, while discouraging arable farming and land drainage, slows runoff and reduces the risk of flooding.
- *Problem*: conflict with landowners who may disagree with these methods of preventing flooding.

Reservoirs and dams
- Dams hold back floodwaters.
- Reservoirs store floodwaters and have other purposes: e.g. water supply, recreation, HEP.
- *Problems*: cost of constructing dams; reservoirs permanently flood valley land.

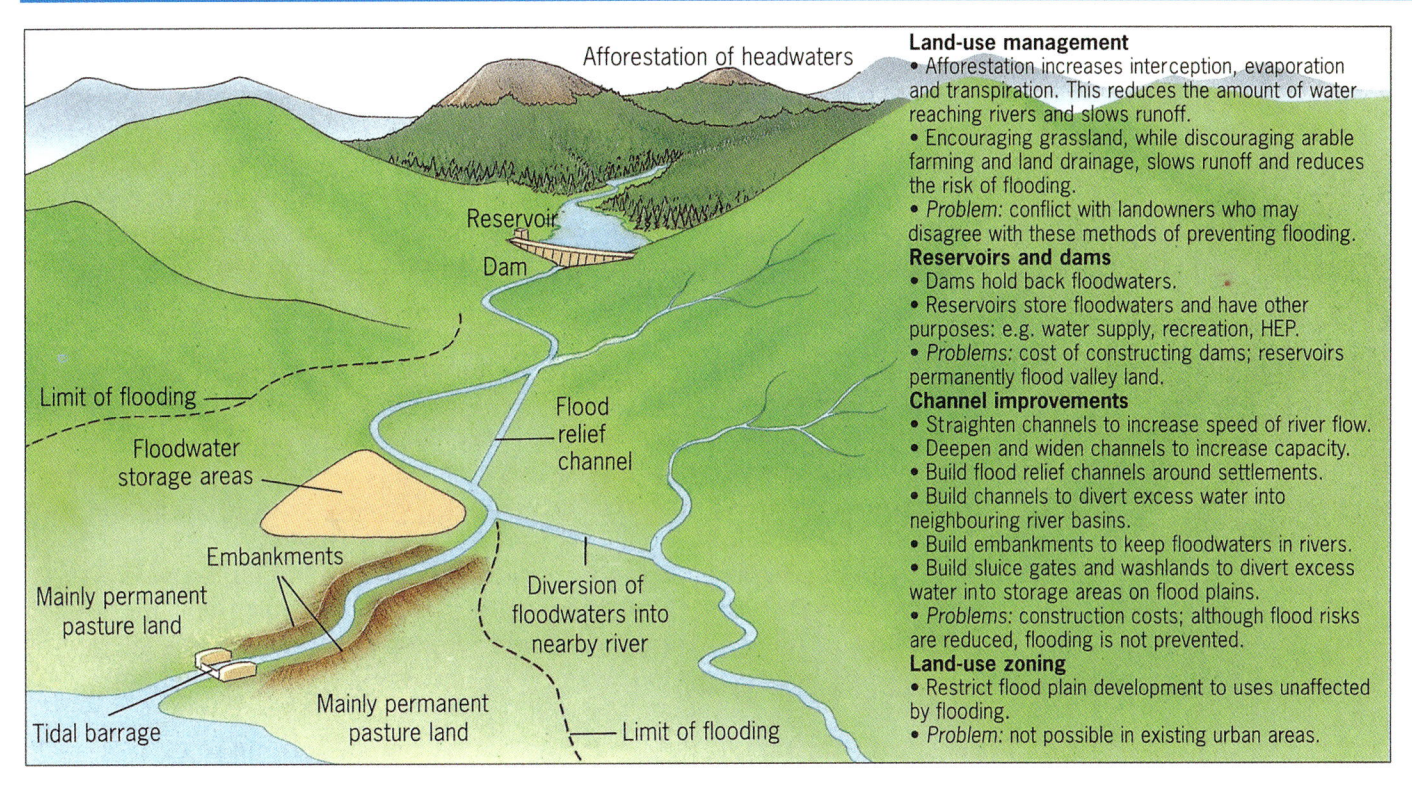

Channel improvements
- Straighten channels to increase speed of river flow.
- Deepen and widen channels to increase capacity.
- Build flood relief channels around settlements.
- Build channels to divert excess water into neighbouring river basins.
- Build embankments to keep floodwaters in rivers.
- Build sluice gates and washlands to divert excess water into storage areas on flood plains.
- *Problems*: construction costs; although flood risks are reduced, flooding is not prevented.

Land-use zoning
- Restrict flood plain development to uses unaffected by flooding.
- *Problem*: not possible in existing urban areas.

Figure 1.33 Flood control and flood prevention

CASE STUDY

1.6 Kielder Water: the UK's first regional water grid

Kielder Water, in the North Tyne valley in Northumberland, is northern Europe's largest artificial lake (Fig. 1.34). Completed in 1982, it was built to meet the growing demand for water from both industry and domestic users in the North-east region. Its effect has been to double the region's water resources.

Geography of water supply and demand in the North-east

There has never been an overall shortage of water in the North-east. The problem of supply is a geographical one. This is because the areas of heaviest demand are also the areas of lowest rainfall. In an average year, the densely populated urban areas of Tyneside, Wearside and Teesside (Fig. 1.36) receive only 650 mm of precipitation. This contrasts with more than 1200 mm in the sparsely populated North Pennines.

Neither Teesdale nor Weardale could provide the extra water needed by the towns and cities along the coast. Teesdale already had three major reservoirs and the River Wear was too small to supply a new reservoir. The River North Tyne had no such disadvantages, though. It drained a large catchment in which there were no existing reservoirs. In addition, the North Tyne valley was sparsely settled, so there were few people to object to flooding of the area.

Figure 1.34 Aerial view of Kielder Water

1 River landscapes and processes

> **EXERCISES**
>
> **18a** Suggest possible advantages and disadvantages of transferring water by river rather than by tunnel or pipe.
> **b** How might Kielder Water assist with flood control on the River Tyne?

Figure 1.35 Kielder Water

Kielder scheme

The Kielder scheme is the first regional water grid in the UK. Its network of tunnels and rivers allows water from Kielder to supply areas throughout the region and up to 120 km away. The grid is made up of 43 km of tunnels linking the River Tyne with the Derwent, Wear and Tees rivers (Fig. 1.36). To supply the grid, water taken from the River Tyne near Corbridge is transferred southwards by tunnel. Outlets on the Derwent, Wear and Tees then top up the flow of these rivers. These rivers then carry the water eastwards before its final abstraction and treatment.

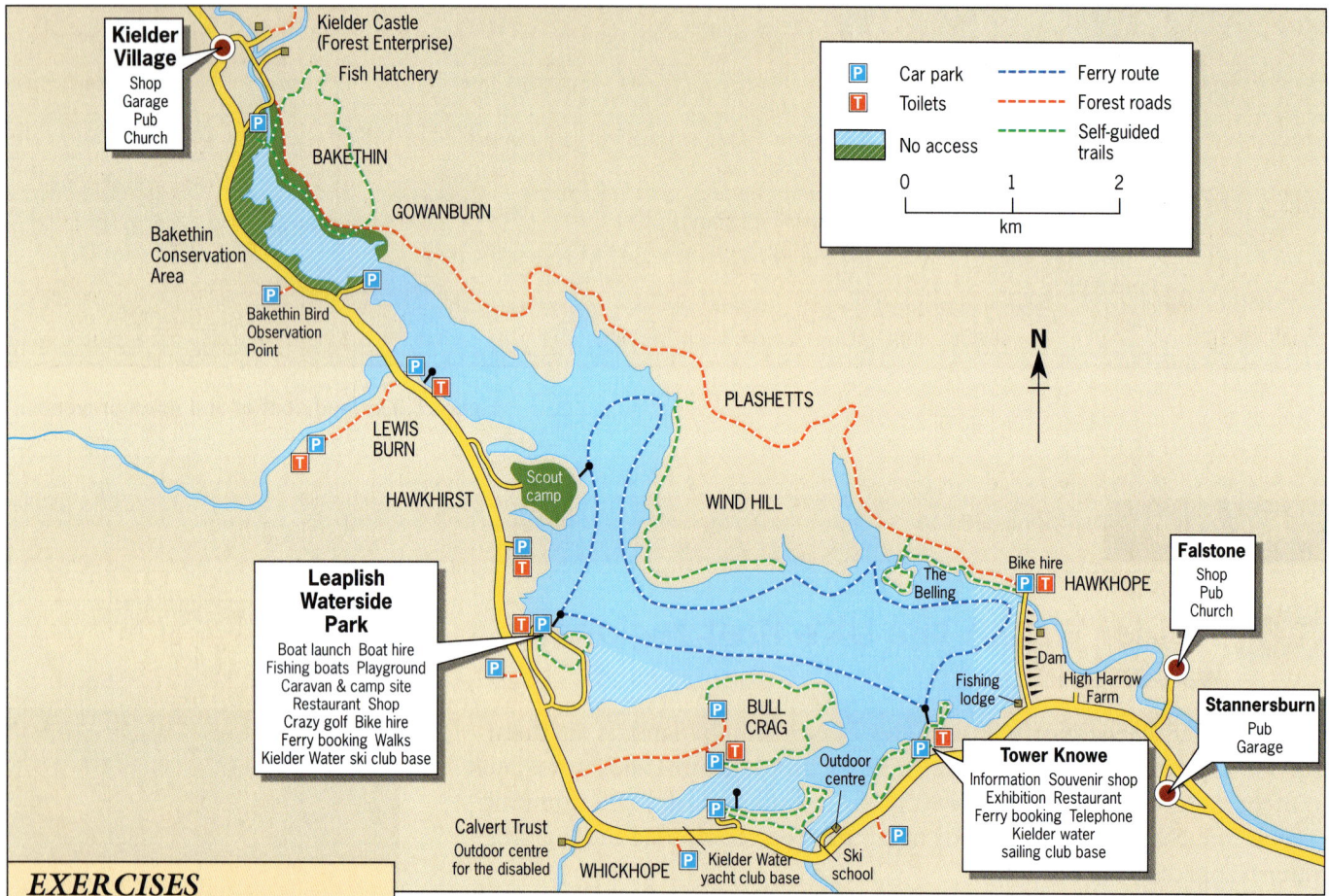

Water security in the North-east

Thanks to Kielder, the North-east has greater water security than any other region in the UK. Furthermore, its water resources are likely to be more than adequate until well into the 21st century. As a bonus, the quality of the region's water and the security of its supply helped to persuade giant electronic companies such as Fujitsu, Siemens and Samsung to invest heavily in the North-east during the 1990s.

The Kielder Water scheme also created jobs in the North Tyne valley. Kielder is a multi-purpose scheme. This means that as well as supplying water, it has also become a major centre for recreation and leisure activities, attracting 500 000 visitors a year (Fig. 1.35). It even generates small amounts of hydroelectric power. All this has given the local people, who previously depended almost entirely on hill farming and forestry, much-needed employment.

> **EXERCISES**
>
> **19** Study Figure 1.35
> **a** Describe how Kielder Water generates employment in the North Tyne valley.
> **b** Make a list of some of the recreational and leisure activities at Kielder that might conflict with each other (e.g. water skiing and angling).
> **c** State briefly why the people involved in the different activities might conflict.
> **d** Explain how some of these conflicts have been resolved. (Fig. 1.35).

1 River landscapes and processes

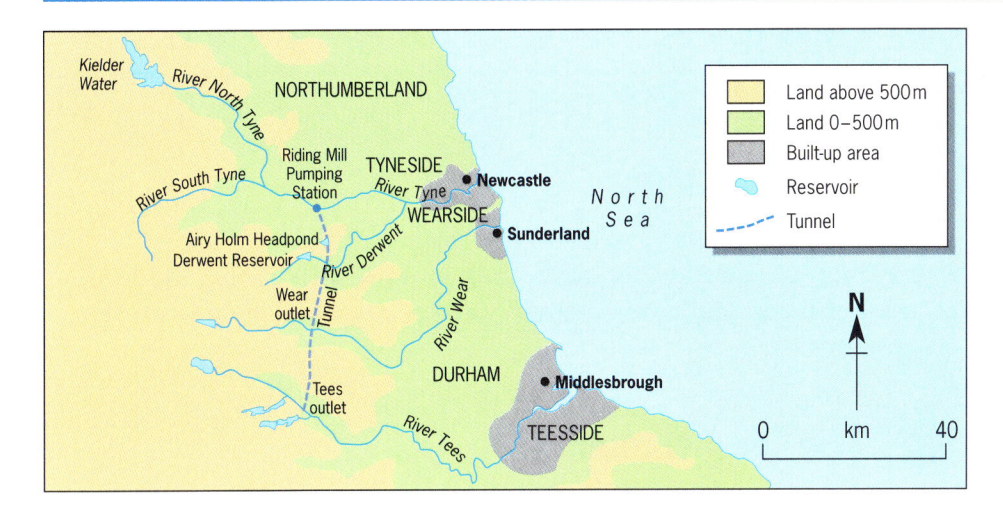

Figure 1.36 The Kielder scheme

REMEMBER
Rivers have multiple uses as resources. They may be used:
- as a source of water supply;
- for recreation;
- for navigation;
- for generating hydroelectricity.

1.7 Summary: River landscapes and processes

KEY SKILLS OPPORTUNITIES
C1.2: Ex. 1, 16; **C1/2.3:** Ex. 4b, 5b; **C2.2:** Ex. 8a, 12b, 18a; **N1/2.1:** Ex. 3, 4b, 4d, 4e, 7b, 7c, 8b, 8c, 8d, 8e, 11, 12b, 13, 15, 17a, 19; **N1/2.2:** Ex. 14, 15c; **IT1/2.1:** Ex. 2a, 2b.

Key ideas	Generalisations and detail
Rivers are important land-shaping agents.	• Rivers erode the land, transport sediment, and deposit it elsewhere. In the process, they create new landforms.
A variety of processes are responsible for erosion and transport of sediment by rivers.	• Rivers erode land when they have more energy than needed to transport water and sediment. • There are three erosional processes: abrasion, hydraulic action and solution. • A river's load is the sediment it transports. Coarse material rolls and slides along the river bed (bedload); sand is bounced along (saltation); silt and clay are moved in suspension; and rocks like chalk and limestone are dissolved and transported in solution.
A river's course, from source to mouth, comprises upland and lowland stages.	• Different processes dominate in each stage. These give rise to distinctive landforms. • Vertical erosion dominates in the uplands, producing V-shaped valleys, interlocking spurs, waterfalls and rapids. • In the lowlands, lateral erosion and deposition result in flood plains and terraces. • Deposition in the lowlands produces levées, estuaries and deltas.
Rivers play an important part in the water cycle.	• Rivers and their tributaries drain an area known as a drainage basin. They transfer water from the land to seas and oceans.
River flow varies over time.	• Rivers have a seasonal pattern of flow, or regime. This is influenced mainly by climate. In the British Isles, rivers have their maximum flow in winter. • The response of a river to short periods of precipitation is recorded on a flood hydrograph. The characteristics of the hydrograph reflect the amount and intensity of precipitation and the characteristics of the drainage basin (geology, slopes, vegetation cover, etc.).
Rivers are natural hazards.	• Flooding by rivers causes damage, loss of life and damage to property. The probability of flooding depends on the level of peak discharge (determined by precipitation and drainage basin characteristics), the relief of an area, and the location of settlement. • Various flood prevention and flood control measures are used to stop floods or minimise their impact. These measures vary in their effectiveness and cost.
Rivers are valuable resources.	• Rivers supply water for domestic use, industry and commerce. They provide recreation and leisure opportunities; they support a rich and varied wildlife; they are an important means of disposing of sewage and industrial effluent; they can be used for transport.

2 Glacial landscapes and processes

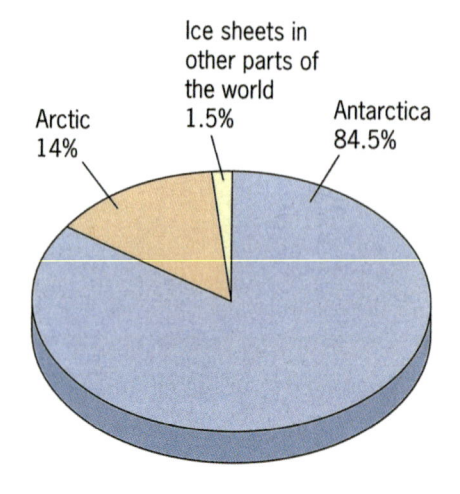

Figure 2.1 Global distribution of ice sheets and glaciers

2.1 Introduction

Ice covers nearly 15 million sq km of the Earth's surface (Fig. 2.1). Although this is only a fraction of the area covered by ice during the last ice age, three-quarters of the world's fresh water is today locked up in glaciers and ice sheets.

2.2 The long winter

The ice age

Imagine that you are a time traveller. You are in the British Isles and it is 20 000 years before present (BP) (Fig. 2.2). As you step out of your time machine, what do you see? In northern Britain, there is a vast sheet of ice stretching to the horizon (Fig. 2.3). Meanwhile, the scene in southern Britain is hardly more inviting. Although the landscape is free of ice, the ground is frozen solid. There are no trees and few plants and animals.

Figure 2.2 Glacial limits and areas of greatest erosion in the last ice age

EXERCISES

1a Describe and explain the distribution of ice sheets and glaciers (Fig. 2.1).
b In what type of environment are ice sheets and glaciers likely to be found outside the polar areas?
c Apart from ice, suggest three other sources of the Earth's fresh water.

EXERCISES

2a Find out where your school is located on Figure 2.2. Was its location covered by ice or was it ice-free in the last ice age?
b* What clues would you look for to find out if the area around your school was glaciated in the last ice age?

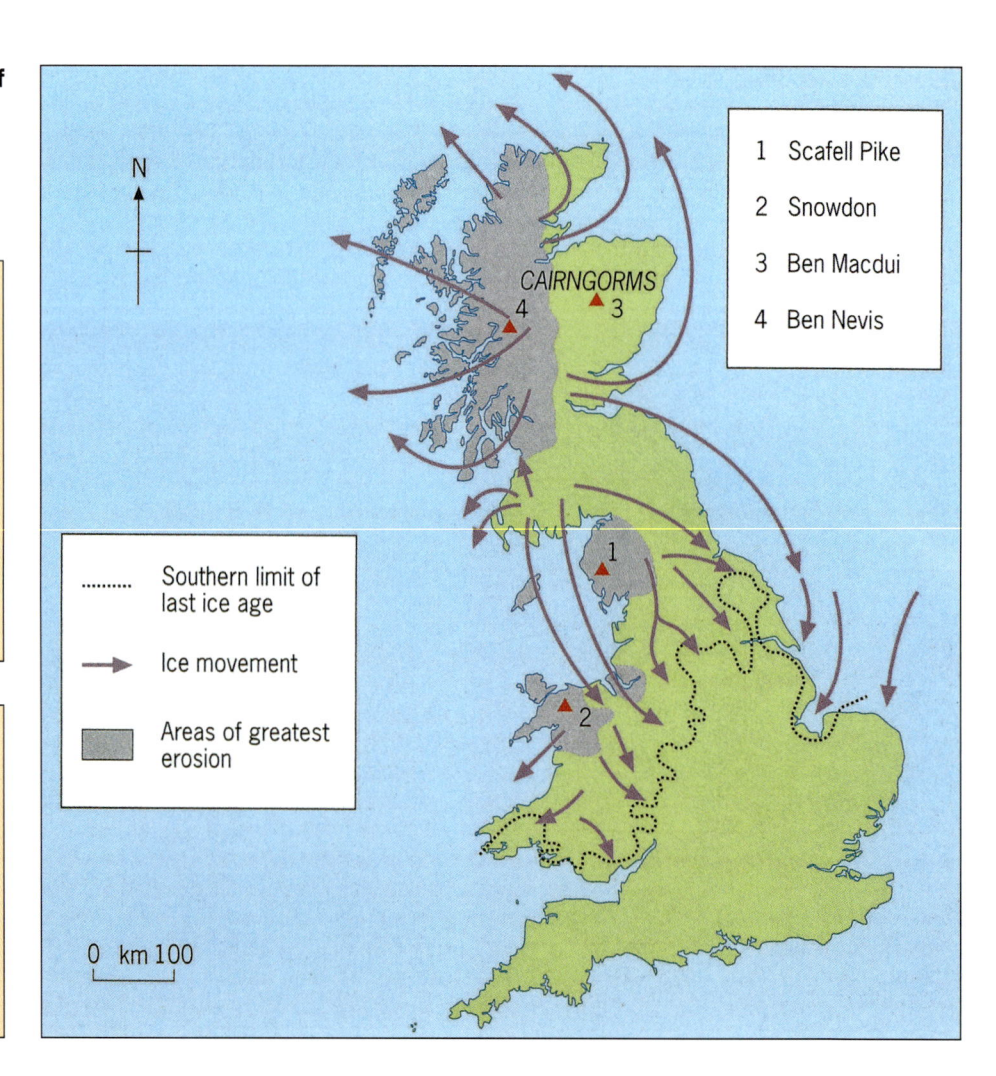

2 Glacial landscapes and processes

20 000 years ago Britain was in the grip of an ice age: a big freeze, which had already lasted for 80 000 years!

The retreat of the glaciers

About 12 000 years ago, the ice age finally ended. A brief cold snap, lasting just 1000 years (Fig. 2.3), saw the return of small glaciers to the uplands of Scotland, the Lake District and North Wales. But by 10 000 BP, these had also melted. This was the last time that glaciers existed in the British Isles. Today, only a few snow patches survive the summer on Ben Nevis and in the Cairngorms (Fig. 2.4). None the less, they remind us that it would need only a small drop in temperature for glaciers to return to Scotland.

During the last two million years there may have been up to 20 ice ages. In other words, this long period of time has been one of almost continuous ice and cold. In contrast, we live today in a warm climate, or **interglacial**, and it is this that is unusual. But with interglacials normally lasting about 10 000 years (Fig. 2.3), the next ice age could be just around the corner!

Figure 2.4 Permanent snow patch on a mountain top in mid August, the Cairngorms, Scotland

Figure 2.3 Timeline: climate change and glaciation in the last 140 000 years

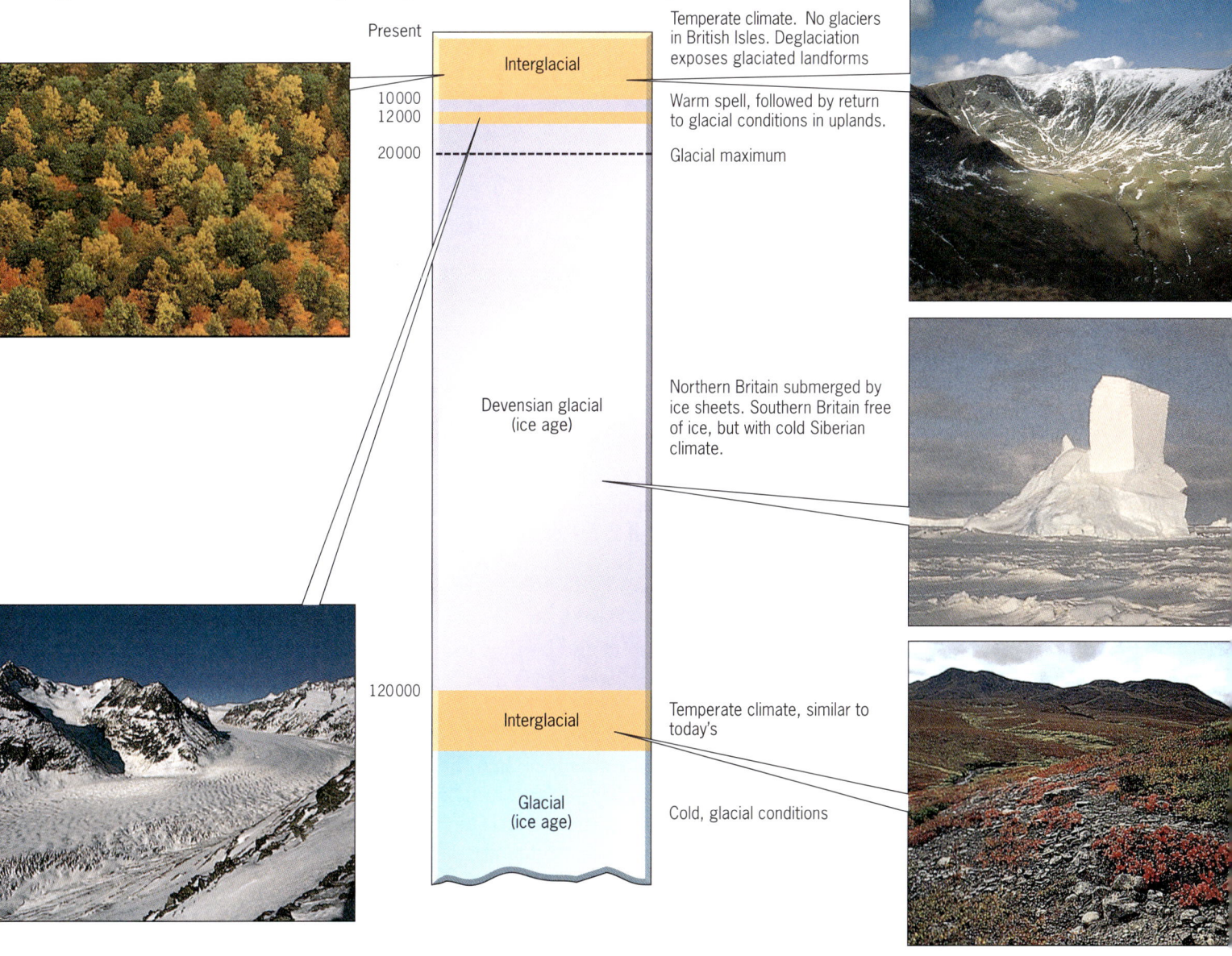

Present — Interglacial — Temperate climate. No glaciers in British Isles. Deglaciation exposes glaciated landforms
10 000
12 000 — Warm spell, followed by return to glacial conditions in uplands.
20 000 — Glacial maximum

Devensian glacial (ice age) — Northern Britain submerged by ice sheets. Southern Britain free of ice, but with cold Siberian climate.

120 000 — Interglacial — Temperate climate, similar to today's

Glacial (ice age) — Cold, glacial conditions

2 Glacial landscapes and processes

EXERCISES

3 Study Figure 2.5.
a Estimate how far Ben Nevis, Ben Macdui, Scafell Pike and Snowdon are below the snowline.
b* If temperatures fall by 6.5°C for every kilometre above sea level, by how much would the climate need to cool for glaciers to develop again on the highest mountains in Scotland?

Figure 2.5 Theoretical snowline for the British Isles

EXERCISES

4a* Use an atlas to find the latitude and height of the following mountains: Aconcagua, Etna, Everest, Erebus, Galdhopiggen, Hekla, Kebnekaise, Kilimanjaro, Mont Blanc, Mount Cook, Mount Kosciusko, Mount McKinley, Tibesti Plateau, and Vesuvius.
b Refer to Table 2.1 and decide which mountains are likely to have permanent snowfields and glaciers.

Table 2.1 Height of snowline

Latitude	Altitude (m)
0°	5000
45°	3000
60°	1500

c Present your results as a table, with columns for • the name of the mountain • its height • its latitude • whether or not the mountain has snowfields and glaciers.
d* Apart from temperature, suggest one other factor that might influence the height of the snowline.

2.3 Glaciers as systems

Glacier ice forms in upland areas above the **snowline**. This is because here, the accumulation of snow (the input) exceeds the rate at which it melts (the glacier's output). This creates a permanent cover of snow and ice. In the British Isles, even the highest mountains are below the snowline (Fig. 2.5). For example, in Snowdonia, North Wales, the snowline is close to 2000 m – almost twice as high as the tallest mountains. However, in the last ice age the snowline was much lower, falling to about 500 m.

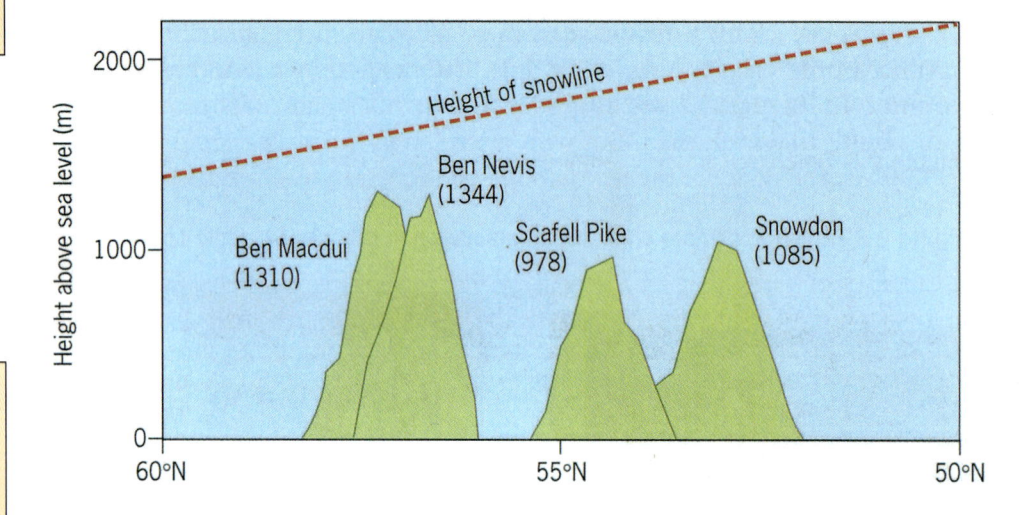

Formation of glacier ice

Permanent snowfields are **accumulation zones** for the ice that feeds glaciers (Fig. 2.6). It takes many years for snow to become glacier ice because it is light: 90 per cent of its volume is empty space. But as it accumulates it is compressed. At the same time, melting and re-freezing helps to convert it to ice. Eventually, after 30 or 40 years, a hard, dense ice forms and starts to flow downslope under its own weight. We call these 'rivers' of ice, glaciers.

Advance and retreat of glaciers

If the climate cools and there are lower temperatures or heavier snowfalls, glaciers advance. If the climate gets warmer or drier, glaciers are likely to 'retreat'. In the last 100 years, most glaciers have been shrinking rapidly.

How glaciers move

Despite being solid and hard, glacier ice can flow like thick custard. This happens in two ways:
◆ When individual ice crystals slip across each other.
◆ When, in warmer conditions, whole glaciers, or parts of them, move by sliding. This is possible when there is water at the base of the glacier. The water acts as a lubricant, allowing movement to occur.
　Glaciers vary in their speed of movement, from 3 m to 300 m a year. Speed depends not just on temperature at the base of the ice, but also on gradient and the amount of ice produced in the accumulation zone. In addition, different parts of a glacier move at different rates. One effect of this is to wrinkle the surface of the ice into great cracks, or **crevasses**.

2 Glacial landscapes and processes

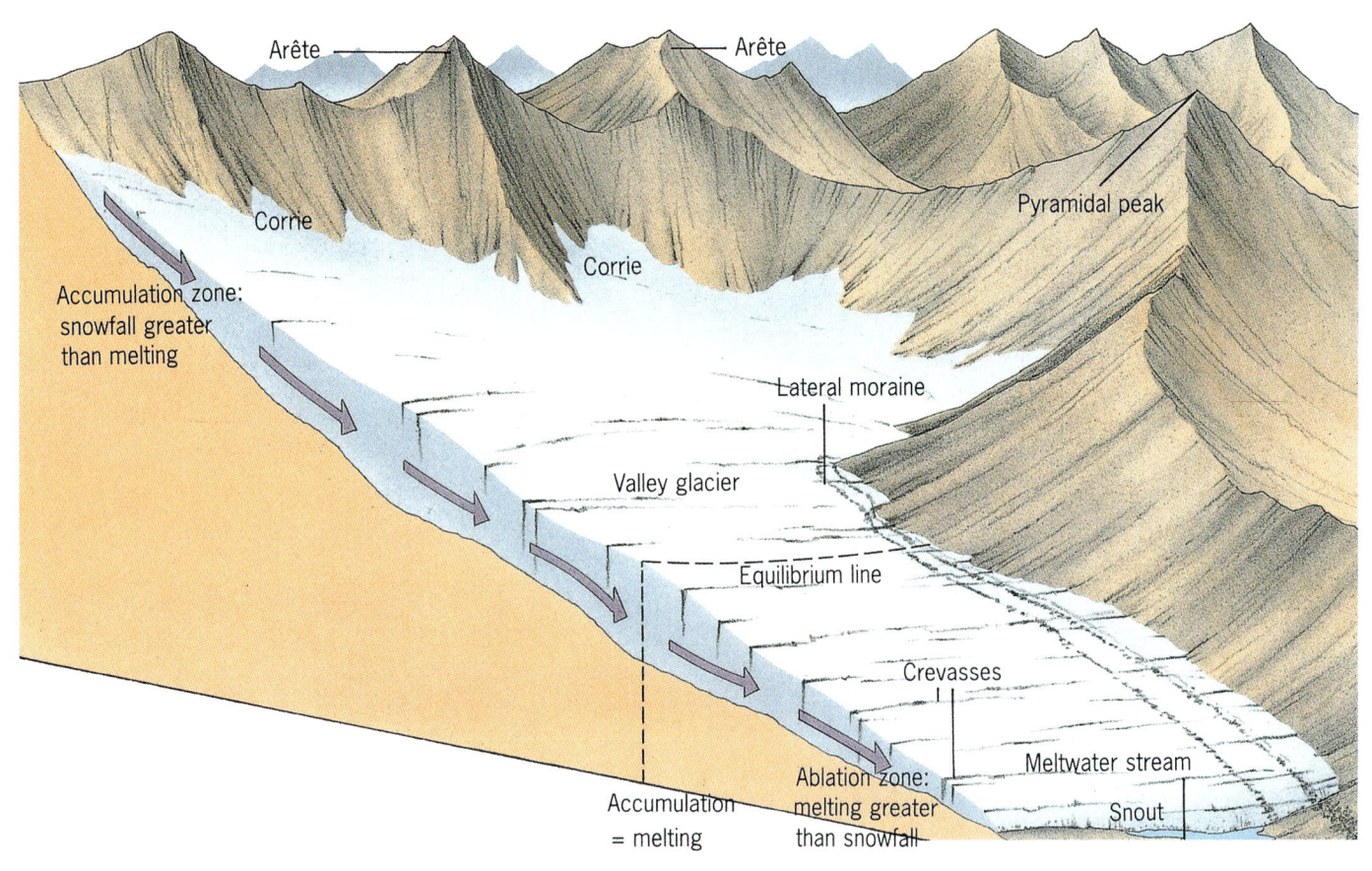

Figure 2.6 (above) A valley glacier

Figure 2.7 Rhône glacier, Switzerland, c. 1880

Figure 2.8 Rhône glacier, Switzerland, 1995

2.4 Glaciers as land-shaping agents

Today we live in an interglacial between ice ages. But throughout northern Europe, in both uplands and lowlands, the imprint of the last ice age is all around us. Sometimes the evidence is so clear that it's as if the ice melted only yesterday. This section is about that evidence: the landforms of glacial erosion and glacial deposition.

EXERCISES

5 Study Figures 2.7 and 2.8.
a Describe how the Rhône glacier has retreated over the last 100 years.
b* Suggest possible reasons for this retreat.

2 Glacial landscapes and processes

> **REMEMBER**
> Glaciers advance when input of snow exceeds the rate of melting. They retreat when melting exceeds snowfall. However, it often takes many years before a glacier responds to changes in inputs and outputs, and advances or retreats.

Of all the forces that shape the landscape, none has a more dramatic effect than glacier ice. Glaciers are like powerful earth-moving machines. They excavate rocks in the uplands, transport rock debris like giant conveyor belts and finally dump it in the lowlands. By such activities, glaciers change the landscape. The processes that cause such change are listed in Table 2.2.

Table 2.2 Processes of weathering and erosion in glaciated uplands

Example	Process
Freeze-thaw weathering	Water collects in rock crevices and joints. As the water freezes, it increases in volume and puts pressure on the rock joints. Continuous freezing and thawing of the ice causes pieces of rock to break off from the valley slopes.
Abrasion	Rock fragments carried in the glacier grind against the valley sides and floor as the glacier moves. This wears away and erodes the land surface.
Quarrying/plucking	Ice freezes on to rock outcrops. As the glacier moves forwards it pulls away pieces of rock.

CASE STUDY

2.5 Landscapes of glacial erosion: Snowdonia

Snowdonia (Fig. 2.9) in North Wales is a rugged upland made of hard volcanic rocks. During the last two million years ice sheets and glaciers have carved this upland into spectacular shapes. Landforms of glacial erosion, including corries, arêtes, glacial troughs (**U-shaped valleys**), hanging valleys and ribbon lakes dominate the mountain scenery.

Corries

Glaslyn, nestling below the summit of Snowdon (Figs 2.10, 2.11) is a perfect example of a **corrie** (also known as a cirque, or a cwm in Wales). A corrie is rock basin surrounded on three sides by steep rock walls. Today, a small circular lake fills Glaslyn. However, for most of the last 100 000 years,

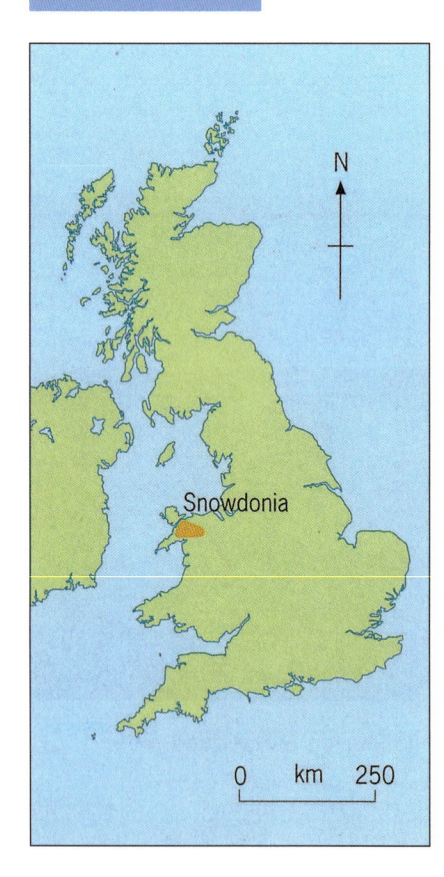

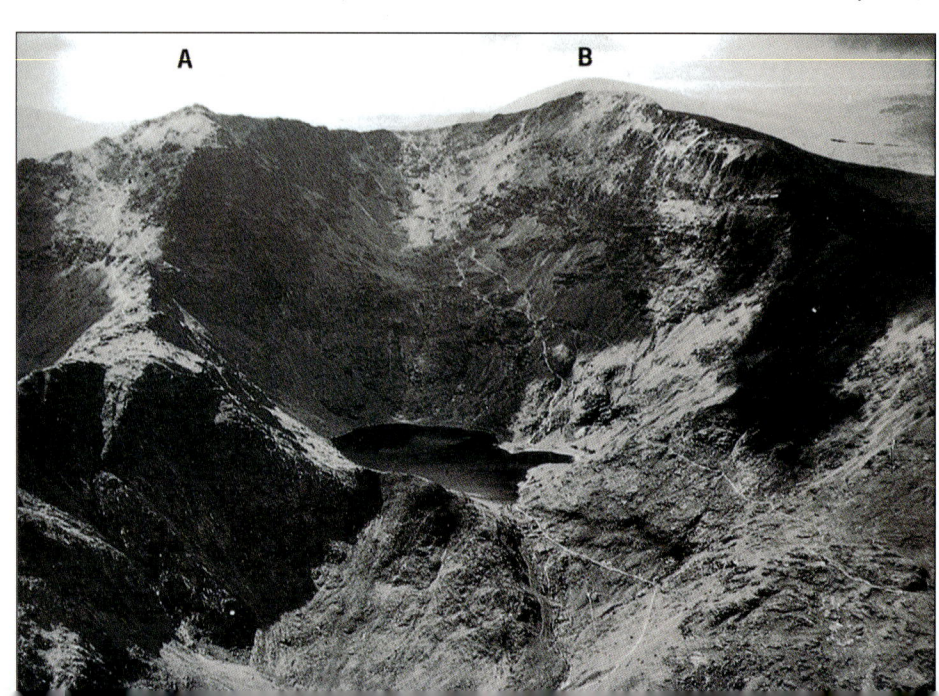

Figure 2.9 (above) Location of Snowdonia

Figure 2.10 (right) Aerial view of Glaslyn and Snowdon, Wales

2 Glacial landscapes and processes

Figure 2.11 Snowdonia area, Wales
© Crown copyright

Table 2.3 OS skills: how to describe a glaciated upland

1. What is the average altitude of the highest ground?
2. What is the average height range between the summit ridges and the valley floors?
3. How steep are the slopes?
4. Is there evidence of cliffs (or crags) and rocky outcrops on the valley slopes?
5. In which direction do the main glacial valleys run?
6. How straight are the glacial valleys?
7. Do the glacial valleys have hanging tributary valleys, ribbon lakes and truncated spurs?
8. Do the valley heads contain corries and corrie lakes?
9. Do the corries face in a preferred direction?
10. Are there any knife-edged ridges (arêtes) separating the glacial corries?

EXERCISES

6 Study Figures 2.10 and 2.11.
a Give the six-figure grid reference of the camera and the direction it was facing.
b Name the summits A and B (Fig. 2.10) and the three corries surrounding the summit of Snowdon (Fig. 2.11).
c* Draw a sketch map of the area covered by eastings 59–64 and northings 52–56 (Fig. 2.11). Use symbols and labels to show these features: corries, corrie lakes, arêtes, steep rocky outcrops.
d* Write a general description of the glacial upland in Fig. 2.11 by answering the questions in Table 2.3.

---

a glacier filled the bowl of Glaslyn. It was this glacier that gave the corrie its distinctive shape.

Formation of corries

During the ice age, prevailing westerly winds swept snow from Snowdon's summit and piled it up on the eastern (leeward) slopes. Over many winters, the snow gradually turned to ice, which then formed a small glacier. As the glacier grew, it began to slide and flow downslope. Rock fragments frozen

2 Glacial landscapes and processes

into the ice scraped and eroded the sides of the mountain – a process called **abrasion** (Fig. 2.12, Table 2.2.). The Glaslyn glacier also eroded its basin by **quarrying** (plucking). This occurred as meltwater at the base of the glacier froze on to rocks. As the glacier moved forwards it dragged, or 'quarried' loose blocks from the corrie floor. Slowly, the processes of abrasion and quarrying enlarged the Glaslyn hollow to its distinctive circular shape (Fig. 2.10).

Meanwhile, on the exposed rock walls above the glacier, freeze-thaw weathering shattered the rocks and helped the corrie to eat into the Snowdon ridge. Then, as weathered rock fragments showered on to the glacier, they became new tools for further abrasion.

Pyramidal peaks and arêtes

Corries also developed on Snowdon's southern and western slopes. These, too, cut back into the mountain and helped re-shape Snowdon's summit into a low **pyramidal peak**. Similar changes occurred on the two ridges – Crib Goch (Fig. 2.14) and Y Lliwedd – leading up to Snowdon. Glacial

Figure 2.12 (above) Rocks smoothed by ice abrasion, at the snout of the Athabasca Glacier, Alberta, USA

Figure 2.13 (right) Cross–section through a corrie during the ice age and today

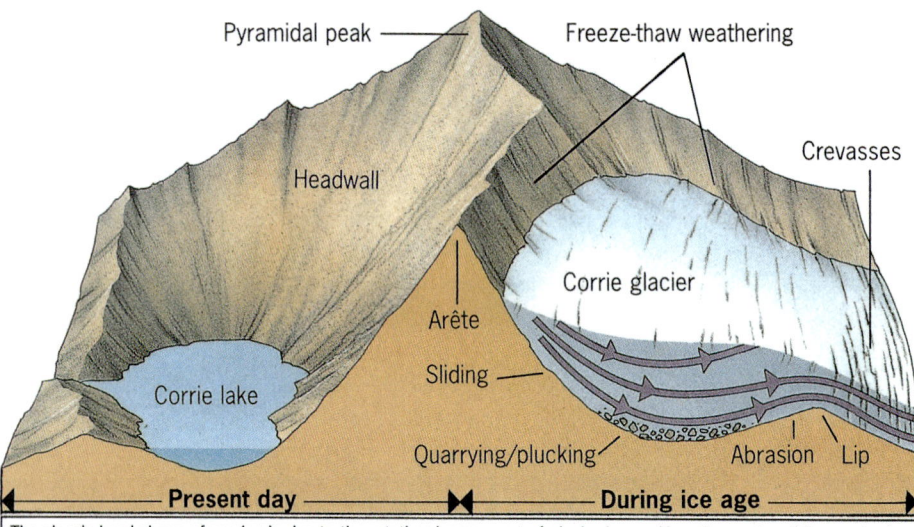

Figure 2.14 (below left) Crib Goch, Snowdonia

Figure 2.15 (below right) Blea Water Tarn, Cumbria – the largest corrie in the English Lake District

The classic bowl shape of corries is due to the rotational movement of glacier ice: • Near the centre of the corrie, ice flow is directed downwards to the rock surface. This gives maximum erosion by plucking and abrasion and leads to over-deepening. • Near the lip, the flow is towards the glacier surface. Erosion is less important here. • Plucking causes retreat of the headwall, which is steepened by freeze-thaw weathering.

2 Glacial landscapes and processes

erosion and freeze-thaw weathering reduced them to narrow, sharp-edged ridges known as **arêtes**.

Glacial valleys

As the Glaslyn glacier grew, it spilled out of its corrie and flowed downslope. It joined ice from the adjacent Llydaw corrie to form a valley glacier. The glacier then flowed around Crib Goch to merge with ice moving north through Nant Peris (Fig. 2.11).

Nant Peris has all the features of a classic glacial valley:

◆ A U-shaped cross-section, with a flat floor and steep sides. Before the ice age, Nant Peris was a V-shaped river valley (Fig. 2.16). Glacial erosion widened, deepened and transformed the valley to its present U-shape.
◆ Hanging tributary valleys. Glacial erosion was so effective that smaller tributary valleys were left high above Nant Peris. Many of these **hanging valleys** (Fig. 2.16) contain streams, which tumble down to Nant Peris in a series of spectacular waterfalls.
◆ A straight plan-form. The Nant Peris glacier sliced through interlocking spurs like a giant bulldozer, leaving them beheaded or **truncated** (Fig. 2.16).
◆ An uneven long profile. Where the valley floor steepened, or where the main glacier was joined by tributary glaciers, erosion increased. As a result, the glacier carved rock basins on the valley floor. The **ribbon lakes** of Llyn Padarn and Lyn Peris occupy these rock basins today.

Roches moutonnées

Roche moutonnées are common erosional features in glacial valleys. They are small outcrops of resistant rock that have been smoothed and steepened by glaciers (Fig. 2.17). The slope facing up-valley has been smoothed by abrasion. In contrast, the steeper down-valley slope has been formed by plucking and the removal of rocks along joints and bedding planes.

Figure 2.16 The development of glacially eroded landscapes

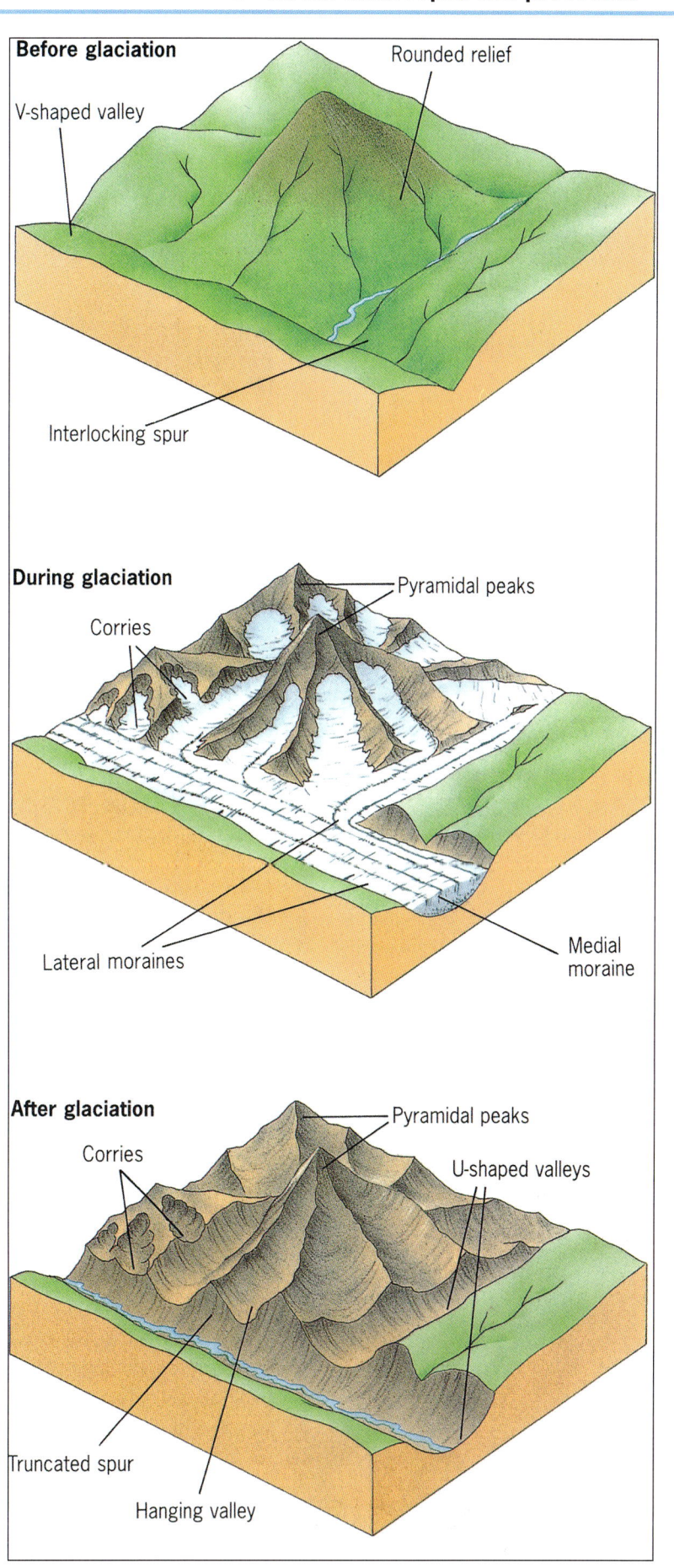

2 Glacial landscapes and processes

EXERCISES

7 Study Figure 2.11.
a Using evidence from Figure 2.11, make a table to describe the differences between a glacial valley (Nant Peris or Pass of Llanberis) and an upland river valley (Fig. 1.11). In your table use the following headings: size; plan; cross-sectional shape; long profile; tributaries.
b* Suggest two reasons why glaciers produce different shaped valleys to rivers.
c Make a list of the features in Figure 2.11 that suggest that Snowdonia is important for recreational activities.
d* Explain how these recreational activities might be linked to the glaciation of Snowdonia.

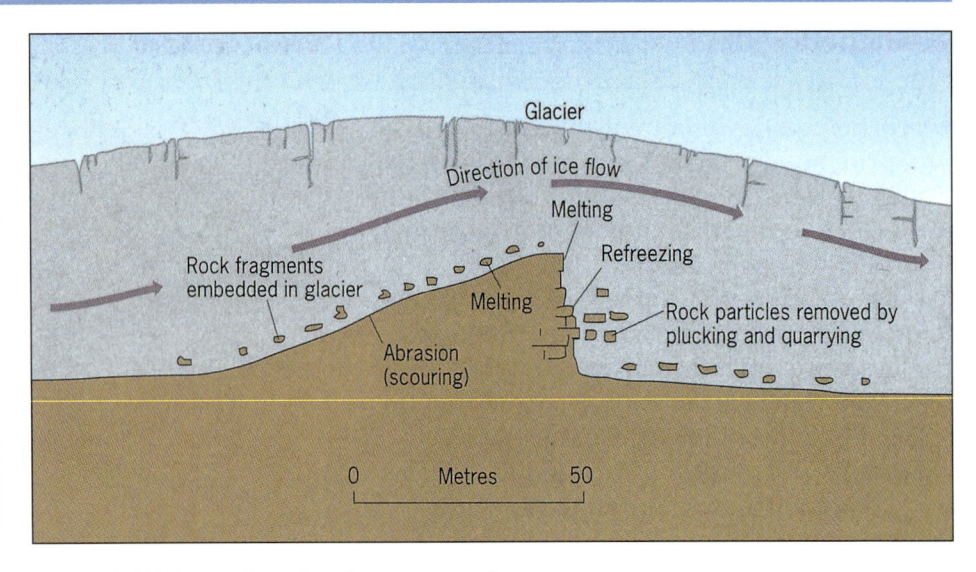

Figure 2.17 Formation of roches moutonnées

REMEMBER
Rock debris deposited by glaciers is a random mixture of boulders, sand and clay known as till. In contrast, debris deposited by meltwater is sorted by size in layers.

2.6 Landscapes of glacial deposition

Glaciers and ice sheets transport huge amounts of debris ranging from boulders as big as houses to the finest rock particles. Where does all this debris come from? There are two main sources. Some comes from glacial erosion: the rest from rockfalls. Freeze-thaw weathering on steep valley slopes feeds glaciers with a continuous supply of rock fragments. Meanwhile, rock avalanches, caused by glaciers undercutting valley slopes (although less frequent) also provide much rock debris.

Moraines

Glaciers and ice sheets eventually dump all the rock debris they carry. We refer to this material as **till** or boulder clay. It is easy to recognise by its unsorted appearance: a random mixture of boulders, rocks, sand and clay (Fig. 2.18).

Most glacial deposition takes place in lowland areas. In the Arfon Lowlands in North Wales (Fig. 2.19), ice sheets from Snowdonia and the Irish Sea plastered the landscape with a thick layer of till – so thick that it has completely buried the old landscape. **Till plains**, like those at Arfon, produce fertile soils and are valuable for both arable and livestock farming.

Till also forms the smaller scale features of mounds and ridges, which we call moraines.

- **Lateral moraines** are carried on the surface of glaciers (Fig. 2.20). They result from rockfalls from valley slopes. However, if a glacier shrinks and

Figure 2.18 Till at Dungeon Ghyll, Cumbria

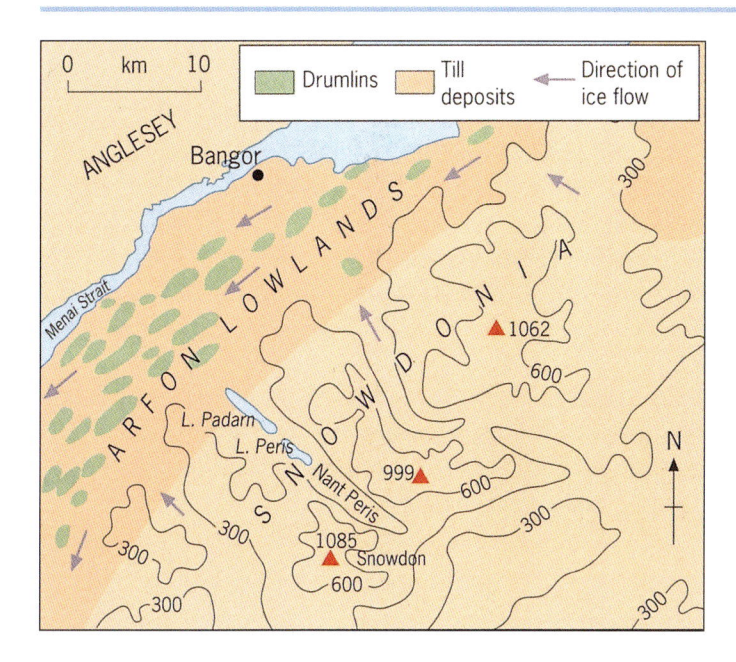

2 Glacial landscapes and processes

Figure 2.19 (above left) Glacial deposition in the Arfon Lowlands, North Wales

Figure 2.20 (above right) La Mer de Glace, France (note the well-developed lateral and medial moraines)

retreats back up its valley, these lateral moraines are left as low ridges along the valley side (Fig. 2.21).
- **Medial moraines** develop when two valley glaciers meet and their lateral moraines join together.
- **Terminal moraines** form from material deposited at the front of an ice sheet or glacier. Ice sheets can build up huge terminal moraines. For example, at the end of the last ice age, the great Scandinavian ice sheet remained stationary for several years over central Jutland, Denmark. The result was a terminal moraine up to 180 m high and 100 km long! However, terminal moraines deposited by valley glaciers are much smaller features. They often form arc-like ridges that stretch from one side of a glacial valley to the other (Fig. 2.22).

Figure 2.21 (left) Lateral moraine near the snout of the Nisqually Glacier, Washington, USA. The moraine forms a well defined ridge running along the right side of the valley.

Figure 2.22 (above) Deposits at the snout of a valley glacier

33

2 Glacial landscapes and processes

Drumlins

Drumlins are elongated hills made of till, which usually occur in large numbers or 'swarms' (Fig. 2.23). They are rounded and blunt at one end, while the other end is longer and tapering (Fig. 2.24). This gives them the shape of half an egg in cross-section. Their streamlined shape suggests that they were formed by moving ice. The drumlins of the Arfon Lowlands have a north-east to south-west alignment, which is parallel to the direction of ice flow (see Fig. 2.19).

Erratics

Ice sheets often transport rocks hundreds of kilometres and deposit them in completely different geological areas. These rocks, which stand out as being 'foreign', are called erratics (Fig. 2.25). For example, we can find erratics made from a granite that crops out only on Ailsa Craig in western Scotland, as far away as West Wales and South-east Ireland (Fig. 2.26).

Figure 2.23 Drumlins, Brampton, Cumbria

Figure 2.24 (right) Plan and profile of a drumlin

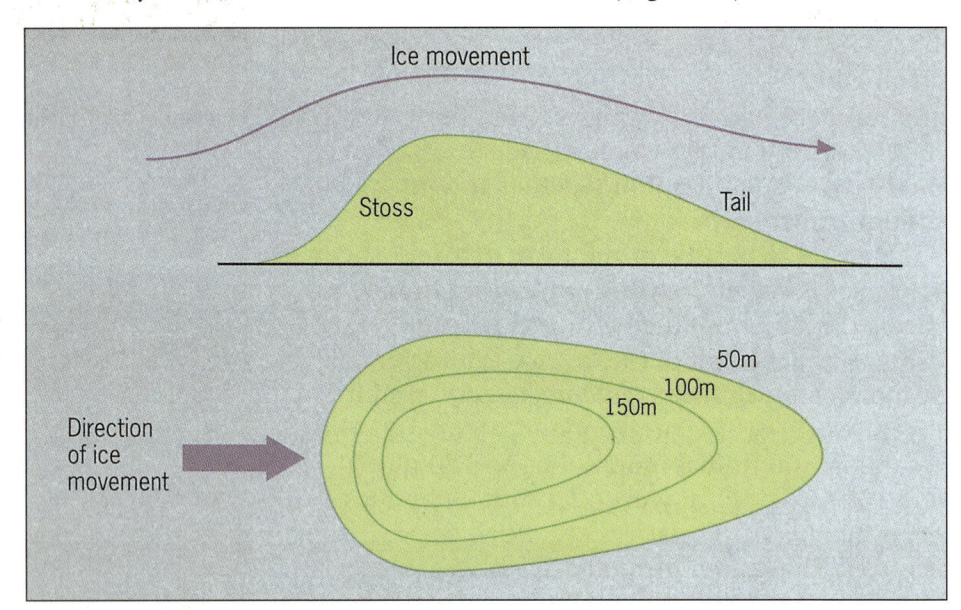

EXERCISES

8 Study Figure 2.23.
a In which direction did the ice sheet flow to form the drumlin? Explain your answer.
b* How can the study of erratics help geographers to understand the movement of glaciers and ice sheets?

Figure 2.25 (below right) Glacially transported limestone boulders, Kingsdale, North Yorkshire

Figure 2.26 (below left) Distribution of Ailsa Craig erratics

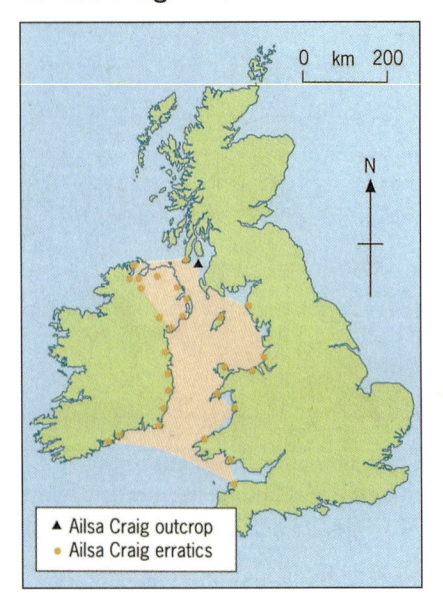

2 Glacial landscapes and processes

Meltwater deposits

As the ice age drew to a close, glaciers and ice sheets began to melt rapidly. Powerful streams swollen by meltwater flowed within, beneath and on the surface of glaciers. These streams could transport huge loads of boulders, gravel and sand. As a result, when the streams deposited these sediments, they created new landforms. We call these **meltwater deposits**, and they include outwash plains, kames, kame terraces and eskers (Fig. 2.27).

Except for outwash plains, all of these meltwater deposits were originally sediments piled up against the walls of glaciers and ice sheets. When the ice finally melted, and the deposits lost their support, they slumped to the valley floor. Their appearance today is largely due to this slumping process.

2.7 The human use of glaciated uplands

Table 2.4 The human use of glaciated uplands

Type of use	Description and examples
Recreation and tourism	Spectacular glaciated scenery e.g. corries, arêtes, U-shaped valleys, hanging valleys, waterfalls, ribbon lakes, etc. Opportunities for sight-seeing, rock climbing, scrambling, walking, mountain biking, etc. Winter tourism in the Alps and Scotland based on skiing. Summer tourism based on sight-seeing and water-based recreation on the many ribbon lakes. Many glaciated uplands in the UK, Europe and North America are protected as national parks.
Hydroelectric power	Glaciated uplands are important for hydroelectric power (HEP) in France, Switzerland, Austria, Norway and Sweden. Glaciated landforms favour HEP generation (e.g. over-deepened U-shaped valleys and hanging valleys provide high 'heads' of water and massive HEP potential). Ribbon lakes and corrie lakes are deep, natural storage reservoirs. High precipitation and permanent snowfields provide plenty of water for river flow all year round.
Farming	Steep slopes, cold and damp climate, and thin soils severely limit farming. Only extensive livestock farming (hill sheep) is viable in the glaciated uplands of the British Isles.
Water catchment	Heavy relief precipitation, deep glacial lakes for water storage, and low population densities give ideal conditions for water catchment. In the English Lake District, Haweswater and Thirlmere (both ribbon lakes) provide water for towns and cities in North-west England.

Glaciated uplands provide opportunities and impose limits for human activities (Table 2.4). On the one hand, their spectacular scenery, landforms and snow make them increasingly popular for recreation and tourism and ideal for hydroelectric power and water catchment. On the other hand, the low population densities and sparse settlement of glaciated uplands tell another story. With their harsh climate, thin soils and rugged relief, they provide few resources for farming.

> **EXERCISES**
> 9a What are the differences between moraines and meltwater deposits?
> b* Why do you think that meltwater streams were so powerful and able to transport such large amounts of sediment?

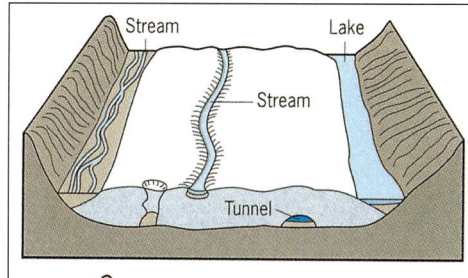

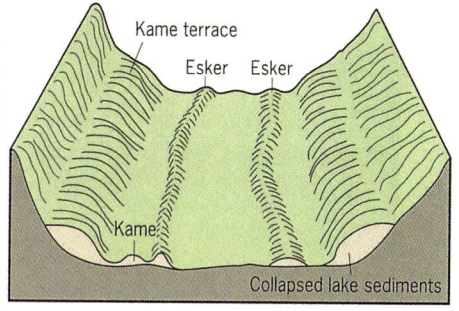

Kame terrace: sediments deposited between the edge of a glacier and valley side by meltwater streams.

Esker: long winding ridges of meltwater sediment (often many kilometres long), laid down in channels of meltwater streams flowing within and beneath the ice.

Kame: small, isolated mound formed from meltwater sediments which filled crevasses and caverns in the ice.

Collapsed lake sediments form Kame terraces.

> Outwash plains: extensive lowlands covered with coarse sand and gravel deposited by meltwater streams flowing from a large ice sheet.

Figure 2.27 Meltwater deposits

> **REMEMBER**
> Most glaciated uplands form spectacular landscapes. Conflicts often arise in glaciated uplands between economic activities (e.g. tourism, HEP) and conservation.

2 Glacial landscapes and processes

> **EXERCISES**
>
> **10a** How many months a year is the temperature below freezing on Ben Nevis (Fig. 2.29)?
> **b** Imagine that you were planning a skiing holiday in Scotland. From the evidence of Figure 2.29 which month would you choose? Explain your answer.
> **c*** Suggest two reasons why snow cover on mountains in the British Isles is often unreliable.

2.8 Skiing in Scotland

Scotland's first ski resort opened in the Cairngorms in 1961. Since then, skiing in Scotland has become big business. Today, there are five ski resorts (Fig. 2.28), which attract about 500 000 skiers a year. Skiing is important to the economy of the Scottish Highlands: it creates about 3000 jobs and is worth £30 million a year.

Commercial skiing in the British Isles has developed only in the Highlands of Scotland. Why is this? The simple answer is climate (see Figs. 2.29 and 2.5). Only in Scotland are there mountains high enough to give a reliable snow cover in winter. Even there, the ski slopes must be sited close to the summits (between 1000 m and 1200 m above sea-level): rarely do they descend much below 800 m.

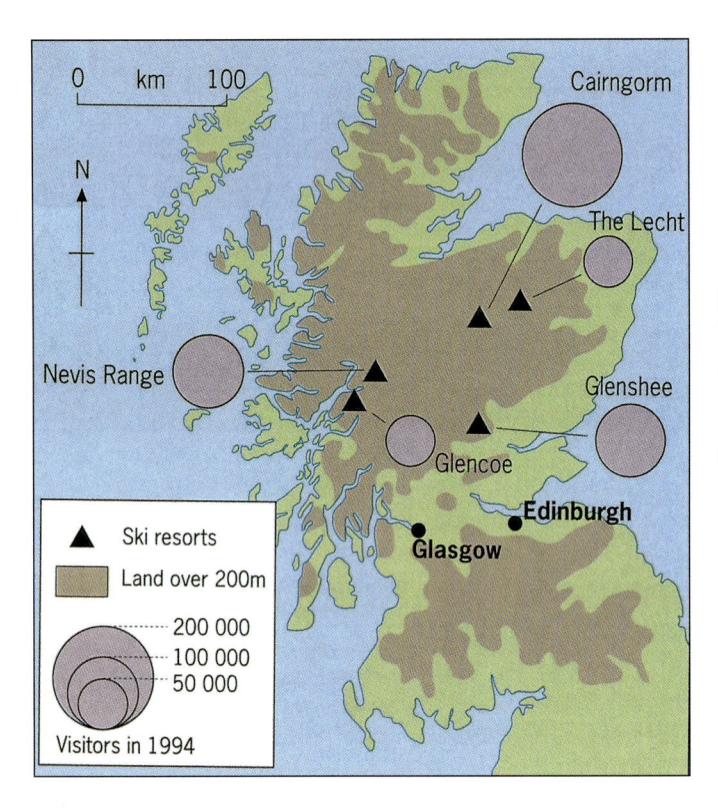

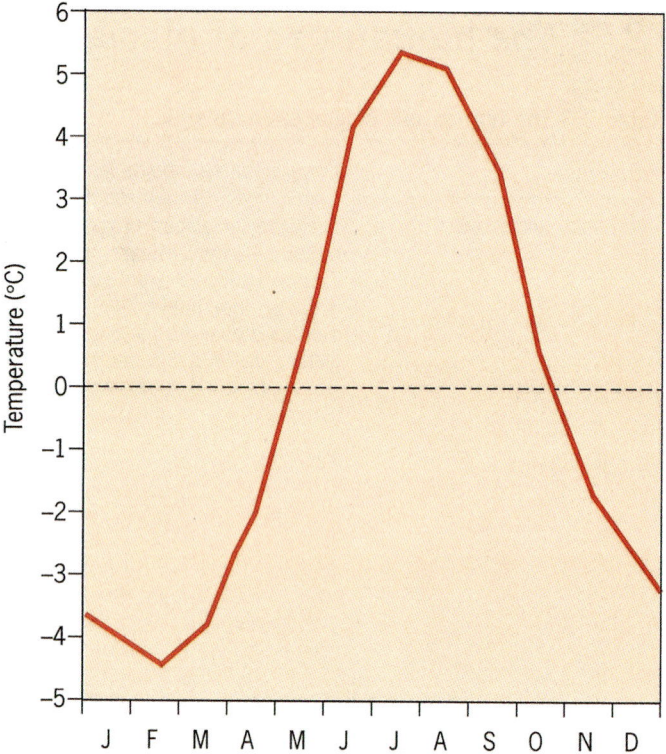

Figure 2.28 Scotland's skiing industry

Figure 2.29 Mean monthly temperatures on Ben Nevis

2.9 Skiing in the Cairngorms

Cairn Gorm is Scotland's oldest and most popular skiing centre (Fig. 2.28). Skiing began at Coire Cas in 1961 when developers built a new access road from Aviemore through Glen More to Coire Cas (Fig. 2.30). This opened up the Cairngorms' northern corries for commercial skiing. A second centre at nearby Coire na Ciste followed in 1974. Undoubtedly, Cairn Gorm ski centre has been highly successful. With more than 50 km of downhill runs, it attracts 200 000 skiers a year.

2 Glacial landscapes and processes

Aviemore (Figs 2.31 and 2.32), located 15 km from Coire Cas, has grown as the main resort for Cairn Gorm. During the 1960s and 1970s, a range of services and facilities, including hotels, guest houses, chalets, a swimming pool and a new shopping centre were built for tourists. As a result, Aviemore became one of the few year-round resorts in the UK. While winter visitors flock to the ski slopes, summer visitors come to roam the hills and ancient pine forests, and admire the scenery and wildlife (Figs 2.33–2.37).

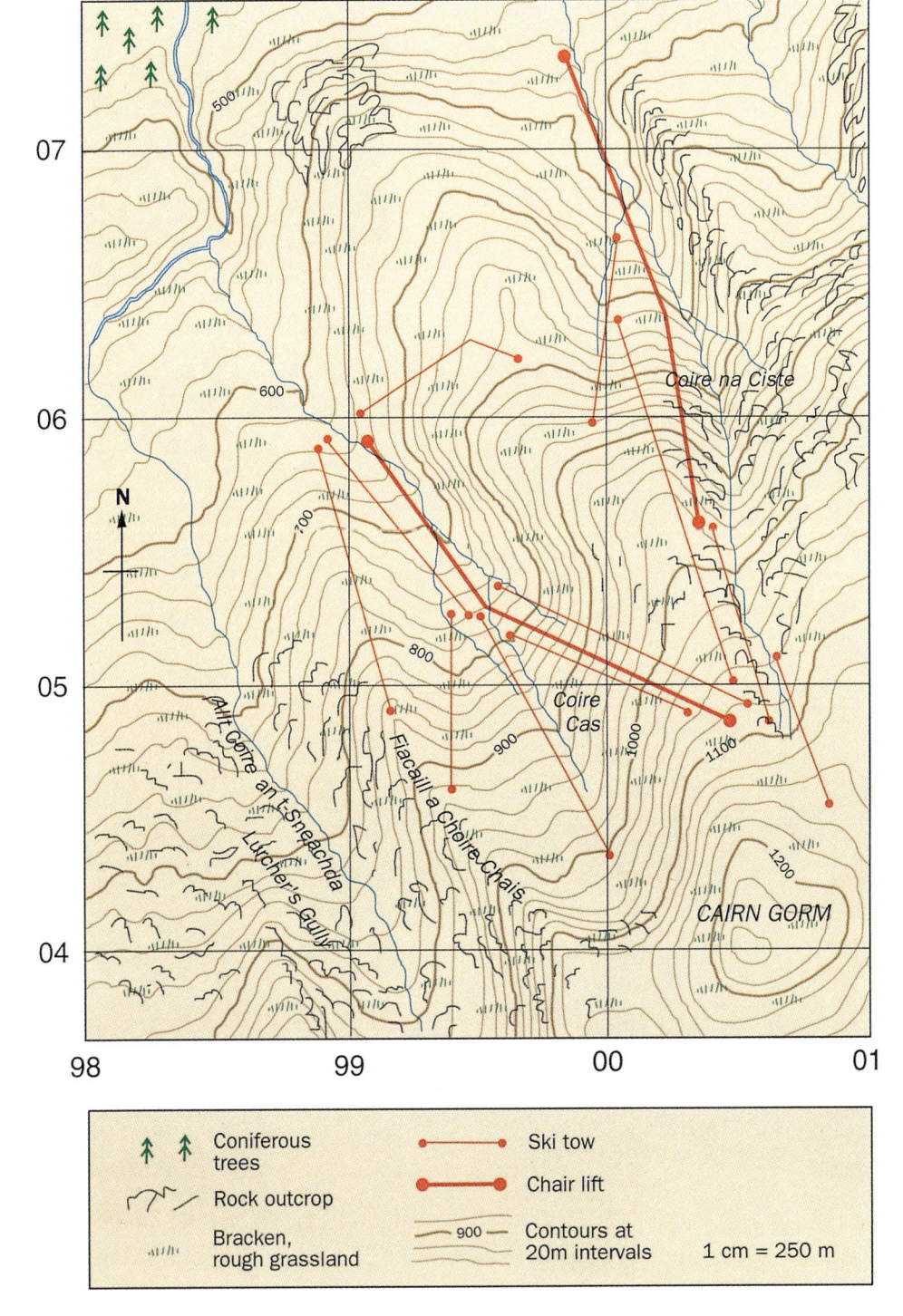

EXERCISES

11 With reference to Figure 2.30:
a Draw a cross-section between 000044 and 989059. Describe the shape of the slope and say how it favours downhill skiing.
b At approximately what altitude do • the highest ski runs start? • the ski runs end? Suggest reasons for the length of the main ski runs.
c Suggest how Aviemore's geographical situation might have contributed to its growth as a resort (Fig. 2.32).

REMEMBER
Glaciated uplands often support fragile ecosystems. Tourism can do permanent damage to these ecosystems and is often unsustainable.

Figure 2.30 (left) Cairn Gorm area (redrawn from Crown copyright, with permission)

Figure 2.31 (below) Aviemore

2 Glacial landscapes and processes

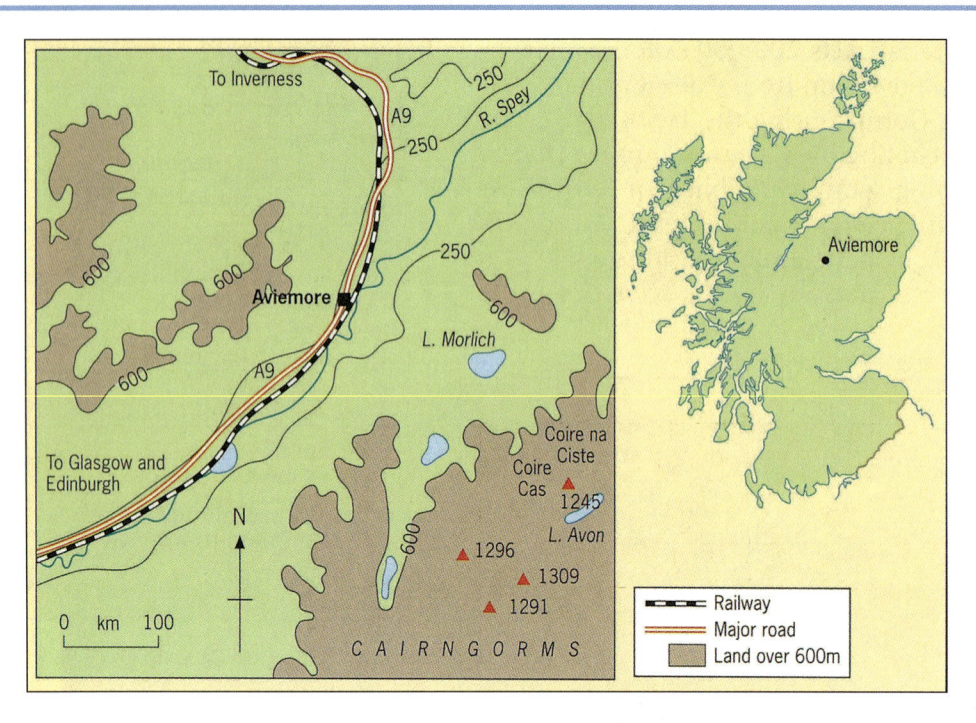

Figure 2.32 Aviemore's situation

Figure 2.33 (below left) Cairngorm plateau

Figure 2.34 Loch Avon: a ribbon lake in the Cairngorms

Figure 2.35 (below right) Cushion of moss campion

Figure 2.36 (bottom left) Ptarmigan

Figure 2.37 (bottom right) Mountain hare

38

2 Glacial landscapes and processes

Glaciated uplands in crisis

Glaciated uplands like the Cairngorms are under pressure. Such areas, which were once remote, are today threatened by the enormous increase in recreation and tourism. Hill walkers, climbers, mountain bikers, downhill skiers and others compete to use the upland environment. In the process, these people come into direct conflict with conservationists. In this section, we shall assess the impact of skiing and other recreational activities on the environment of the Cairngorms.

> **EXERCISES**
>
> **12** Study Figure 2.38 and describe the effects of skiing on the landscape.

Conflict in the Cairngorms

Skiing and conservation confront each other head on in the Cairngorms.

Conservationists argue that the effects of skiing, although localised, have been disastrous. The environment around the ski slopes has been badly damaged. At Coire Cas, builders used heavy machines to install uplift facilities (chairlift and ski tows). To do this, they had to bulldoze access roads, which have permanently scarred the landscape.

FACTFILE
- *The Cairngorm plateau, at 1100 m, is the largest area of high ground in the British Isles.*
- *The Cairngorms contain four of the UK's five highest mountains.*
- *The plateau has a sub-arctic climate, unique glacial landforms, and many rare plants and birds.*
- *The Cairngorms have special status as a conservation area: they are protected as a National Scenic Area and a National Nature Reserve.*
- *The international importance of the Cairngorms is recognised in their proposed designation as a World Heritage Site.*

Figure 2.38 Ski slopes at Coire Cas in summer

Figure 2.39 Cairn Gorm – the UK's fifth-highest mountain and site of the controversial funicular railway.

39

2 Glacial landscapes and processes

> **EXERCISES**
> **13** Read the paragraphs about 'Conflict in the Cairngorms' and 'The Cairn Gorm funicular railway', and Table 2.5. The Scottish Secretary has invited interest groups to submit their views on the planned Cairn Gorm funicular project. On behalf of either a conservation group or a group supporting the project (Table 2.5) write a letter to the Scottish Secretary giving and explaining your opinion on the project.

In addition, skiing in Scotland depends on drifted snow. To allow deep drifts to form, developers built snow fences and dug bulldozed tracks at an angle into the hillside. All of this spoils the area's natural beauty and causes damage to the environment (Fig. 2.38).

The Cairn Gorm funicular railway

In the late 1990s, hostilities between conservation and the skiing industry in the Cairngorms resurfaced. The Cairngorm Lift Company put forward a controversial plan to replace its ageing chairlift to the summit of Cairn Gorm, with a new funicular railway. The development also included a new visitor centre and restaurant. The cost of the scheme was £17 million, which included £9 million from public funds. The Scottish Secretary gave strong backing to the plan, which was finally approved in 1998. The arguments for and against the funicular are summarised in Table 2.5.

Environmental management

Damage to the environment of the Cairngorms is not caused only by skiing. The chairlift at Coire Cas reaches to within 200 m of the summit of Cairn Gorm. As a result, 60 times more people visit the summit of Cairn Gorm today than before the chairlift was built. This improved access led to people trampling vegetation, widening footpaths and causing unsightly erosion (Fig. 2.40).

Table 2.5 The issue of the Cairn Gorm funicular

Groups supporting the funicular	Arguments supporting funicular development	Groups against the funicular	Arguments against funicular development
Chairlift company Highland and Islands Enterprise Highland Regional Council Scottish National Heritage Scottish sports Council Scottish Tourist Board	• Ski areas cover only 1 per cent of the Cairngorms total area. • The funicular will double the number of skiers who can be transported to the summit of Cairn Gorm. • New infrastructure will increase visitor demand to an estimated 200 000 visitors in winter and 165 000 in summer. • Cairn Gorm will become an all-year-round resort. • 2500 jobs on Speyside indirectly depend on tourism. The funicular will generate new jobs and secure existing ones. • Speyside is one of the poorest regions in the UK. It is one of the few UK regions that qualifies for EU funding. New investment is essential to the region's economic future.	Worldwide Fund for Nature RSPB Scottish Ramblers' Association	• The further development of tourism is unsustainable. It will put added pressure on the environment and result in degradation. • In order to secure planning permission for the funicular, estimates of visitor numbers are wildly exaggerated. • The Chairlift company will provide only 112 full-time jobs. • The cost of creating each new job is £107 000 – ten times the usual cost. • The Cairngorms are a fragile arctic-alpine environment which is highly sensitive to visitor pressure. • The Cairngorms are unique in the British Isles, supporting many rare plants (e.g. arctic saxifrage) and animals (e.g. ptarmigan, dotterel, arctic hares). • The Cairngorms environment is of international importance: it is a candidate for World Heritage status (like the Grand Canyon and Yellowstone Park in the USA).

2 Glacial landscapes and processes

Sustainable tourism

If tourism and recreation in the Cairngorms are to be sustainable and not degrade the environment, careful management is needed. One possibility is zoning. This means that planners could set aside the most accessible areas for intensive recreation (e.g. skiing, camping, forest trails, picnic sites, information centres). Then, most visitors who want to walk in the hills could be directed to specially managed areas where badly eroded paths have been repaired, refuge huts have been closed and where there is some signposting. However, in the central core of the plateau there would be little or no management. Here, conservation would have absolute priority. Only a small number of visitors seeking the 'wilderness' experience would come this far.

Figure 2.40 Visitor pressure and environmental damage

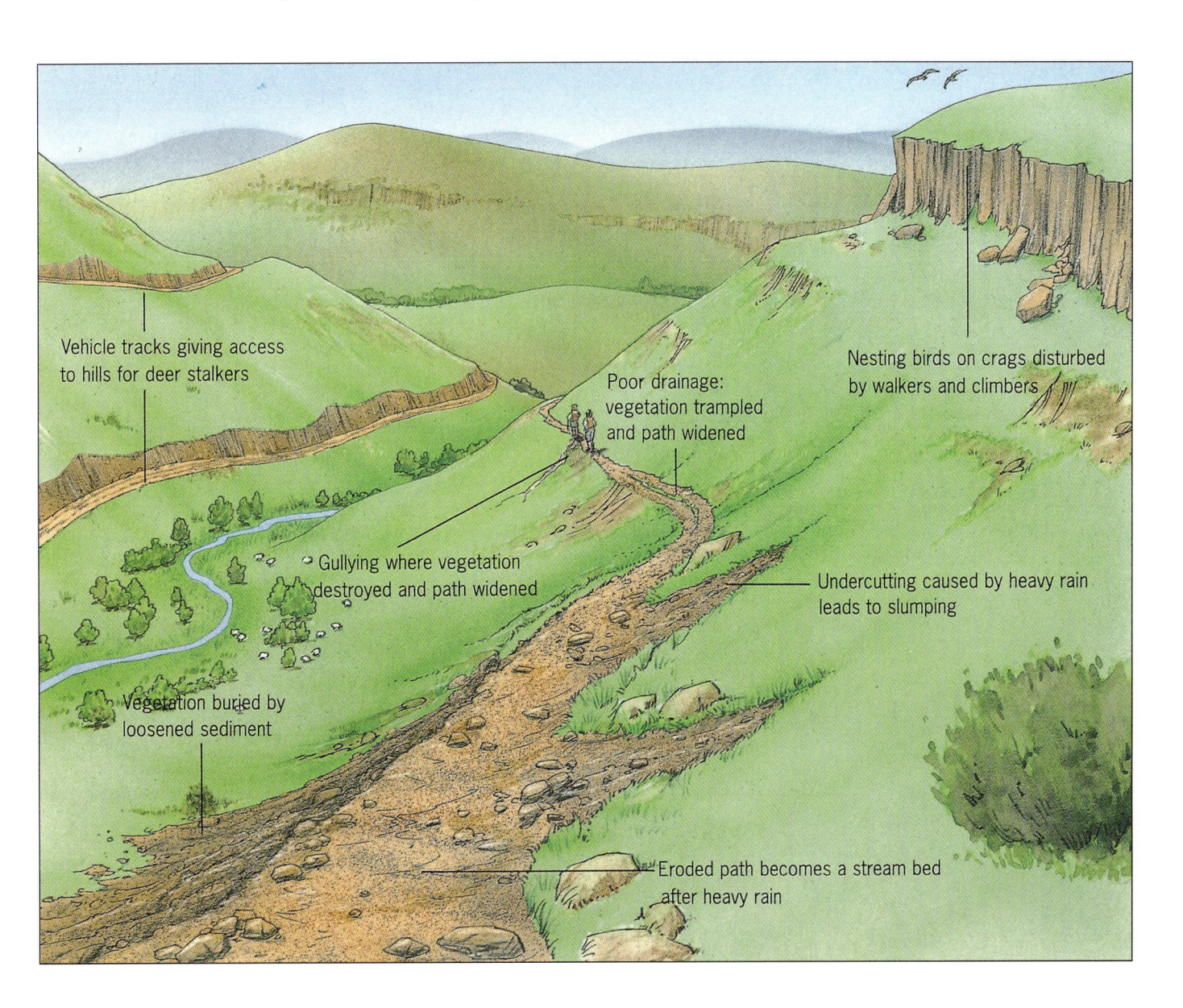

2 Glacial landscapes and processes

2.10 Summary: Glacial landscapes and processes

> **KEY SKILLS OPPORTUNITIES**
> **C1.2:** Ex. 6a, 6b, 7a, 9;
> **C1/2.3:** Ex. 6c, 6d, 7a, 12, 13; **C2.2:** Ex. 13; **N1/2.1:** Ex. 1, 2a, 5a, 8a, 10b, 11b, 11c; **N1/2.2:** Ex. 3, 4b, 10a; **N1/2.3:** 4c.

Key ideas	Generalisations and detail
For most of the last two million years, northern Britain and Ireland have been covered by ice sheets and glaciers.	• The last glaciers in the British Isles finally melted 10 000 years ago. This ended an ice age that had lasted for nearly 100 000 years. • The last ice age had an enormous effect on both upland and lowland landscapes in the British Isles.
Glaciers are systems.	• Glaciers are systems with inputs of ice from snowfall, and outputs through melting. • Glacier ice forms in the accumulation zone above the snowline, where snowfall exceeds the rate of melting. Any increase in accumulation of snow and ice causes a glacier to advance. Glaciers retreat when accumulation decreases. Currently, glaciers are retreating everywhere.
Glaciers erode the landscape and create new landforms.	• Glaciers erode by abrasion and quarrying (plucking). • Erosion is concentrated in upland areas. • Landforms of glacial erosion include corries, arêtes, U-shaped valleys, hanging valleys and ribbon lakes. They are best developed in uplands formed from hard, resistant rocks, e.g. Snowdonia.
Glaciers transport huge amounts of rock debris which they eventually deposit to create new landforms.	• The rock debris transported by glaciers is called till. • Deposition of till often forms low mounds and ridges known as moraine. • Landforms of glacial deposition include till plains, moraines (lateral, medial, terminal) and drumlins. These features are best developed in lowlands, e.g. Arfon Lowlands.
Glacial meltwater produces a variety of depositional landforms.	• Powerful meltwater streams flow on, within, and under ice sheets and glaciers. • Meltwater streams transport large amounts of rock debris. • Deposition of this material results in meltwater features: outwash plains, kames, kame terraces and eskers.
Glaciated uplands provide resources for a range of human activities.	• The main human activities in glaciated uplands are: recreation and tourism; hydroelectric power; hill sheep farming; and water catchment. • Recreation and tourism in the uplands are growing rapidly. • Farming in the uplands is declining.
Use of glaciated uplands can lead to conflict.	• Glaciated uplands are fragile environments with unique landforms and wildlife. • Recreation and tourism increasingly conflict with conservation in the uplands, (e.g. skiing causes significant, though localised, environmental damage). Apart from unsightly uplift facilities, skiing (and hill walking and climbing) destroys vegetation, causes erosion and footpath widening, and disturbs wildlife, e.g. in the Cairngorms.

Fig 2.41 Snout of a glacier fed by the Vatnajökull ice cap, Iceland

3 Coastal processes and landforms

3.1 Introduction

Coasts are dynamic places: they change rapidly. Think of your favourite beach. In summer, it's probably steep and sandy. But in winter, it will be battered by storm waves, which flatten beaches and remove most of the sand, leaving instead a rubble of cobbles and shingle. Such rapid change reminds us that waves differ from rivers and glaciers as land-shaping agents.

- Rivers do most of their work during floods, which may occur just once or twice a year.
- Glaciers have enormous power to erode and transport, but only when the climate is in deep freeze.
- Waves erode land and shift sediment along the coast continuously.

Coasts and people

Waves are responsible for most of the landforms on the coast. Often these landforms are important resources for people and wildlife (Figs 3.1–3.4). Coastlines are a magnet for settlement, and many of the world's largest cities occupy coastal sites. But coasts, like flood plains, are also hazardous places. This is especially evident where rapid erosion is taking place and where coastal lowlands are at risk from flooding.

Figure 3.1 Paignton, Devon

EXERCISES

1a Describe how people use coasts in Figures 3.1 to 3.4.
b Suggest any problems that might arise from these different uses.
c Using an atlas, find the location of the following major world cities: Buenos Aires, Rio de Janeiro, Mexico City, New York, Los Angeles, London, Paris, Moscow, Cairo, Johannesburg, Calcutta, Bombay, Delhi, Beijing, Hong Kong, Singapore, Tokyo, Seoul, Sydney.
d List these cities in a table and tick those that have a coastal location.

Figure 3.2 (above left) Milford Haven oil terminal, Wales

Figure 3.3 (bottom left) Heavy industries on the Mersey Estuary, Cheshire

Figure 3.4 (bottom right) Extracting sand and gravel from ancient beach deposits, Dungeness, Kent

3 Coastal processes and landforms

Figure 3.5 The coastal system

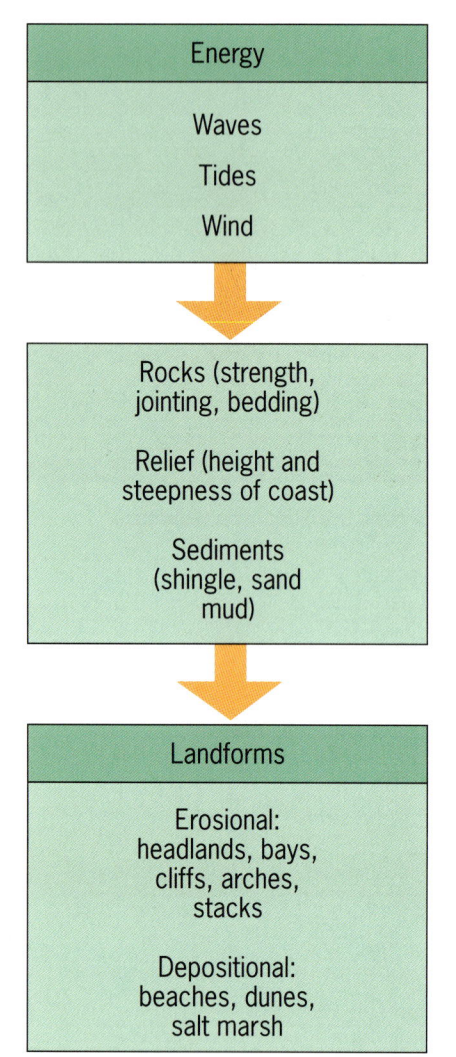

Energy
Waves
Tides
Wind

↓

Rocks (strength, jointing, bedding)
Relief (height and steepness of coast)
Sediments (shingle, sand mud)

↓

Landforms
Erosional: headlands, bays, cliffs, arches, stacks
Depositional: beaches, dunes, salt marsh

3.2 The coastal system

The coast is the narrow zone where land meets sea. We can understand the coastline better if we think of it as a system (Fig. 3.5). A system is simply a group of objects that are linked together by flows of energy and materials (see Book 1, Chapter 4). Waves, tides and wind (the inputs) provide the energy that drives the coastal system. This energy does two things: it erodes land and transports sand, shingle and mud. The end result (or output) from the coastal system is landforms such as headlands, cliffs, beaches and dunes.

The first part of this chapter is about coastal landforms. But first we need to look more closely at the main energy input to the coastal system: waves.

3.3 Waves

Waves are movements of energy through water. They are caused by the wind. As wind blows across the sea, friction (between the wind and the sea's surface) leads to turbulence. Downward gusts of wind depress the sea surface to form wave troughs. In contrast, upward air movements allow the surface to rise to form wave crests (Fig. 3.6).

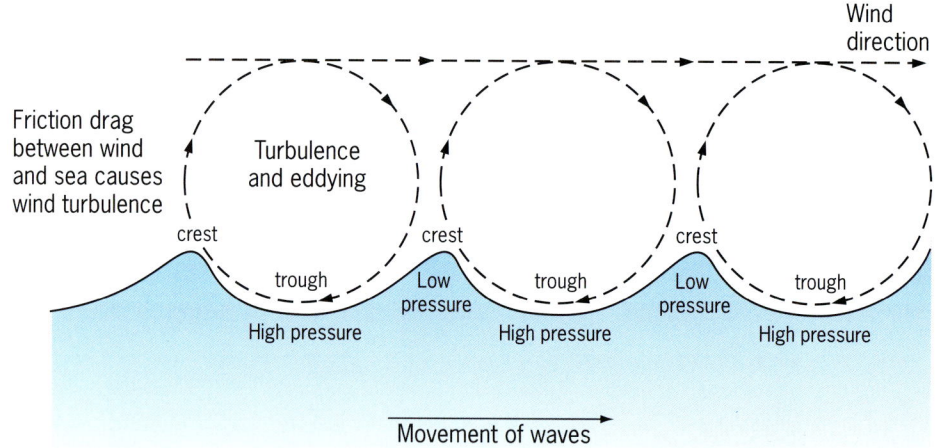

Figure 3.6 Formation of waves

> **REMEMBER**
> Waves on sand and shingle beaches may be constructive or destructive. Constructive waves add sediment to beaches; destructive waves remove sediment.

Wave height and length

The amount of energy in waves depends on their height. Three factors influence wave height and wave energy.
- Wind speed. When the weather is very windy, damaging storm waves crash against the coastline.
- Wind duration – the length of time a wind blows. The wind does not have to be strong to generate large waves. A moderate wind blowing steadily for several days can also produce high waves.
- Fetch – the distance of open sea over which a wind blows. Where the fetch is very long (e.g. in a westerly direction from the coast of North Cornwall), large waves can occur. Where it is short, or the coastline is sheltered (e.g. Morecambe Bay) waves are generally much smaller.

3 Coastal processes and landforms

Movement of waves

If you watch a wave travelling across the sea's surface, it is easy to think that the water in the wave is moving forwards. In fact, what you are seeing is a movement of energy. The water particles, rather than moving forwards, follow a circular orbit (Fig. 3.6). However, when waves reach the coast things start to change. In shallow water, waves begin to 'feel' the sea bed and slow down. This causes an increase in wave height, until eventually the wave becomes unstable and breaks. Only then does the water itself move forwards as it surges up the beach as the **swash**, and returns to the sea as **backwash**.

Table 3.1 Wind direction, fetch and wave height

	Wind direction	Max. fetch (km)	Wave height (m)
Blackpool	W	207	5.2
Bridlington	E		
Dover	S		
Cape Wrath	NW		
Aberdeen	NE		

Constructive and destructive waves

Breaking waves may be either **constructive** or **destructive**.

- Constructive waves are only a metre or so high and have low energy. They are most common in summer, or in spells of calm weather. These waves gently push sand and shingle onshore and form steep beaches.
- Destructive waves are quite different. They are powerful storm waves, which may be five or six metres high. They move sand and shingle offshore and create flat beaches. Along cliffed coastlines, destructive waves have enough power to erode even the toughest rocks.

EXERCISES

2 We can calculate the effect of fetch on wave height using the following formula:
$$H = (\sqrt{F}) \times 0.36$$
where: H = wave height in metres.
F = fetch in kilometres.
a Use an atlas to find • the fetch • the wave height at each location in Table 3.1.
b Study an atlas map of the British Isles and North Atlantic. Then, on an outline map of the British Isles, shade in the coastlines that, on the basis of fetch, are likely to have high, medium and low wave energy.
c* Suggest how the distribution of wave energy might affect landforms on different stretches of coastline.

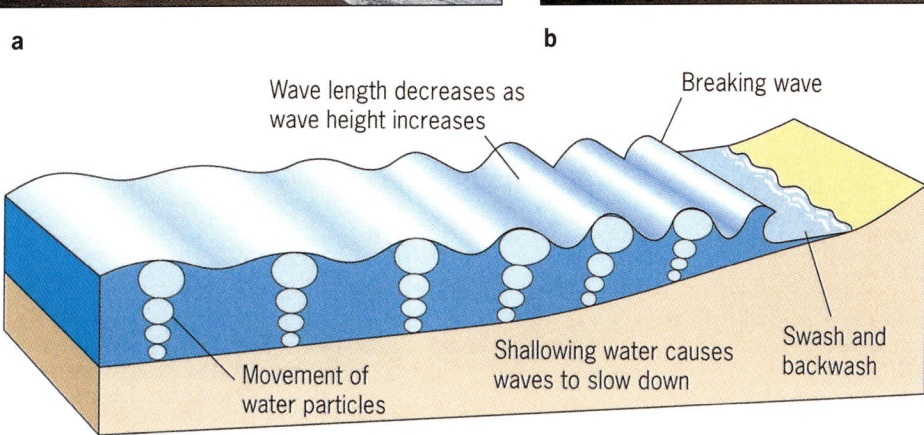

a

b

Figure 3.7 (above) Swash (a) and backwash (b) on the Californian coast at Point Reyes

Figure 3.8 Waves in shallow water

45

3 Coastal processes and landforms

3.4 Coastal features: South Devon and Dorset

Figure 3.10 Cliffs in Lyme Bay, Dorset

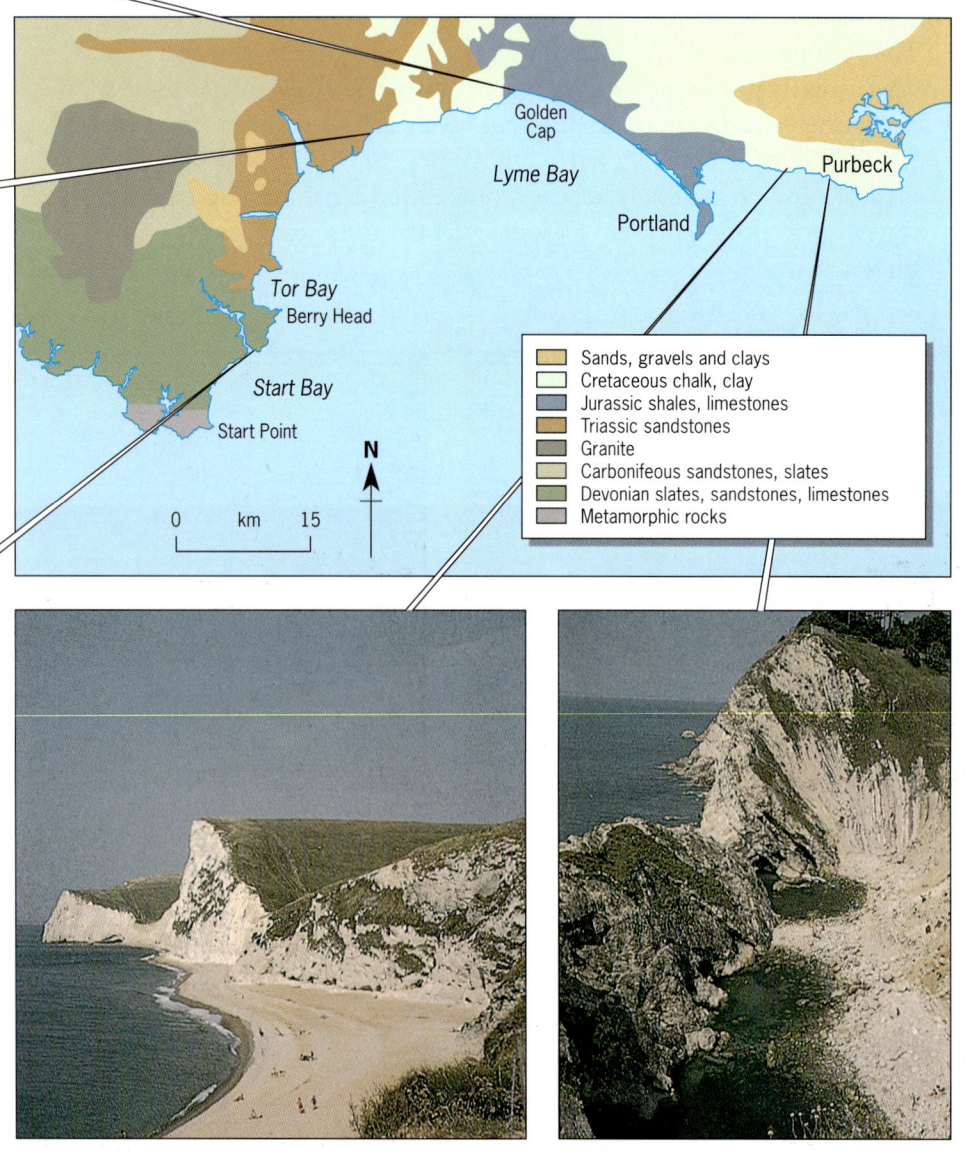

Rock type and coastal landforms

The coastline of southern England, between Start Point in Devon and the Isle of Purbeck in Dorset, has some of the most spectacular scenery in Britain (Figs 3.9–3.14). This coastal scenery owes much to the rocks and relief of the area. For example, hard rocks such as schist at Start Point and limestone at Berry Head, form rugged **headlands**. In contrast, where less-resistant rocks meet the coast, the sea has eroded broad bays like Lyme Bay in Dorset. Much of the Devon and Dorset coastline is upland. Erosion of these upland coasts creates dramatic **cliffs** like Golden Cap in Dorset, the highest on the south coast.

Figure 3.9 Coastline of south Devon and Dorset

- Sands, gravels and clays
- Cretaceous chalk, clay
- Jurassic shales, limestones
- Triassic sandstones
- Granite
- Carboniferous sandstones, slates
- Devonian slates, sandstones, limestones
- Metamorphic rocks

Figure 3.11 (above centre) Cliffs at Sidmouth, Devon

Figure 3.12 (above) Cliffs in Start Bay, Devon

Figure 3.13 Cliffs at Bat's Head, Dorset

Figure 3.14 Cliffs at Stair Hole, Dorset

3 Coastal processes and landforms

Wave energy and coastal landforms

Rocks and relief are not the only influences on coastal scenery. Wave energy along the south coast is particularly high. This is because waves approaching from the south-west have a fetch of several thousand kilometres. Such powerful waves erode the more-exposed stretches of coast to form features such as cliffs, **stacks** and **shore platforms** (**wave-cut platforms**). Elsewhere, in sheltered bays and estuaries, waves transport and deposit sediment. The result is a variety of sand and shingle beaches.

> **EXERCISES**
>
> **3** Study Figure 3.16 and draw sketches to show the cliff profiles in Figures 3.12, 3.13 and 3.14. Add notes to explain how these cliff profiles might have formed.

Erosional landforms

The main erosional processes along coastlines are described in Table 3.2. Together, these processes give rise to distinctive landforms.

Cliffs

Cliffs are common features on upland coasts. There are two reasons for this.
◆ The coast is constantly undercut by wave action.
◆ The rocks eroded from cliffs are quickly removed by waves.

Table 3.2 Processes of coastal erosion

Process	Description
Abrasion/corrasion	Sand and shingle picked up by waves scour the rocks along coastlines. Abrasion cuts a notch at the base of cliffs and is responsible for cliff retreat and shore platforms.
Hydraulic action	The pounding effect of water on the coast during storms. Also, water and air (compressed by wave action in cracks, joints and bedding planes) and marine organisms weaken rocks and lead to rockfall.
Corrosion	The solution of limestones by sea water.
Attrition	The wearing away and rounding of sediments (to form sand and shingle) by abrasion and by rubbing against each other.

Cliff formation

Wave erosion, concentrated at the high water mark, cuts a **notch** at the base of cliffs. This undermining eventually causes cliffs to collapse and retreat inland (Fig. 3.15). The speed of cliff retreat depends on rock strength (see Fig. 3.9). Thus cliffs made from weaker rocks, like the shales of Lyme Bay, may on average retreat several metres a year (Fig. 3.10). Compare this with the slaty rocks of Start Bay (Fig. 3.12), where thousands of years of wave action have made little impact on the cliffs.

Cliff profiles

Cliffs vary in their cross-sectional shape, or profile (Fig. 3.16). Vertical profiles occur when sedimentary rocks are either horizontally bedded or dip inland. Such cliffs, undercut by waves, retreat inland and retain their steep face. However, if the rocks dip seawards, loosened rocks can easily slide into the sea along the bedding planes. In these circumstances, cliff profiles have the same angle as the angle of dip of the rocks.

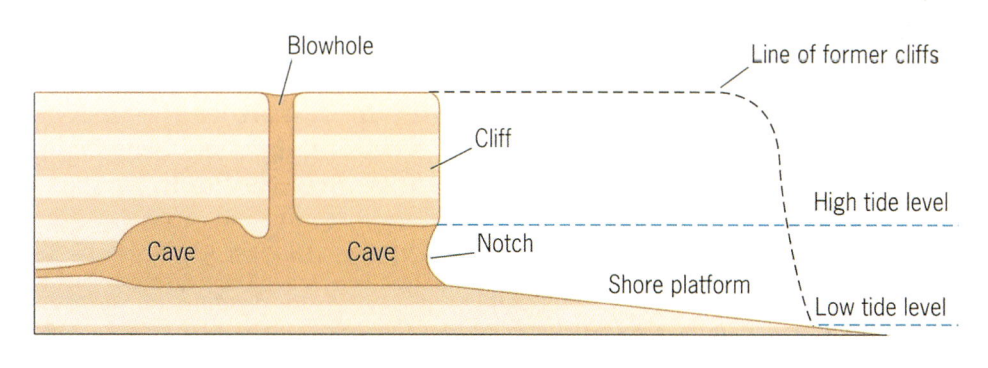

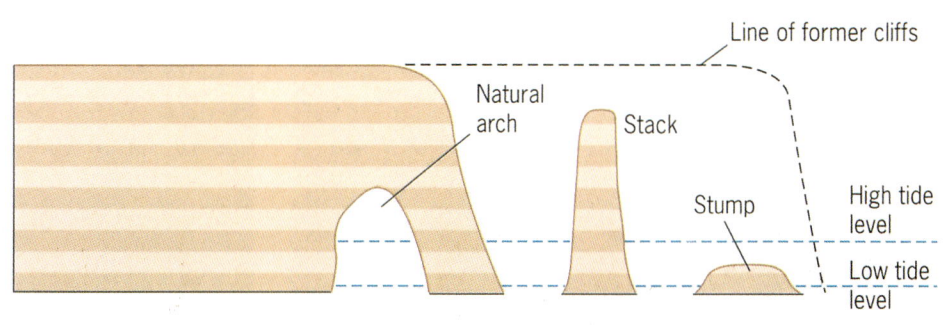

Figure 3.15 Formation of cliffs

3 Coastal processes and landforms

Horizontal strata

Strata dipping inland

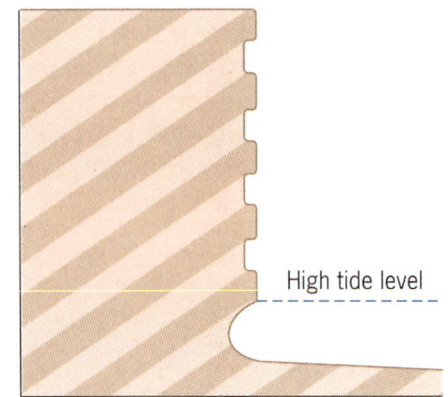

Strata dipping towards the sea

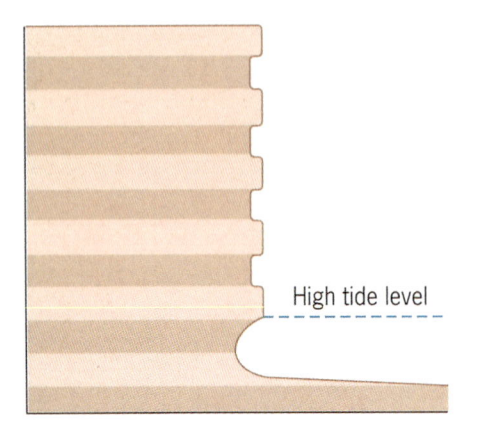

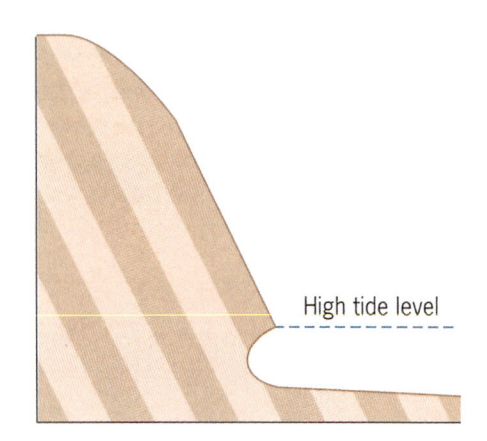

High tide level

High tide level

High tide level

Figure 3.16 Cliff profiles

Figure 3.17 Durdle Door, Dorset

Caves, arches and stacks

Caves form at the base of cliffs (Fig. 3.15). They are simply joints, faults and bedding planes that have been enlarged by erosion. Some caves are linked to the cliff top by a vertical shaft or **blowhole**. Blowholes form where rocks have collapsed into a cave along a major joint. At high tide, or in stormy conditions, spectacular jets of water spray on to cliff tops through blow holes.

Headlands are good places to find caves. This is because wave erosion is particularly strong here. Sometimes, when caves develop on opposite sides of a headland they join up to form a **natural arch**. Durdle Door, carved out of the limestone of the Dorset coast, is a particularly graceful natural arch (Fig. 3.17). Impressive though they are, natural arches are only temporary features: as erosion continues, the arch eventually collapses (Figs 3.18 and 3.19) leaving behind isolated rock pinnacles or stacks.

Arches and stacks are merely stages in the destruction of cliffs. As cliffs recede inland, they leave behind a shore platform (wave-cut platform) (see Fig. 3.15). Shore platforms are the 'roots' of the old cliffs, scoured and quarried by the waves at high tide, and weathered when exposed at low tide.

Figure 3.18 Marsden Rock, Tyne and Wear (before collapse)

Figure 3.19 Marsden Rock (after collapse, 10 Feb 1996)

48

3 Coastal processes and landforms

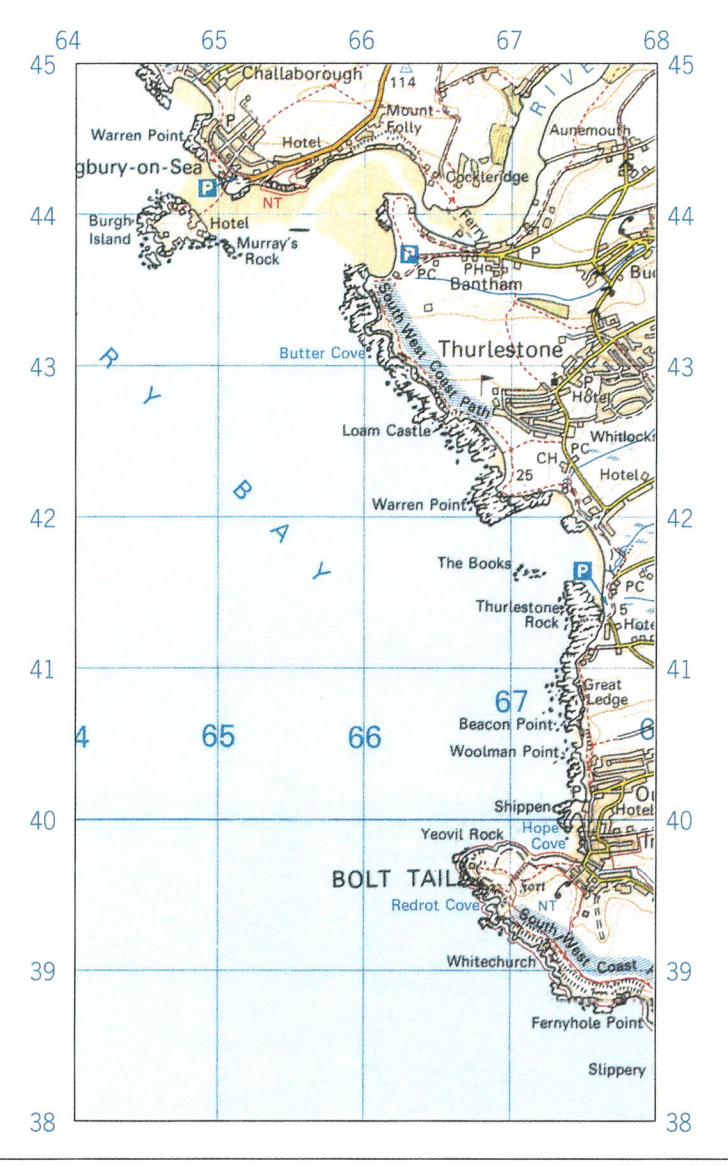

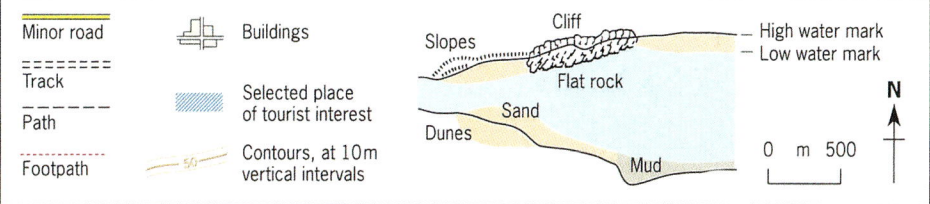

> **EXERCISES**
>
> **4** Study Figure 3.20.
> **a** Describe the upland coastline in Fig. 3.20 between Warren Point (667421) and Whitechurch (673391) by answering the questions in Table 3.3.
> **b** Name and locate (using 6-figure grid references) three features of coastal erosion.
> **c** Explain briefly how each of the features you have named was formed.
> **d*** Suggest two reasons why there are cliffs along this stretch of coastline.
> **5** Draw and label a series of sketches to show how erosion creates a headland.

Table 3.3 OS map skills: how to describe an upland coast

1. In which direction does the coastline run?
2. How high is the coast?
3. Is the coastline straight, or indented with headlands and bays?
4. Does the coastline show any evidence of submergence, with fjords or rias (see page 53)?
5. Are there any erosional landforms such as cliffs, stacks, caves, arches, rocks and shore platforms?

Figure 3.20 South Devon coastline
© Crown copyright

Depositional landforms

Depositional landforms are most common along lowland coastlines. They include various types of beach, mud flats, salt marshes and sand dunes.

Beaches

Beaches are shoreline accumulations of sand and shingle deposited by both waves and currents. Have you ever wondered where this beach material comes from? The obvious answer is from cliff erosion. Yet only a tiny proportion of sand and shingle is from this source. In fact, most beach sediments have been brought down to the coast by rivers.

> **REMEMBER**
> Erosion by waves on upland coasts produces a sequence of landforms. Notches and caves cause cliffs to collapse. As the cliffs retreat inland, they leave behind arches, stacks and ultimately shore platforms.

49

3 Coastal processes and landforms

Spit: Dawlish Warren

Barrier beach: Slapton Sands

Tombolo: Chesil beach

Legend:
- Low-water mark
- Salt marsh
- Mud flats
- Sand/Shingle
- land 0-100m

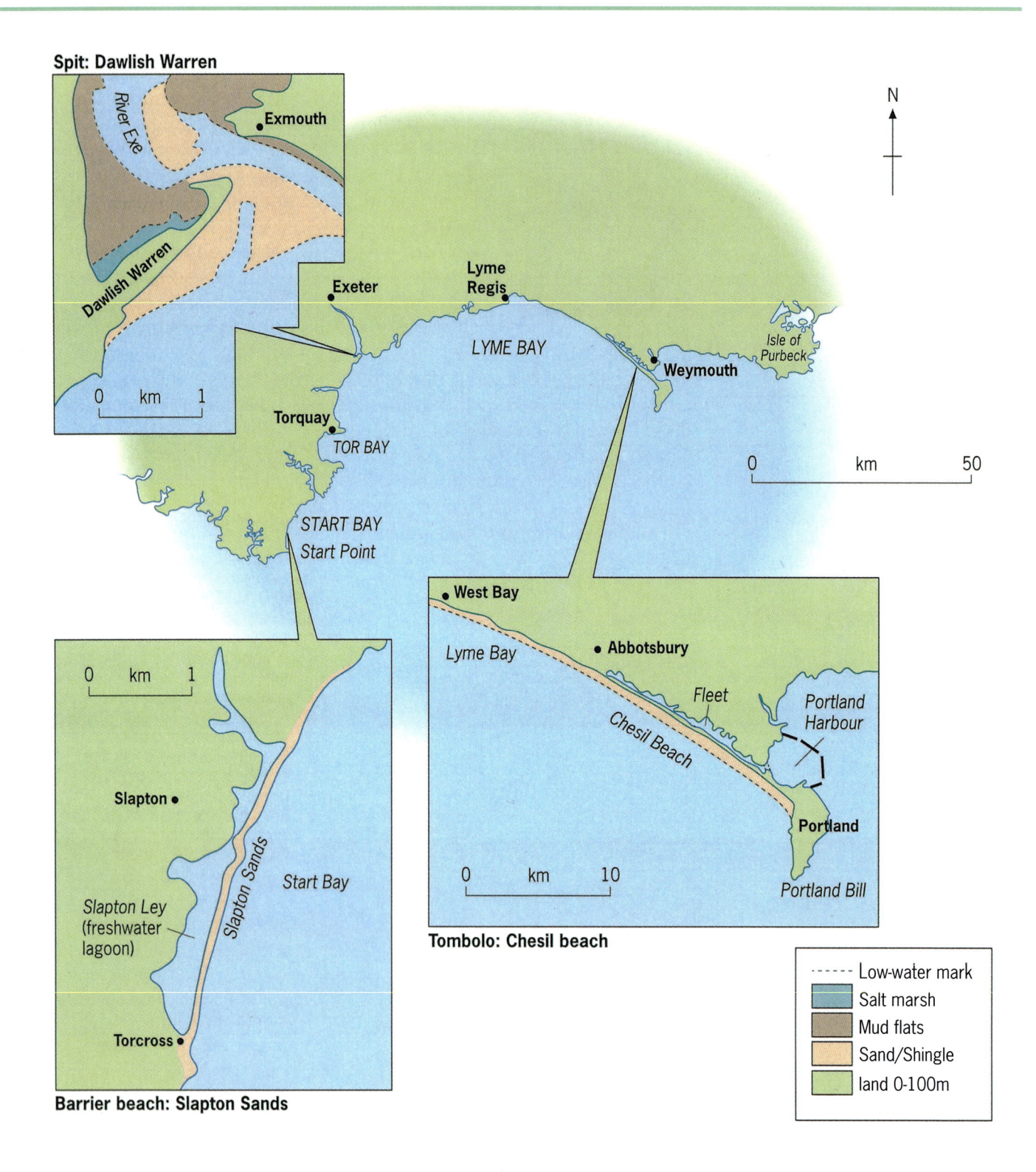

Figure 3.21 Classic depositional landforms of south Devon and Dorset

Bayhead beaches

Beaches have several distinctive shapes. Along indented coastlines like South Devon, crescent-like beaches occupy sheltered bays, coves and other inlets (Fig. 3.20). We call these **bayhead beaches**. They form where waves reach the coast parallel to the shoreline. This means that the swash and backwash move sand and shingle along the same path up and down the beach.

3 Coastal processes and landforms

Figure 3.22 Longshore drift

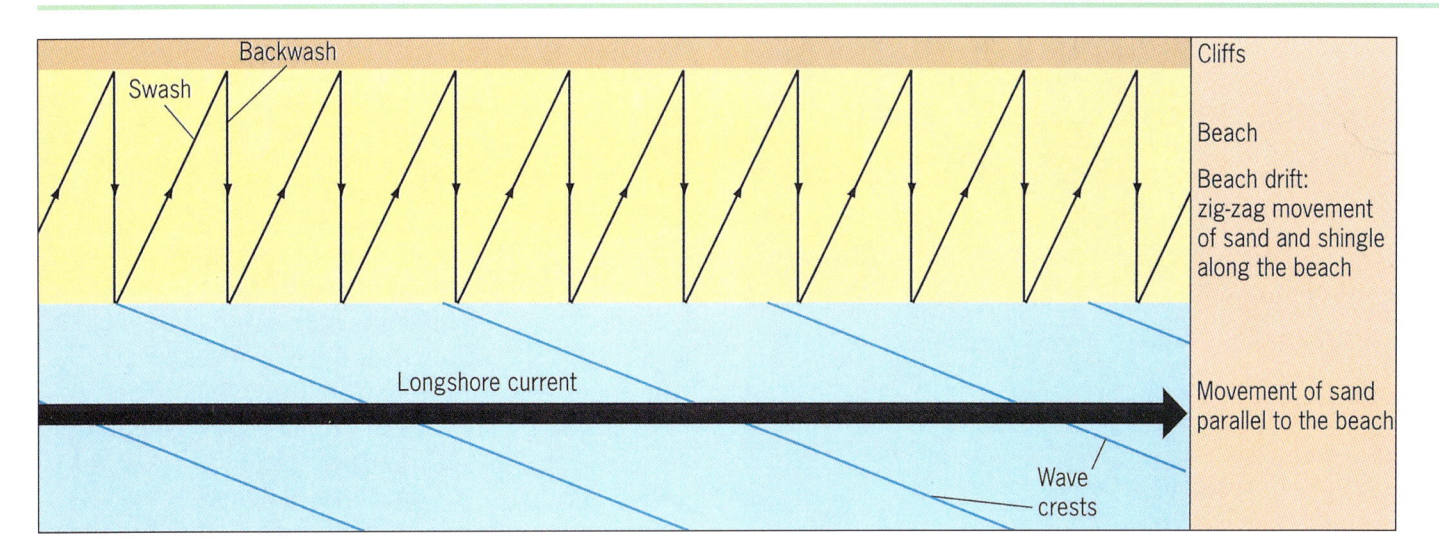

Spits

Less common are beaches known as **spits**. They are easily recognised because they are joined to the land at just one end. Most spits form along coasts where the waves break at an oblique angle (i.e. not parallel to) the shoreline. As a result, the swash pushes sediment up and along the beach, while the backwash drags it down the beach at a 90° angle to the shore. This produces a 'saw-tooth' movement of sediment along the beach known as **longshore drift** (Fig. 3.22).

Table 3.4 OS map skills: how to describe a lowland coast

1. In which direction does the coast run?
2. What is the average height of the coast?
3. What type of beaches (if any) are found along the coast (e.g. bayhead beaches, spits, tombolos, bay bars etc.)?
4. Are the beaches made of sand or shingle or both?
5. Is there evidence of longshore drift? If so, what is the direction of longshore movement?
6. Are there any areas of sand dunes?
7. Are there any areas of mud flat, sand flat and salt marsh?

EXERCISES

6a Draw a sketch of the spit at Dawlish Warren (Fig. 3.23). Add notes to your sketch to show the direction of longshore drift, the recurves (hooks), the mouth of the River Exe, and areas of mud flat and salt marsh.

b* Describe in your own words how longshore drift forms spits.

c Describe the main features of deposition in Fig. 3.20 between Thurlestone Rock (675415) and GR 645447 by answering the questions in Table 3.4.

Figure 3.23 Aerial view of Dawlish Warren

3 Coastal processes and landforms

Dawlish Warren spit

Dawlish Warren (Figs. 3.21 and 3.23), at the mouth of the River Exe, is the only example of a spit between Start Point and Purbeck. Powerful waves from the south-west have created a longshore movement, which has driven sand and shingle across the mouth of the River Exe. In Figure 3.23 you can see a number of recurves, or hooks, at the end of the spit. They mark stages in the spit's growth and tell us that Dawlish Warren has grown by longshore drift. Spits often develop across river mouths or where there are abrupt changes in the direction of the coastline. Tucked away behind spits like Dawlish Warren are quiet backwaters, where waves cannot reach. Here, tidal currents deposit fine silt and form **mud flats** and **salt marshes**.

Chesil Beach tombolo

Chesil Beach in Dorset is the longest shingle ridge in the British Isles (Fig. 3.24). It is 18 km long and runs from Abbotsbury in the west to the Isle of Portland in the east. Beaches like Chesil, which join an island to the mainland, are called **tombolos**. Chesil did not form by longshore drift. It started thousands of years ago as a shingle bar out in the English Channel. After the ice age (see section 2.2), as the climate warmed and sea level rose, waves gradually rolled Chesil onshore. Eventually, about 6000 years ago, it reached its present position.

Slapton Sands barrier beach

Slapton Sands in South Devon (Fig. 3.21) formed in a similar way to Chesil. But Slapton is a **barrier beach** not a tombolo. This is because it extends across a shallow bay, rather than joining an island to the mainland.

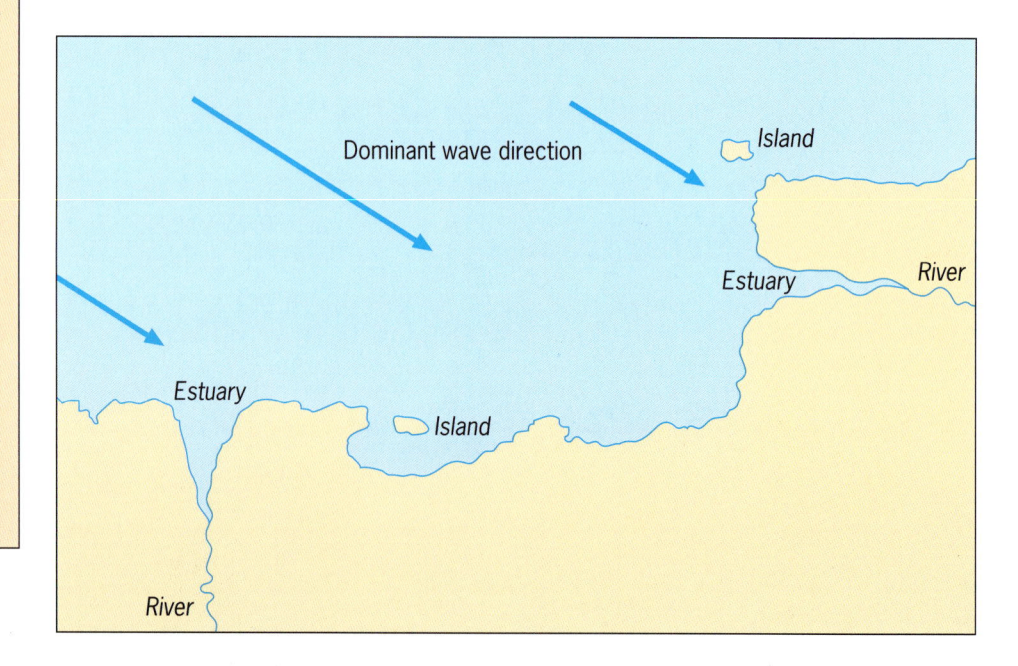

Figure 3.24 Aerial view of Chesil Beach

EXERCISES

7 Study Figure 3.20.
a Name and locate (using 6-figure grid references) two types of beach.
b* Suggest how the beach linking Burgh Island to the mainland might have formed.
c* Find the area covered by the OS map extract in an atlas. Study the atlas map and suggest why there are relatively few beaches along the coastline in Fig. 3.20.
8a Draw a copy of Fig. 3.25 and mark the probable direction of longshore drift.
b Show the possible location where a • spit • tombolo • barrier beach • bayhead beach could develop.
c* Explain your choice of location for these types of beach.
d* Study Fig. 3.24 and explain why it is unlikely that Chesil Beach formed by longshore drift.

Figure 3.25 Longshore drift and depositional landforms

3 Coastal processes and landforms

3.5 Sea-level changes

Worldwide changes

During the last ice age, so much water was transferred from the oceans to great ice sheets and glaciers, that sea level fell by 100 m. At this time, the English Channel, Irish Sea and most of the North Sea were dry land. But as the ice age drew to a close and sea level started to rise, Britain became an island once again.

Landforms of sea level change

The main effect of rising sea level was the submergence of large stretches of Britain's coastline. In South Devon and South Cornwall, deep river valleys were drowned by the sea. We call these drowned valleys, and their tributaries **rias** (Fig. 3.26). Lowland rivers, meandering across broad flood plains were also drowned. They formed wide, shallow inlets or estuaries (see section 1.3), such as the Thames and the Humber. In western Scotland, a different type of valley was flooded. Here we find drowned glacial valleys or **fjords** (Fig. 3.27).

Localised changes

Not all sea-level changes were worldwide. Some changes were caused by vertical movements of the land and affected only small areas. For example, during the ice age, huge masses of ice covered the Highlands of Scotland. So thick was this ice that it depressed the height of the land by hundreds of metres. Then, thousands of years later, when the ice melted, the land began to rise. It is still rising today. As a result of this **isostatic change**, ancient beaches, cliffs, caves and shore platforms have been lifted out of the sea. These **raised beaches** are common around the coast of western Scotland (Fig. 3.28).

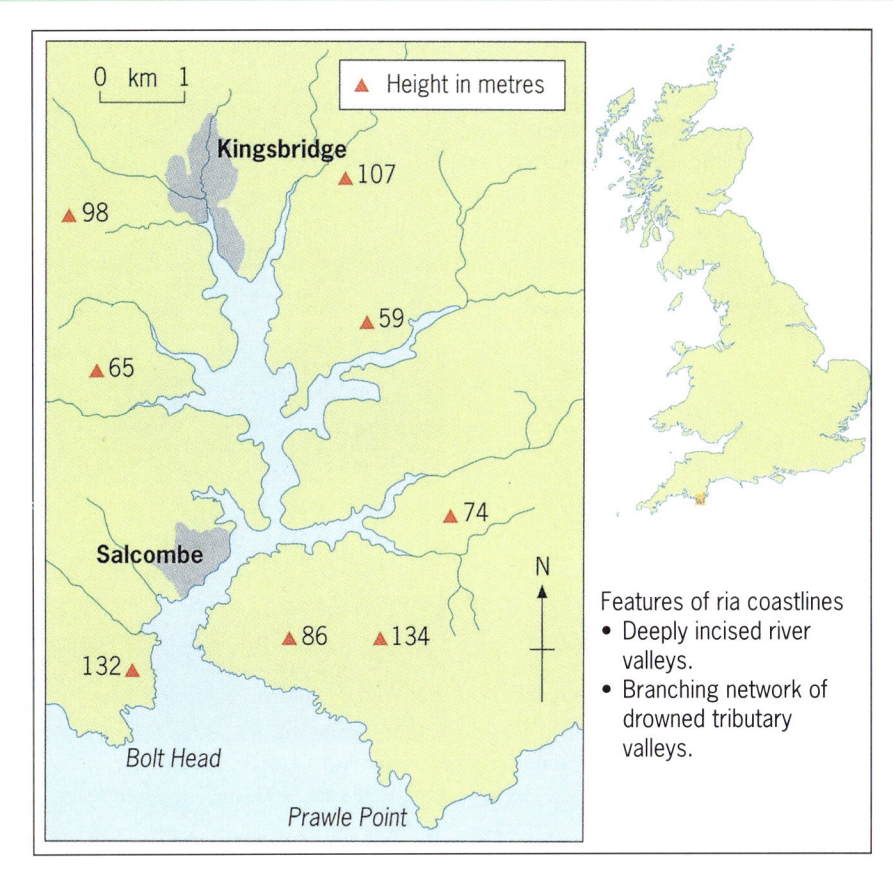

Figure 3.26 Ria at Kingsbridge estuary, Devon

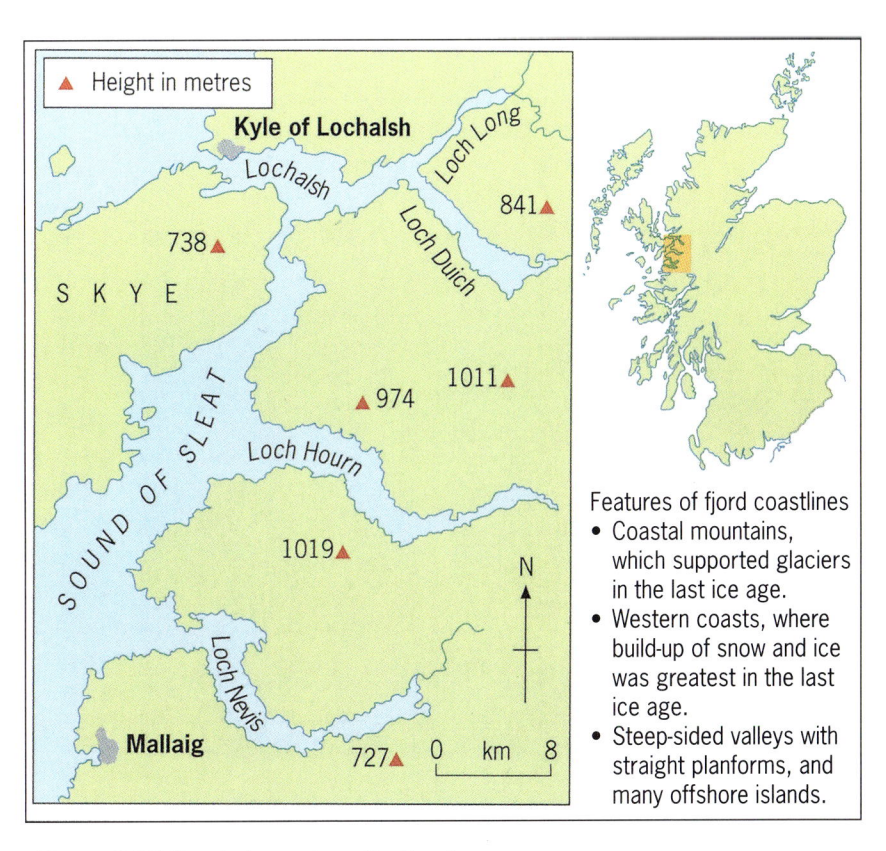

Features of ria coastlines
- Deeply incised river valleys.
- Branching network of drowned tributary valleys.

Figure 3.27 Fjords in western Scotland

Features of fjord coastlines
- Coastal mountains, which supported glaciers in the last ice age.
- Western coasts, where build-up of snow and ice was greatest in the last ice age.
- Steep-sided valleys with straight planforms, and many offshore islands.

3 Coastal processes and landforms

> **EXERCISES**
> 9 Draw a sketch of Figure 3.28. Add notes to your sketch that describe and explain (1) the main features of raised beaches; (2) their formation.

Figure 3.28 Aerial view of raised beaches on Jura, Scotland

CASE STUDY

FACTFILE
- Milford Haven occupies a sheltered, deep-water harbour, which can accommodate oil tankers of up to 280 000 tonnes.
- Thanks to its deep-water harbour, Milford Haven is a major centre for oil refining in the UK. Texaco and Elf operate refineries and the former Gulf refinery is used for oil storage.
- 2000 large vessels use Milford Haven's harbour every year.
- The coast of south-west Wales is of international importance for marine life.
- The islands of Skomer and Skokholm are among the most important seabird breeding sites in Europe. The world's largest population of Manx shearwaters breeds on Skomer. There are large populations of puffins, razorbills, guillemots, cormorants, fulmars and kittiwakes.
- Skomer has one of the two marine nature reserves in the UK. There are important populations of marine mammals, including grey seals, porpoises and dolphins.

3.6 The *Sea Empress* oil disaster, Milford Haven

On 15 February 1996, the supertanker *Sea Empress* hit rocks at the entrance to Milford Haven in Pembrokeshire (Fig. 3.29). Over the next six days, 70 000 tonnes of crude oil leaked from the vessel. The oil polluted beaches, mud flats and salt marshes in South-west Wales, and as far away as North Devon and Lundy.

Oil refining at Milford Haven

Milford Haven is a drowned river valley, or ria. Its deep, sheltered waters provide a superb natural harbour where supertankers discharge their cargoes of crude oil. As a result, both Texaco and Gulf have built oil refineries here. But deep water is only one of the attractions of Milford Haven to these companies. The other is its location in the southern part of the UK. From Milford Haven, it is relatively easy to serve the large markets for refined oil products in South Wales, the Midlands and southern England.

Figure 3.29 The *Sea Empress* run aground and leaking oil, St Ann's Head, Pembrokeshire

3 Coastal processes and landforms

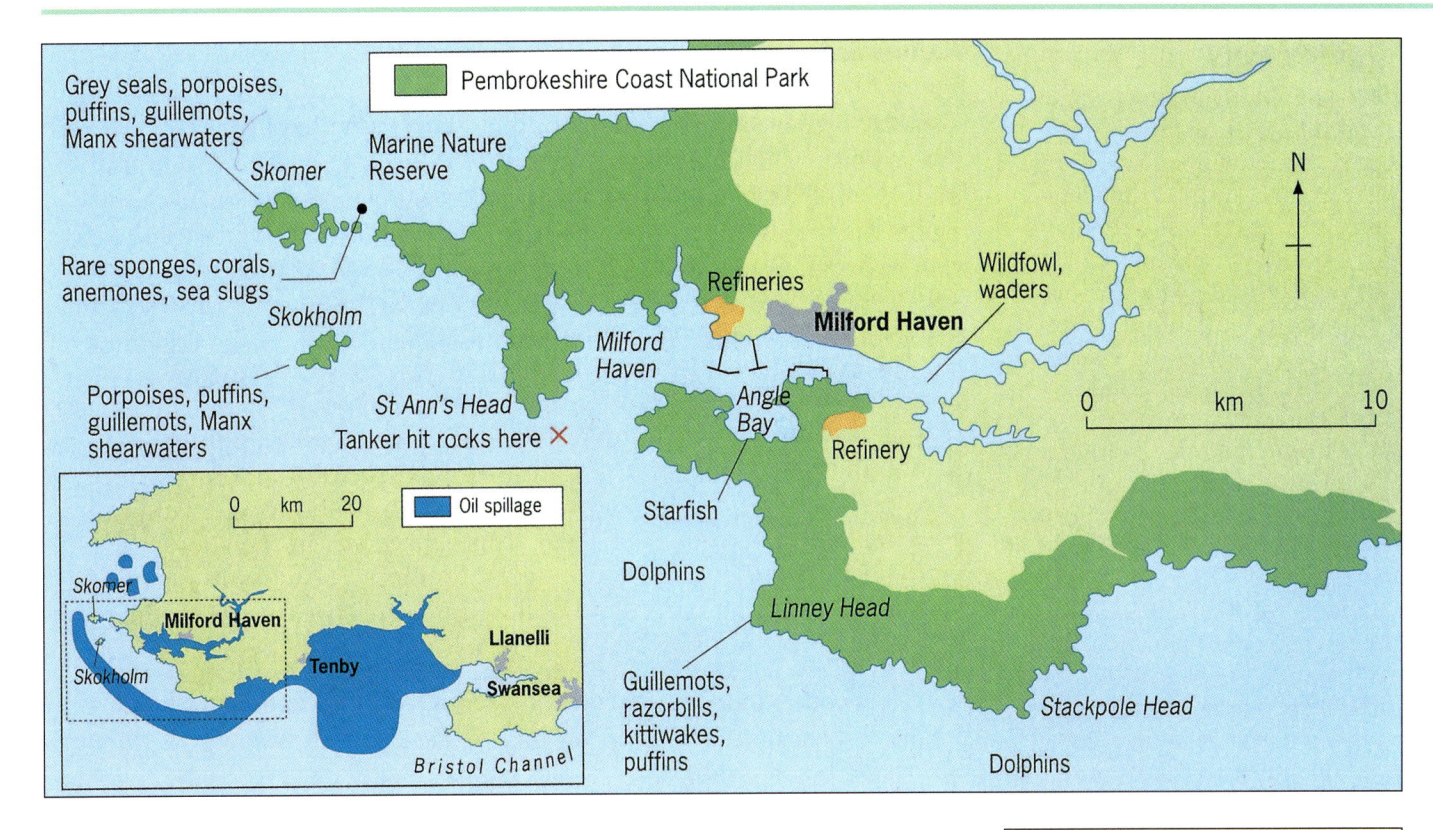

Figure 3.30 The *Sea Empress* oil spill and its effect on wildlife

Environmental resources

Environmentally, Milford Haven is not an ideal place for oil refining. Much of the coastline of south-west Wales is protected by the Pembrokeshire Coast National Park, and there is an underwater nature reserve within a few kilometres of Milford Haven. Most important, is the area's status as an international refuge for marine life. Half a million sea birds breed along the coast and there are colonies of grey seals, dolphins and porpoises (Fig. 3.30).

The impact of the oil spillage

Any major oil spillage close inshore spells ecological disaster for nearby coastlines (Fig. 3.31). But in a small area like South-west Wales, which has so much marine life, the effects are potentially catastrophic. In addition, such an oil spill results in economic problems.

Environmental impact

As soon as oil started to leak from the *Sea Empress*, aeroplanes sprayed it with dispersants, and booms were used to contain the oil slicks. This didn't stop large amounts of oil from being washed up on beaches. Here, road-cleaning vehicles sucked up the oil, and task forces sprayed rocky shorelines.

In spite of the efforts of the special task forces, the impact of the spillage was disastrous.

- More than 200 km of coastline were polluted.
- Although conservationists rescued and cleaned thousands of oiled sea birds, at least 50 000 (mainly guillemots and razorbills) died. The spillage occurred just as many sea birds were returning to Pembrokeshire to breed, but before the winter migrants had left.

EXERCISES

10 Fjords offer excellent deep-water anchorages for large oil tankers. However, few have been developed. Study the distribution of fjords in the UK (i.e. North-west Scotland) and suggest a possible reason for this.

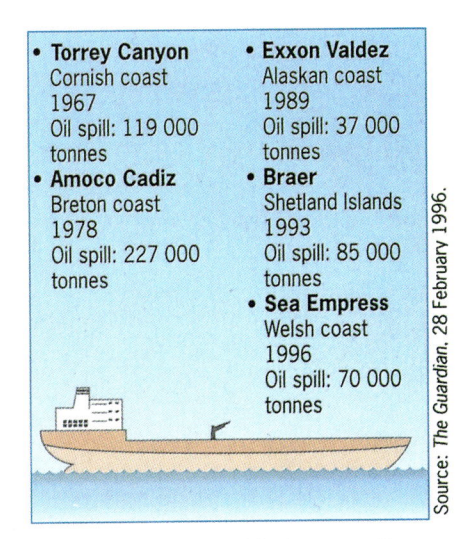

- **Torrey Canyon**
  Cornish coast
  1967
  Oil spill: 119 000 tonnes
- **Amoco Cadiz**
  Breton coast
  1978
  Oil spill: 227 000 tonnes
- **Exxon Valdez**
  Alaskan coast
  1989
  Oil spill: 37 000 tonnes
- **Braer**
  Shetland Islands
  1993
  Oil spill: 85 000 tonnes
- **Sea Empress**
  Welsh coast
  1996
  Oil spill: 70 000 tonnes

Source: *The Guardian*, 28 February 1996.

Figure 3.31 The world's largest oil spills

3 Coastal processes and landforms

EXERCISES

11 Should major oil terminals and refineries be located in environmentally sensitive areas like Pembrokeshire? Imagine a meeting between the residents of a coastal village in Pembrokeshire and an oil executive from the Milford Haven refineries. Among the villagers are a fisherman, an hotelier and a conservationist.

a Assume the role of either the fisherman, the hotelier or the conservationist. Write a short statement, addressed to the oil executive, giving your attitude towards the issue and the reasons for your viewpoint.

b* Assume the role of the oil executive and respond to this statement by explaining your attitude towards the issue.

Economic impact

Fishing

Worst hit by the economic effects of the spillage were local fishermen and hotel owners. The local fishing industry (for shellfish, crabs, lobsters and fish), worth £20 million a year, was a major source of employment in the region. Shellfish were exported to Japan and South Korea; crabs and lobsters went to Spain, France and Italy. Following the accident, foreign buyers cancelled their orders, and the government banned fishing in a 750 sq km exclusion zone around the coast. Without compensation, many fishermen faced bankruptcy.

Tourism

The life-blood of tourism in south-west Wales is the region's rocky coastline and unspoilt sandy beaches. The beaches at Tenby, the biggest seaside resort on the Pembrokeshire coast, were badly polluted. Although two months after the disaster few traces of oil remained on Tenby's beaches, most hoteliers were pessimistic about achieving full bookings for the summer of 1996. This was because many visitors were discouraged by the sight of polluted beaches in newspapers and on television. Cleaning up polluted beaches is easy: persuading holidaymakers that their favourite beach will not be covered by oil during the summer is a more difficult task.

CASE STUDY

3.7 The crumbling cliffs of Holderness

FACTFILE

- The Holderness coast is the most rapidly eroding coastline in Europe – average rates of erosion are 2 m per year.
- Since Roman times, more than 30 villages have disappeared beneath the North Sea.
- The coastline comprises soft, easily eroded till or boulder clay.
- Rapid erosion is also due to the absence of beaches and to powerful storm waves from the north-east.

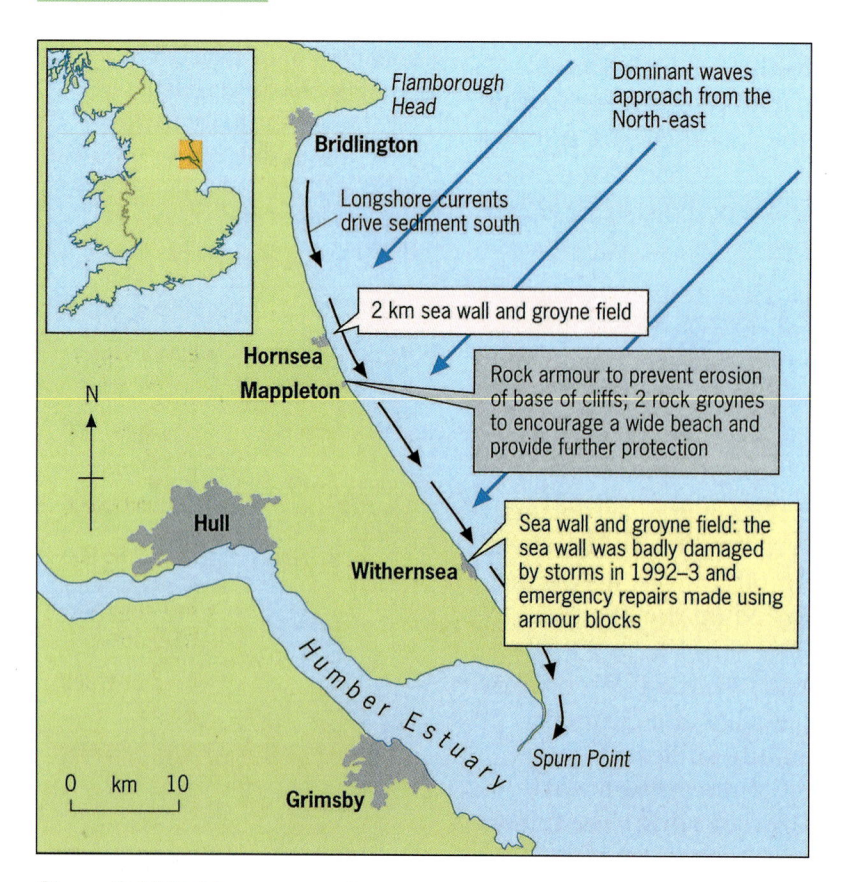

Figure 3.32 Holderness coast

3 Coastal processes and landforms

Figure 3.33 Cliff erosion at Holderness threatens caravans as the ground beneath steadily disappears

Figure 3.34 Villages of Holderness lost to erosion since Roman times

Causes of rapid erosion

There are three causes of rapid erosion at Holderness: soft material that forms the cliffs; high energy waves; and the absence of beaches.

- The Holderness coast is made of soft glacial till. This material, deposited by glaciers in the last ice age, is easily eroded by waves (Fig. 3.35).
- Holderness is an exposed coast with little protection from waves from the north-east. These waves have a long fetch (see section 3.3) and are very powerful.
- A lack of well-developed beaches. Beaches protect cliffs by stopping waves. At Holderness, the beaches are so narrow and thin that they provide little protection from wave attack.

Two factors explain the absence of beaches. First, because the Holderness cliffs are mostly clay (Table 3.5), there is little sand and shingle to make beaches. And second, longshore drift carries what sand and shingle is available south to Spurn Point. Taking all of these factors together, it is little wonder that erosion is so rapid along the Holderness coast.

Table 3.5 Composition of the till cliffs at Holderness

Mud	72%
Sand	27%
Boulders	1%

EXERCISES

12a Show the data in Table 3.5 as a pie chart.

b* Suggest what is likely to happen to the mud eroded from cliffs at Holderness. Explain your answer.

c* Erosion is less rapid in the most northerly section of the Holderness coast. Study Figure 3.32 and suggest a reason why.

Figure 3.35 Cliff erosion at Holderness

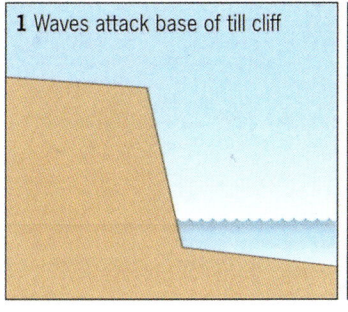

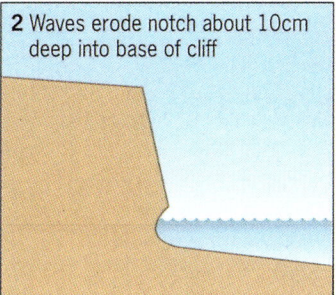

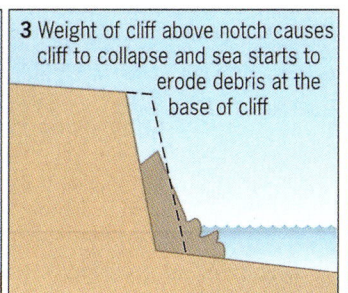

| 1 Waves attack base of till cliff | 2 Waves erode notch about 10cm deep into base of cliff | 3 Weight of cliff above notch causes cliff to collapse and sea starts to erode debris at the base of cliff | 4 Longshore currents and wave action erode debris at base of cliff exposing cliff to further erosion |

3 Coastal processes and landforms

> **EXERCISES**
>
> **13** Study Figure 3.36. Many experts think that sea walls are not the answer to the problem of cliff erosion. Suggest reasons for their point of view.

Managing the Holderness coast

Many settlements dot the Holderness coast. Most of these are isolated farms, but there are also villages, such as Atwick and Mappleton, and small towns, such as Hornsea and Withernsea (Fig. 3.32). Coastal erosion directly threatens some of these settlements. Without protection, many buildings would fall into the sea.

Coastal protection measures

A variety of measures has been used to protect the coast from erosion (Fig. 3.36). Hornsea and Withernsea have a combination of sea walls and groynes. At Mappleton, both armour blocks and groynes protect the cliffs (Fig. 3.37). Elsewhere in Holderness, there are few sea defences. This means that every year agricultural land and even farms tumble into the sea.

Coastal management issues

Coastal erosion at Holderness is controversial. Many local people believe

Figure 3.36 Types of sea defence

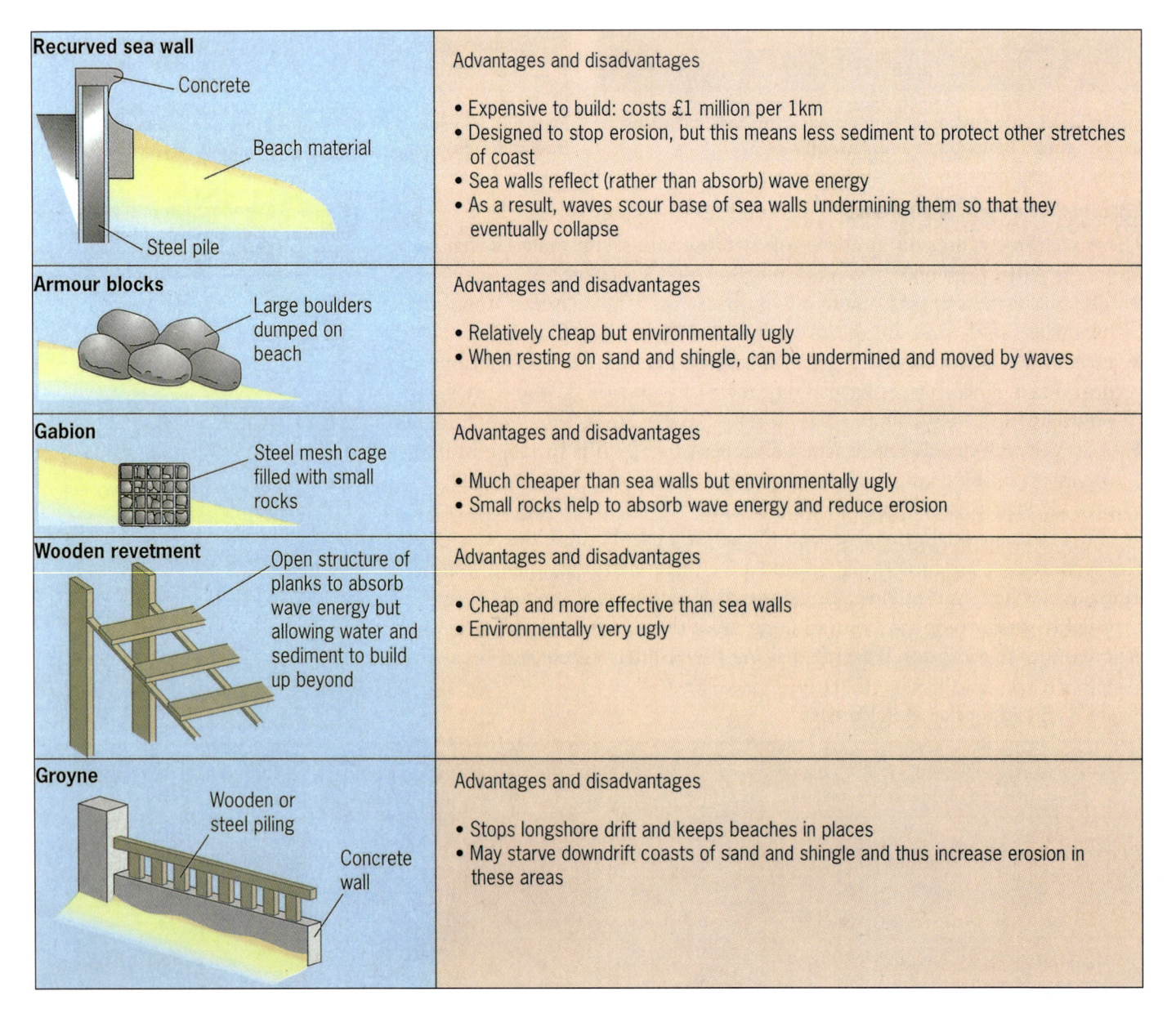

Type	Description	Advantages and disadvantages
Recurved sea wall	Concrete, Beach material, Steel pile	• Expensive to build: costs £1 million per 1km • Designed to stop erosion, but this means less sediment to protect other stretches of coast • Sea walls reflect (rather than absorb) wave energy • As a result, waves scour base of sea walls undermining them so that they eventually collapse
Armour blocks	Large boulders dumped on beach	• Relatively cheap but environmentally ugly • When resting on sand and shingle, can be undermined and moved by waves
Gabion	Steel mesh cage filled with small rocks	• Much cheaper than sea walls but environmentally ugly • Small rocks help to absorb wave energy and reduce erosion
Wooden revetment	Open structure of planks to absorb wave energy but allowing water and sediment to build up beyond	• Cheap and more effective than sea walls • Environmentally very ugly
Groyne	Wooden or steel piling, Concrete wall	• Stops longshore drift and keeps beaches in places • May starve downdrift coasts of sand and shingle and thus increase erosion in these areas

3 Coastal processes and landforms

that not enough is being done to protect Holderness. Geographers take a different view, arguing that it is better to let nature take its course.

Responsibility for sea defences at Holderness rests with local district councils. Their policy is to defend only the larger settlements. The main reason for this is cost: it would simply be too expensive to protect the entire Holderness coastline. And, as the bulk of the land at risk is farmland, which has limited value, the cost of protecting it would be hard to justify.

Erosion and deposition on the east coast

There is an even stronger argument for allowing erosion to continue at Holderness. Cliff erosion at Holderness adds 2.5 million cubic metres of mud to the North Sea every year. Currents carry this mud south to the Humber estuary and to the coast of Lincolnshire. Here, it builds mud flats and salt marshes, which help to protect the low lying coasts of the Humber Estuary and Lincolnshire from flooding.

If we stop erosion at Holderness, we may create an even bigger problem of flooding elsewhere. Remember that nearly 500 000 people live around the Humber estuary in large urban centres, such as Hull and Grimsby. This is ten times the number living in Holderness, where only a small number of people are directly at risk from coastal erosion.

Figure 3.37 Armour blocks protecting the cliffs at Mappleton

> ### EXERCISES
> **14 Either**, imagine that you are the owner of a farm in Holderness, just 50 m from the sea. Write a letter to the local council stating • your views on coastal erosion • the reasons why you hold these views • what action the council should take to tackle the erosion problem.
> **Or**, imagine that you are a planner required to attend a meeting of local people concerned about coastal erosion at Holderness. Write a speech for delivery in class outlining the council's policy on erosion, explaining in full the reasoning behind the policy.
> **15** The problem of erosion at Holderness shows us how the coast operates as a system.
> **a*** Draw a systems diagram (see Fig. 3.5) of the Holderness, Humber Estuary and Lincolnshire coast. Your diagram should include cliffs, sediment, erosion, sediment transport, deposition, mud flats/salt marshes.
> **b*** With reference to Holderness, explain how altering one part of the coastal system leads to changes and problems elsewhere.

Sustainable coastal management

Coastal protection is an important issue not just in East Yorkshire and Lincolnshire, but along most of the coast of south-east Britain. Because of the enormous costs, maintaining existing coastal defences is not sustainable. Instead, a policy called managed retreat has been introduced (Fig. 3.38). This is

Figure 3.38 Managed coastal retreat

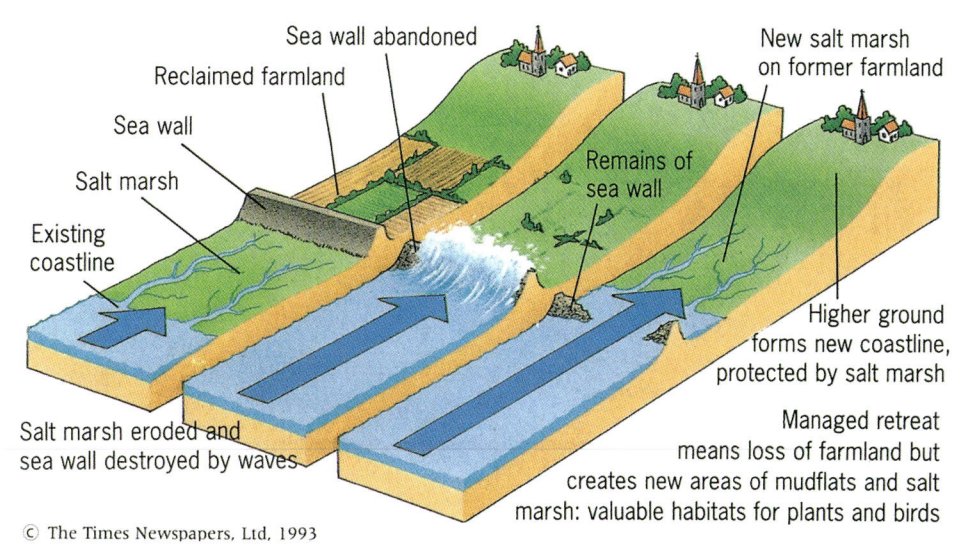

© The Times Newspapers, Ltd, 1993

3 Coastal processes and landforms

> **REMEMBER**
> Protecting coastlines against erosion and flooding, with sea walls, groynes etc., is no longer sustainable. Responses to coastal protection, such as managed realignment, are sustainable as well as being environmentally friendly.

when some sea walls and tidal embankments are not maintained, thus allowing farmland behind these barriers to be flooded by sea water. Eventually, the sea will build its own natural barriers – mud flats, salt marshes and beaches – which will stop further flooding and erosion. Set against the loss of farmland, valuable new marshland habitats will be created for plants and birds. This approach to coastal management is both sustainable and environmentally friendly. Managed retreat works *with* natural systems rather than against them.

CASE STUDY

Figure 3.39 (above) Dunes on the Pembrokeshire coast

Figure 3.40 Human impact on coastal dunes

EXERCISES
16* Study Figure 3.40 and describe the impact of recreational activities on dune coastlines.

3.8 Coastal sand dune erosion and recreation

There are 47 000 hectares of sand dunes around the UK coastline. Sand dunes are rich wildlife habitats for a wide variety of specialised plants and animals. They also have an important role in protecting lowland coastlines from flooding by the sea.

Pressure on Pembrokeshire's sand dunes

The Pembrokeshire Coast National Park in west Wales has 13 dune systems, which vary in size from 2 to 145 hectares. Dunes are fragile environments, easily damaged by unsustainable human activities. The main cause of damage is recreational pressures, including golf courses, horse riding and camping. Dune systems that are easily accessible and close to popular beaches are most vulnerable. Trampling destroys the vegetation cover, exposing the thin soils and sand to erosion by wind and rainfall. Large-scale wind erosion causes 'blow-outs', destroying large areas of dune ridge. In some cases, erosion is so severe that the coast is threatened with encroachment by the sea. Once erosion of soil and sand has occurred, it is very difficult for plants to recolonise the affected areas.

Dune management

The Pembrokeshire Coast National Park has taken steps to control dune erosion and encourage sustainable use of the dunes. It has:
- Planted marram grass. Marram stabilises the dunes; the roots knit the sand and soil together, while the stems trap wind-blown sand.
- Fenced off replanted areas to protect the vegetation from trampling.
- Directed visitors along defined footpaths.
- Restricted vehicular access to car parks.

3.9 Summary: Coastal processes and landforms

KEY SKILLS OPPORTUNITIES
C1.2: Ex. 4b; **C1/2.3:** Ex. 1a, 11, 14; **C2.1:** Ex. 14; **C2.2:** Ex. 4c, 6b, 6c; **N1/2.1:** Ex. 3, 7b, 8d, 10, 12b, 12c; **N1/2.2:** Ex. 2a, 12a; **N1/2.3:** Ex. 2b.

Key ideas	Generalisations and detail
Natural processes cause coastlines to change rapidly.	Beaches respond to changing inputs of wave energy in just a few hours. Rapid erosion (cliff retreat) is taking place in eastern England (e.g. Holderness, East Anglia).
Coasts are systems.	The coastal system is driven by inputs of energy from waves, as well as from tides and winds. This energy interacts with rocks and sediments of the coast to produce distinctive landforms e.g. cliffs, beaches, etc. These are the outputs from the coastal system.
Wave energy is influenced by wind speed, wind duration and fetch.	Gale-force winds produce powerful storm waves. High-energy waves are also produced by moderate winds that blow steadily for several days. Fetch – the extent of open sea over which waves form – also influences wave energy. The longer the fetch, the greater the energy. The direction in which a coast faces determines its fetch.
Waves may be either constructive or destructive.	Constructive waves have low energy and build up sediments to form steep beaches. Destructive waves have high energy. In storm conditions they remove sediments from beaches and erode hard rock coastlines by abrasion/corrasion and hydraulic action.
Geology influences coastal landforms.	Hard, resistant rocks, such as granite and chalk, form headlands. Less-resistant rocks, such as shale and clay, form bays. Horizontally bedded rocks and those dipping landwards tend to form vertical cliffs. Rocks that dip seawards have lower-angled profiles.
Waves erode the coastline by abrasion/corrasion, hydraulic power, corrosion and attrition.	Rock fragments (shingle, cobbles), picked up by waves, wear away the base of cliffs. This is abrasion/corrasion. The pounding of waves against the land weakens rocks by hydraulic action, or quarrying. The rocks disintegrate along lines of weakness i.e. joints and bedding planes. Weathering by solution, salt spray and marine organisms (e.g. molluscs) also leads to cliff destruction. Attrition of beach sediments (by abrasion and by the sediments rubbing together) leads to the formation of sand and shingle.
Wave erosion is most evident on upland coasts where it creates new landforms.	Wave erosion on upland coasts leads to a sequence of landforms: cliffs, caves, arches, stacks, blowholes and shore (wave-cut) platforms. The initial landform is a cliff. The other landforms are simply stages in the destruction of cliffs.
Waves transport and deposit mud, sand and shingle to create new landforms.	Depositional landforms, such as beaches, mud flats, salt marshes and sand dunes, are best developed on lowland coasts. Longshore drift transports sediments laterally along the coast. It gives rise to distinctive beaches known as spits (e.g. Dawlish Warren). Other distinctive beaches formed either by longshore drift or by bars of shingle rolled onshore, are tombolos and barrier beaches.
Rising sea levels in the last 15 000 years have produced distinctive coastlines.	The worldwide rise in sea level following the end of the last ice age flooded many coastlines. Drowned lowland river valleys became estuaries; drowned valleys that were deeply incised became rias; drowned glacial troughs became fjords. Locally (e.g. Scotland), where the land has risen faster than sea level (due to the melting of great ice sheets), raised beaches have formed.
Rapid coastal erosion is a problem along some stretches of coastline.	On the Holderness coast, erosion averages 2 m a year. Agricultural land, farms and even villages are at risk. Sea defences, such as sea walls and armour blocks, are used to protect the larger settlements from wave attack. These defences are costly. With global warming the cost of defending the coastline will rise. The policy of managed retreat recognises that not all coastlines can be defended. It is a sustainable policy.
Human interference in the coastal system often leads to environmental problems.	Sea defences may stop erosion and cut off supplies of sediment to low-lying coasts. These sediments form salt marshes and mud flats, which help to protect these areas (e.g. Lincolnshire) from flooding. The intensive use of sand dunes for recreation often results in environmental degradation.
Deep-water anchorages in rias and fjords offer good industrial sites, but may conflict with conservation.	Milford Haven ria has a deep-water oil terminal and is the location for two oil refineries. The *Sea Empress* oil spillage at the entrance to the ria killed thousands of sea birds, polluted beaches and had severe economic impact on local fishing and tourism.

4 Population: distribution and change

EXERCISES

1 Use the data in Table 4.1 to draw a bar graph showing the world's population by continent.

4.1 Introduction

At the start of the 21st century, two features of the world's population demand our attention: its rapid growth and its uneven geographical distribution.

Population in time

Since the middle of the 20th century, the scale of population increase has been staggering: from 2.5 billion in 1950 to an estimated 6.2 billion in 2001. And this is not the end of the growth: within your lifetime you will almost certainly see a further doubling of the world's population.

Population in space

While the global population has changed rapidly over time, its distribution across the surface of the Earth has stayed much the same. Most remarkable is the geographical unevenness of this distribution (Fig. 4.1). For example, nearly 40 per cent of the world's population lives in China and India, which between them account for less than 10 per cent of the world's land area. At the other extreme, four countries – Australia, Brazil, Canada, and Russia – cover more than 30 per cent of the world's land area and yet have only 6 per cent of the world's total population.

Figure 4.1 World population distribution

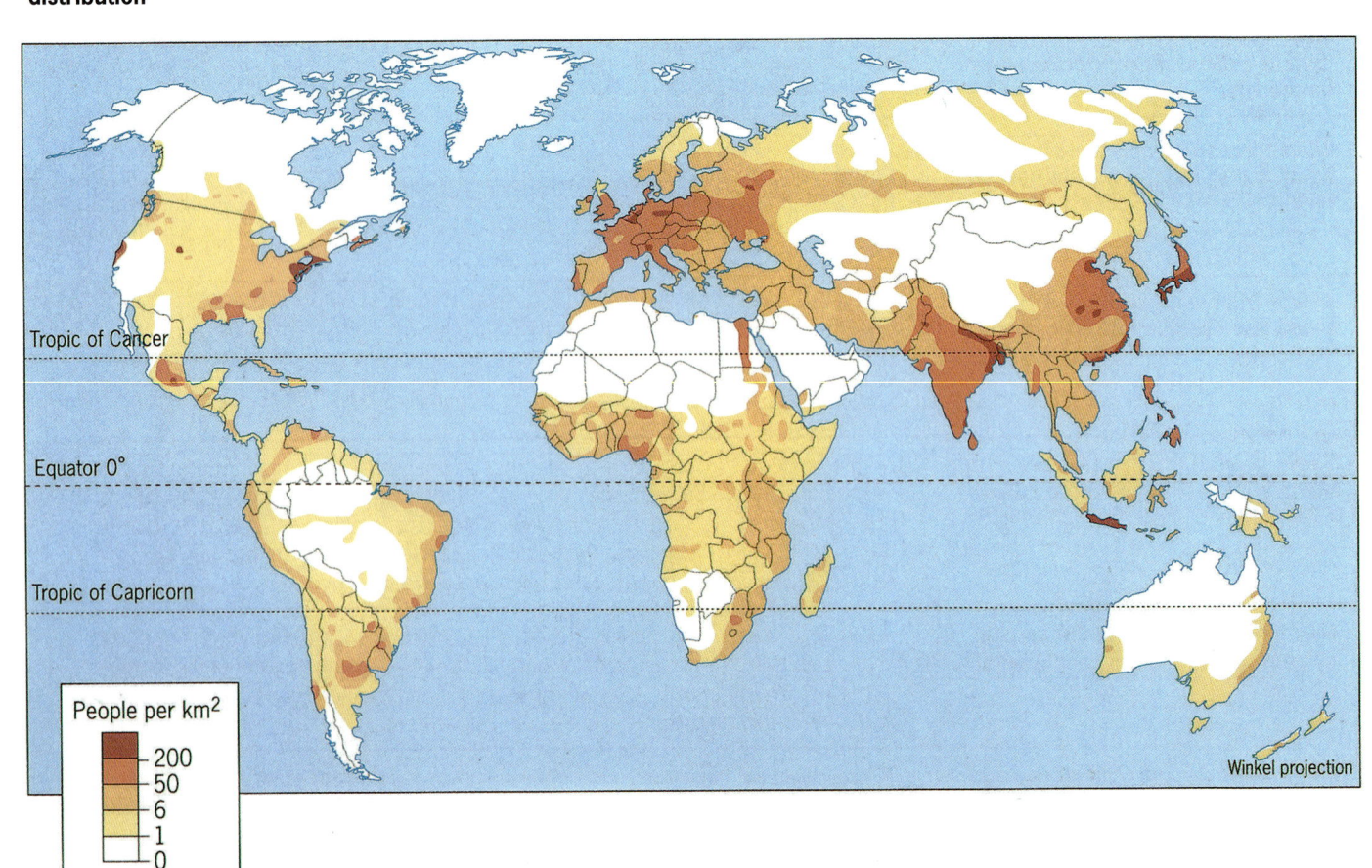

4 Population: distribution and change

Table 4.1 Population of continents

Continent	Population, 2001 (millions)
Africa	823
Asia	3737
North America	310
South America	527
Europe	729
Oceania	31

Table 4.2 Land area of continents

Continent	Land area (000s sq km)
Africa	29 665
Asia	44 609
North America	18 388
South America	20 539
Europe	9181
Oceania	8429

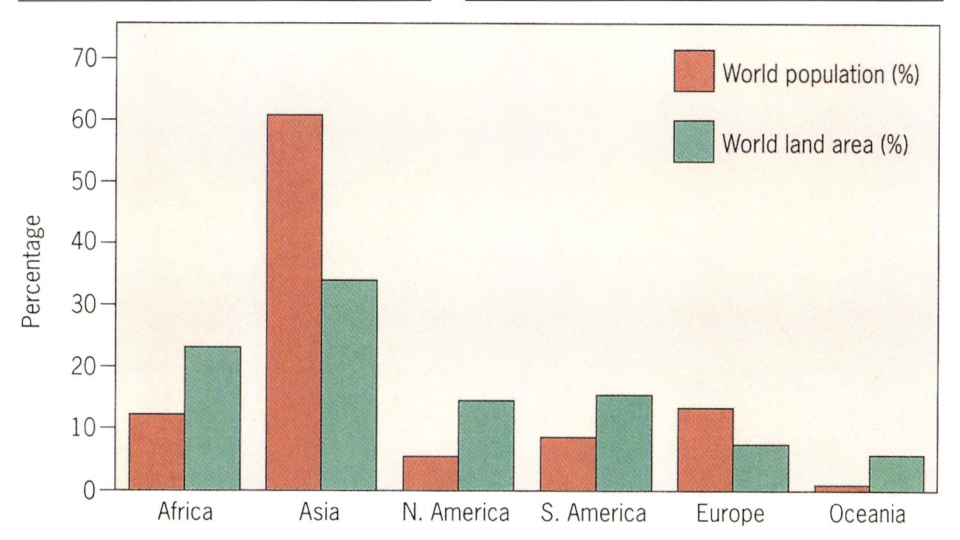

Figure 4.2 Population distribution and land area

4.2 Global population distribution

We often hear people say that we live on an 'overcrowded planet'. If you live in South-east England, Hong Kong, Java or any other densely populated part of the world, it is easy to understand the real meaning of overcrowding. And yet the average population density of the planet (excluding Antarctica) is only 47 people per sq km. This tells us that large areas of the world are very sparsely populated. We can understand the global distribution of population (Fig. 4.1) only by looking at both physical and human factors.

Explaining the global distribution of population: physical factors

Physical factors, especially climate, relief, vegetation and soil, determine the broad outline of population distribution at the global scale. Sometimes, such physical influences combine to produce favourable environments. Then they attract people and settlement. But just as often, physical factors may create harsh environments. These offer few opportunities for economic activities and permanent settlement.

Harsh climates
Large parts of the continents are either too dry or too cold to support many people. The world's hot deserts (Fig. 4.3) are too dry for cultivation. As a result, much of Africa north of the equator (Fig. 4.4), the Arabian peninsula and central Australia are unpopulated. Even temperate deserts, such as the

EXERCISES

2 Study Figure 4.2, then copy the following paragraph and insert the missing words:

The world's population is unevenly distributed. .... supports three-fifths of the world's population on little more than one-third of the world's total land area. The only other continent with more than its average share of population is .... The most sparsely settled continent is ...., followed by ...., ..., and .... Generally the continents in the .... hemisphere are more sparsely settled than those in the .... hemisphere.

3 Use Tables 4.1 and 4.2.
a Calculate the average population density of the continents. (Population density is the total population divided by the total land area in sq km.)
b Which is • the most densely populated continent • the least densely populated continent listed in Table 4.1?
c Compared to the world average population density, which continents have • above average density • below average density?

REMEMBER
The world's population is unevenly distributed. At the global scale, physical factors, such as climate, relief, soils and vegetation, are the most important influences on population distribution.

4 Population: distribution and change

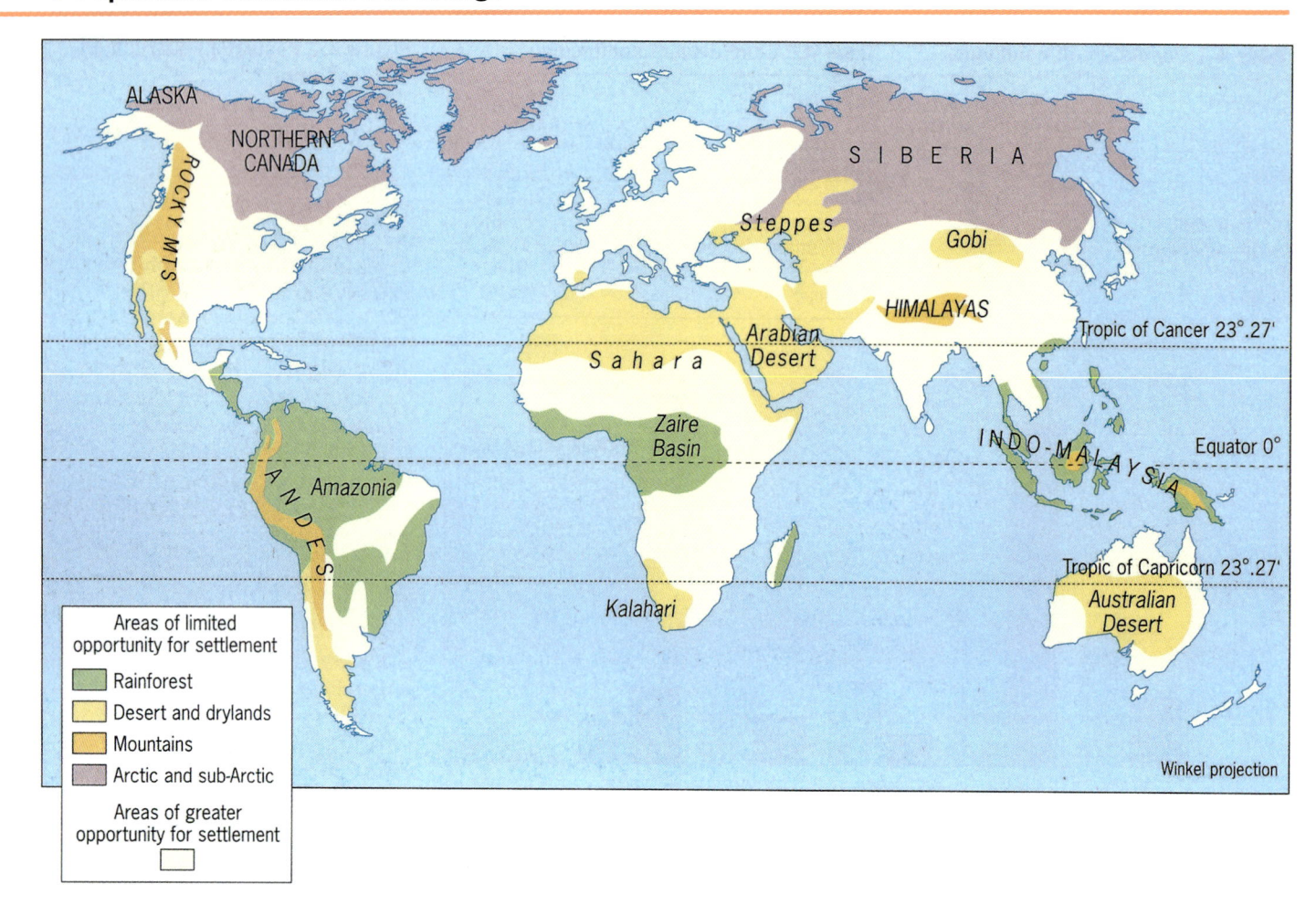

Figure 4.3 Environments of limited opportunity for population and settlement

Figure 4.4 Livestock farming in the drylands of Mali

Gobi in central Asia, are just as sparsely populated. Here, temperatures range from -25°C in January to 32°C in July and there is an average of only 1.4 people per sq km.

In high latitudes, low temperatures deter population and settlement. In Antarctica and Greenland, ice sheets cover an area equal in size to North America. No one lives permanently in these environments of perpetual cold. Less extreme, though, are the areas of tundra (Fig. 4.5) and coniferous forest in Alaska, Canada and Siberia. Even here, the low temperatures and short growing season make it impossible for people to grow crops. Only small groups of hunters and nomadic herders have successfully adapted to these harsh conditions, and they live at very low population densities.

Warm moist climates

Where climate provides a growing season long enough to allow cultivation, population densities are much higher. In humid tropical areas, such as monsoon Asia (South and South-east Asia, Fig. 4.6), high temperatures and abundant rainfall make it possible to grow two or three crops a year. In these areas, wet rice cultivation supports the highest rural population densities in the world. Other favourable regions for agriculture include northern Europe, the Mediterranean, central and eastern USA, eastern China and the seasonally wet tropical areas of Africa and South America. All have higher than average population densities (see Figs. 4.1 and 4.3).

4 Population: distribution and change

Relief

Extensive upland areas usually support few people. Outside the tropics, mountain ranges such as the Himalayas (Fig. 4.7), Rockies and Alps are usually too cold to attract large populations. Even where temperatures are high enough for farming, steep slopes and poor soils often make cultivation impossible. However, within the tropics, mountainous areas sometimes have more favourable climates, and therefore higher population densities than surrounding lowlands. For example, in Ethiopia, the best farming areas are in the highlands. Here, unlike the scorched lowlands, there is enough reliable rainfall for cultivation. Also compared to lowlands, tropical highlands have fewer insect-borne diseases, such as malaria. But in general, because they are flat, lowland areas attract settlement and often support large populations.

Figure 4.5 (above left) The Alaskan tundra

Figure 4.6 (above right) Intensive rice farming, Ganges Plain, India

Soils and vegetation

Soils and vegetation can also influence population distribution. Deltas such as the Nile (see Fig. 1.22), Ganges and Mekong support huge populations because of their fertile alluvial soils. In contrast, the soils of tropical rainforests (Fig. 4.8) are highly infertile. Here, the only sustainable agriculture is shifting cultivation, which supports densities of just 1 or 2 people per sq km. As well as having poor soils, the rainforests (see Fig. 4.3) are home to many tropical diseases, which make it difficult for people to settle large areas of Amazonia, the Zaire Basin and much of Indo-Malaysia.

Figure 4.7 Himalayas, Nepal

4 Population: distribution and change

Figure 4.8 Orinoco River basin, Venezuela

Figure 4.9 Climate, relief and population density

EXERCISES

4 Study Figures 4.4 to 4.8. Describe the environmental conditions in each photograph and suggest how they might influence population density.

5 Study the pattern of mean annual temperature, mean annual precipitation and relief in the hypothetical region in Figure 4.9. Assume that population distribution is influenced only by physical factors:
a For each area (A–F), suggest the likely population density (i.e. high, medium, low).
b Give reasons for your suggested densities.
c Present your results as a table.

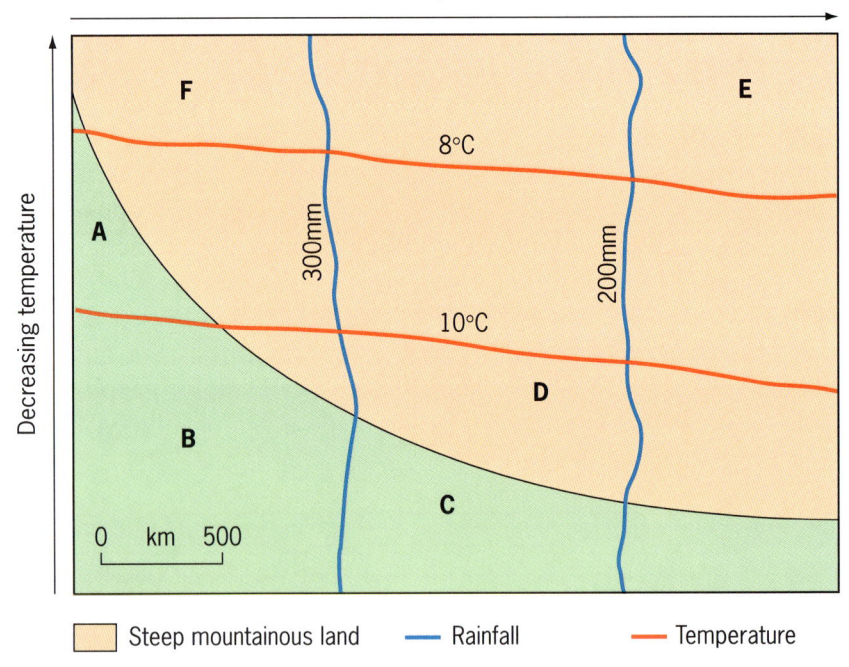

Explaining the global distribution of population: human factors

Physical factors alone cannot explain the global distribution and density of population. Often economic, technological and historical influences have greater importance.

Economic activities

The type of economic activity (farming, industry or services) that dominates a region has a strong control on population density. For example, because farming uses the land as a resource, it takes a large area to support a farming community. On the other hand, industry and services rely on materials,

4 Population: distribution and change

energy and trade brought in from elsewhere. These activities therefore use land for location, not production. In addition, such industries need people to work in them. Thus, they give rise to towns and cities and very high population densities (Fig. 4.10). The world's greatest concentrations of population are vast urbanised areas based on industry, services and trade. Such areas include western Europe, the Pacific coast of Japan, and the North-east and South-west USA (see Fig. 4.1).

Technological development

Closely linked to economic activities are levels of technology, education and skills. In places where the most advanced piece of machinery is a simple scratch plough, peasant farming is likely to dominate. In these conditions, both crop yields and population densities are often low. But in countries with advanced technologies, highly educated workforces and sophisticated communications systems, employment in manufacturing, finance, advertising, tourism and so on are able to support very high densities. We can see these contrasts most clearly in East and South-east Asia. Here, countries such as Cambodia and Vietnam rely heavily on low technology farming. Alongside these are Singapore, Hong Kong and Taiwan, which are some of the most successful and advanced industrial economies in the world (Table 4.3).

You should realise that low population densities in LEDCs do not always indicate a lack of natural resources. Countries like Brazil and Zaire have enormous natural wealth but lack the necessary capital, skills and technology to develop it.

Figure 4.10 Urban sprawl, Tokyo, Japan

> **REMEMBER**
> The growth of large urban populations is often nothing to do with local resources. Cities support huge numbers of jobs in service activities for both national and global markets.

EXERCISES

6a Use the information in Table 4.3 to plot a scattergraph to show the relationship between employment in industry and services (horizontal, x axis) and population density (vertical, y axis).
b Draw a best-fit line through the scatter of points on your graph.
c* Summarise the relationship between employment and population density in a short paragraph.
d* Suggest possible reasons why some countries do not follow the general trend.

Table 4.3 Population density and employment in industry and services in East and South-east Asia

	Employment in industry and services (%)	Population density (sq km)
Burma	36	76
Cambodia	26	55
China	27	135
Indonesia	46	115
Japan	93	340
Laos	24	23
Malaysia	58	67
Philippines	59	258
South Korea	82	469
Taiwan	79	608
Thailand	30	124
Vietnam	32	27

4 Population: distribution and change

Figure 4.11 Yanomami people washing vegetables, Venezeula

Historical influences

Historical influences also affect population distribution. In fact, today's regions of highest density tend to be where people have been settled longest. For example, the high population densities in much of India, eastern China and Europe reflect civilisations that go back thousands of years. Compare this to the continents of North and South America, Africa and Oceania (see Fig. 4.1). Until a few hundred years ago, they had only small numbers of indigenous people. These groups included the Aborigines in Australia and the Native Americans in North and South America (Fig. 4.11).

4.3 World population growth

We have already seen that the world's population is growing rapidly: so rapidly, that the global population increases by three people per second or 210 000 every day! In 2000, 78 million people were added to the world's population, which is a number comparable to the entire population of Germany.

Historically, such rapid rates of population growth are a recent event. They were unknown before the 19th century (Fig. 4.12). Indeed, the human population reached its first billion only in about 1800.

Despite the rapid growth of the world's population, population growth is not occurring everywhere. Ninety-five per cent of current growth is in the economically developing world – in Asia, Africa and South America (Table 4.4). This growth will continue for most of the 21st century. In fact, the latest estimates suggest a world population of 6.8 billion in 2010, rising to 9.1 billion by mid-century. Thereafter, world population should remain level or decline slightly.

Figure 4.12 Growth of the world's population

EXERCISES

7 Use Table 4.4.
a Draw three pie charts to show the proportion of the world's population in MEDCs and LEDCs in 1950, 2001 and 2030.
b Describe how the distribution is likely to change between 1950 and 2030.

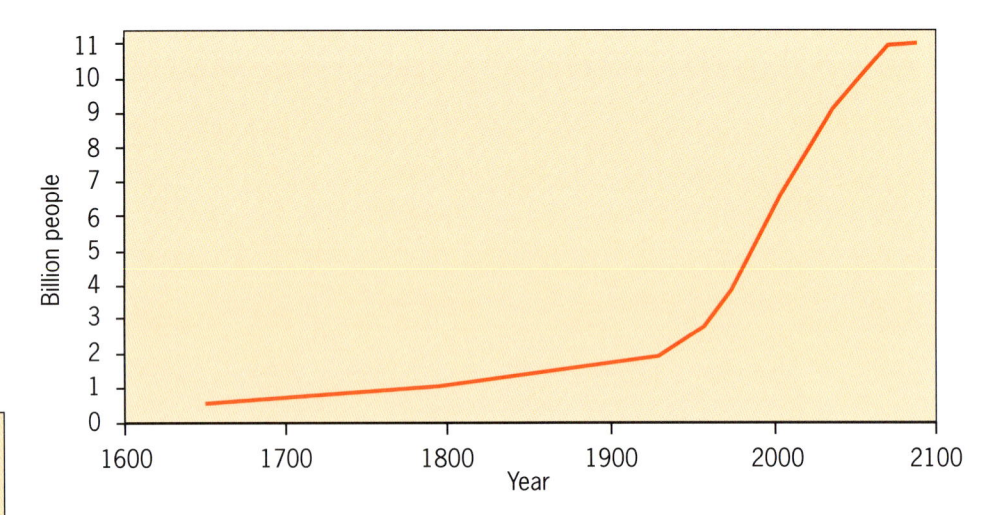

Table 4.4 The changing distribution of the world's population

	MEDCs (millions)	LEDCs (millions)
1950	806	1713
2001	1188	4969
2030	1240	6900

4 Population: distribution and change

4.4 How does population grow?

At a global scale, the population grows when the number of people being born (births) exceeds the number of people dying (deaths). Normally we express the number of births and deaths per 1000 of the population.

- The number of births per 1000 people is known as the **crude birth rate** (CBR).
- The number of deaths per 1000 people is the **crude death rate** (CDR).
- The difference between the crude birth rate and the crude death rate is the **natural population change**, which is usually stated as a percentage per year.

In 2001, the world's crude birth rate was 21 per 1000 and the crude death rate was 9 per 1000. This gave a natural increase of 13 per 1000 or 1.3 per cent and explains the current rapid growth of the world's population.

Globally, there are huge contrasts in birth rates, death rates and rates of natural increase (Fig. 4.13). Birth rates are highest in LEDCs. Here, they averaged 24 in 2001, compared with just 11 in MEDCs. The highest birth rates are in Africa, south of the Sahara Desert, where some countries exceed 50 per 1000. The lowest birth rates in the economically developing world are in China (16 per 1000).

Factors influencing CBR

A large number of social, economic and political factors influence CBRs (see Fig. 4.17). As a result, CBRs are hard to predict. This is the main reason why all forecasts of future population growth contain a degree of uncertainty.

EXERCISES

8a Calculate the natural increase rates for the countries in Table 4.5.

b Which EU countries experienced a natural decrease of their population in 2001?

c* Apart from birth rates and death rates, what other aspect of population might affect population change within a country?

Table 4.5 Crude birth rates and crude death rates in the EU, 2001

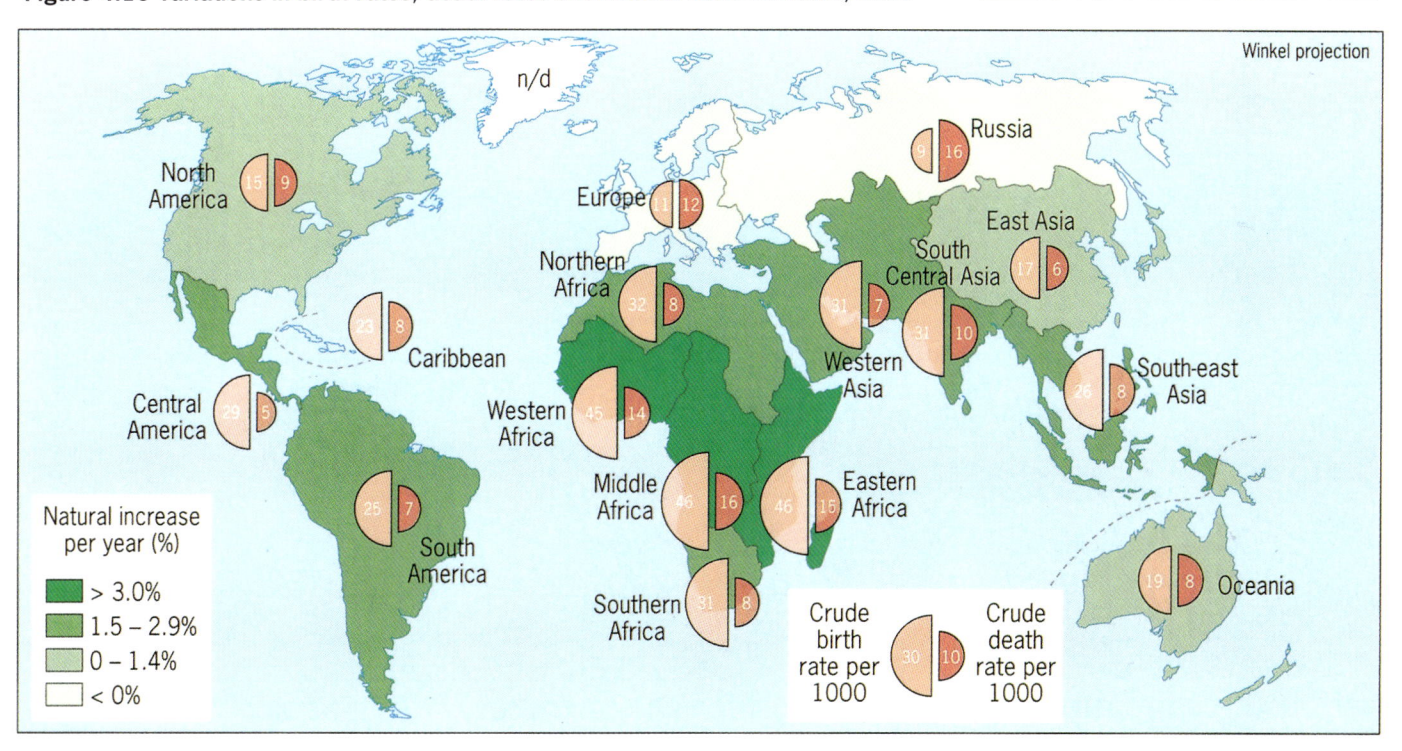

	CBR	CDR
Austria	9.7	9.8
Italy	9.0	10.1
Belgium	10.7	10.1
Luxembourg	12.2	8.9
Denmark	12.0	10.9
Netherlands	11.8	8.7
Finland	10.7	9.8
Portugal	11.5	10.2
France	12.1	9.1
Spain	9.3	9.1
Germany	9.2	10.4
Sweden	9.9	10.6
Greece	9.8	9.7
UK	11.5	10.4
Ireland	14.8	6.1

Figure 4.13 Variations in birth rates, death rates and natural increase rates, 1999

4 Population: distribution and change

EXERCISES

9 Study Figure 4.13 and write a paragraph to describe the global patterns of birth rates, death rates and natural increase rates.

Differences in death rates

Death rates are more variable than birth rates. Some LEDCs with low standards of living and poorly developed medical services continue to have high death rates. For instance, in Malawi and Angola death rates are more than 20 per 1000. Other LEDCs like Mexico and Costa Rica, with higher standards of living and very youthful populations, have current death rates of just 4 or 5. In MEDCs, death rates are fairly steady at about 10 per 1000.

4.5 Population change and development

Sweden's population in 1800 and 2001

FACTFILE

- *In 1800 Sweden was a poor country.*
- *Most people worked in agriculture, were illiterate and had a low standard of living.*
- *Death rates, particularly among children, were high.*
- *High death rates were the result of poverty, diseases such as typhoid and smallpox, and food shortages caused by harvest failure.*
- *Birth rates were high, partly to ensure that some children survived infancy.*

Figure 4.14 Painting of poor people waiting at a police station in Stockholm, Sweden (1857)

Today, the condition of Sweden's population is very different to that of 1800. Economic and social development mean that even the poor are much better off than most people were 200 years ago. The improvement in people's diets, housing, health and levels of education have been accompanied by changes in birth, death and population growth rates. Similar changes have occurred in all MEDCs over the last 200 years. When we view these changes together, we call them the **demographic transition** (Fig. 4.16).

Figure 4.15 (below) Birth and death rates in Sweden, 1737–2000

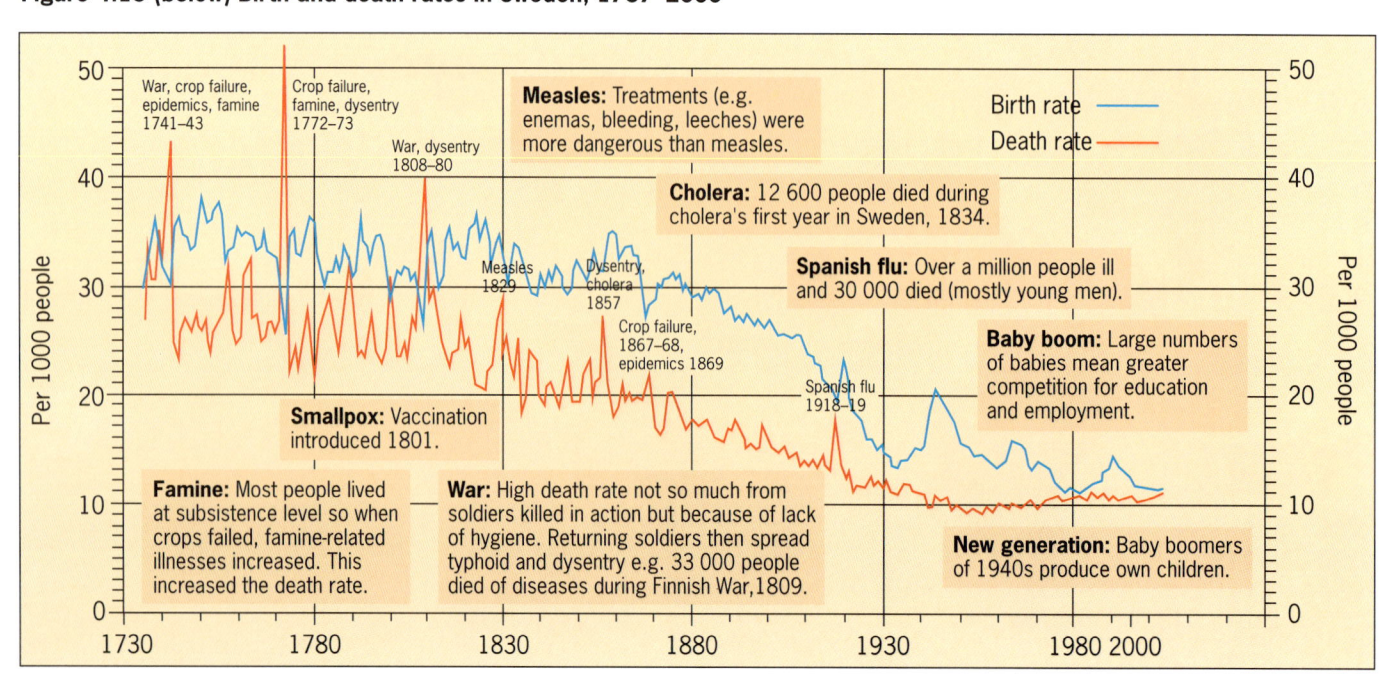

- **War, crop failure, epidemics, famine 1741–43**
- **Crop failure, famine, dysentry 1772–73**
- **War, dysentry 1808–80**
- **Measles:** Treatments (e.g. enemas, bleeding, leeches) were more dangerous than measles.
- **Cholera:** 12 600 people died during cholera's first year in Sweden, 1834.
- **Spanish flu:** Over a million people ill and 30 000 died (mostly young men).
- **Baby boom:** Large numbers of babies mean greater competition for education and employment.
- **Smallpox:** Vaccination introduced 1801.
- **Famine:** Most people lived at subsistence level so when crops failed, famine-related illnesses increased. This increased the death rate.
- **War:** High death rate not so much from soldiers killed in action but because of lack of hygiene. Returning soldiers then spread typhoid and dysentry e.g. 33 000 people died of diseases during Finnish War, 1809.
- **New generation:** Baby boomers of 1940s produce own children.

4 Population: distribution and change

Demographic transition

As its name suggests, the demographic transition is a gradual process. It has four stages:
- Stage 1. Before economic development, birth and death rates are high and there is little or no growth.
- Stage 2. Thanks to improved living conditions and medical advances, death rates begin to fall. However, birth rates remain high and the population increases rapidly.
- Stage 3. Birth rates at last begin to decline. Even so, the gap between births and deaths remains large and the population continues to increase.
- Stage 4. Birth rates fall to the same low level as death rates and population growth comes to an end.

The demographic transition describes what has happened to birth and death rates in MEDCs in the last 200 years. It seems unlikely, though, that today's LEDCs will follow the same transition. Birth rates are currently falling in the economically developing world. However, these falling birth rates are not due to economic development and better standards of living. Instead, they are taking place because of family planning programmes and the use of contraception. Modern mass media such as radio and television have made millions of people in poorer countries aware of the benefits of contraception. In today's MEDCs, contraceptives were neither widely available nor widely accepted until the 20th century.

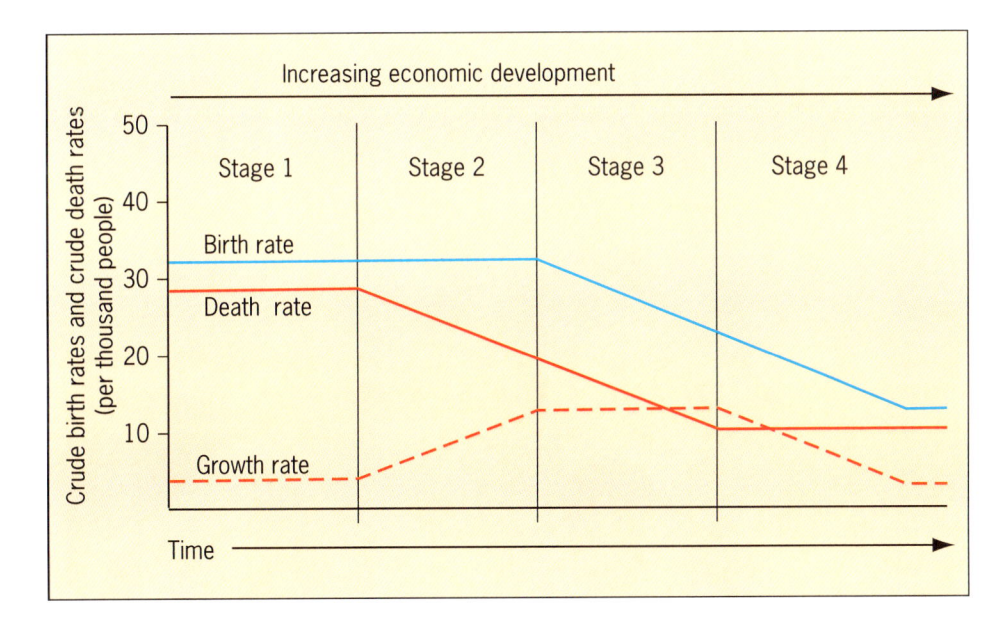

Figure 4.16 The demographic transition

Other reasons for population change

Economic development is only one factor that promotes population change. Other important influences include social, religious, political and environmental factors. These are shown in Figure 4.17.

EXERCISES

10a Make a copy of the demographic transition (Fig. 4.16).
b Identify each stage of the demographic transition from the descriptions below. Add the descriptions as labels to your copy of the demographic transition.

Stage ?
- Falling birth rates; low death rates.
- Population growth begins to slow down.

Stage ?
- High birth rates and high death rates.
- Population is stable.

Stage ?
- Low birth rates and low death rates.
- Population is stable.

Stage ?
- High birth rates and falling death rates.
- Rapid population growth.

c Assuming that all countries follow the demographic transition, on your copy of Figure 4.16, label the countries below at their appropriate stage in 2000.

	CBR	CDR
Chile	17.2	5.5
Japan	10.0	8.1
Ethiopia	45.1	17.6
Pakistan	32.1	9.5
Israel	19.3	6.2
Switzerland	10.4	8.8

EXERCISES

11 Study Figure 4.15.
a At approximately which dates did Sweden enter the second, third and fourth stages of demographic transition?
b For how many years did the death rate exceed the birth rate in Sweden before 1830? What were the causes of these years of high death rate?

4 Population: distribution and change

Figure 4.17 Reasons for variations in birth and death rates

Age structure
A high proportion of women of reproductive age (15–49) increases the birth rate. Similarly, a high proportion of old people in the population increases the death rate.

Diet, housing, living conditions
A balanced diet (protein, carbohydrate, fat, vitamins) and sufficient food intake lower death rates, especially among children. Good housing conditions, with adequate sanitation and clean water supplies, also lowers death rates. Improvements in living conditions mean more children survive. This may help to lower birth rates.

Medicine, health care
The availability of medicines, hospitals and doctors, reduce death rates: people live longer.

Family planning and contraception
Birth rates in the economically developing world are often high because women do not have access to family planning services and contraception. Where family planning is available, birth rates often fall rapidly. These services are more easily available in towns and cities than in the countryside.

Economic conditions
In rural areas of the economically developing world, children are often an economic asset. They work in farming at an early age, so large families make good sense. In urban areas, however, there is less work for children and so they become an economic burden. As a result, birth rates in LEDCs are lower in urban than in rural areas. In MEDCs, children have at least ten or eleven years in full-time education. Children are supported by their parents during this time and so there is little economic advantage to large families.

Social and religious factors
The status of women has an important influence on birth rates in LEDCs. As women become more educated or achieve a higher status, they tend to have fewer children. Religion (or superstition) may forbid contraception (e.g. catholicism). Children may be seen as 'God-given'. In some societies, women marry in their early teens, thus increasing the chance of having large families.

Political factors
Governments may adopt policies either to encourage or to discourage births. For example, governments can encourage population growth by banning abortion or the sale of contraceptives, or by giving financial benefits for children. Currently, most governments in LEDCs have policies aimed at reducing population growth by promoting family planning.

4 Population: distribution and change

4.6 Age–sex structure

We use a special type of graph called a **population pyramid** to show the age–sex structure of a place (Figs. 4.19 and 4.21). In spite of their name, not all population pyramids have a classic triangular shape. Sweden's pyramid (Fig. 4.21) for example, is more rectangular than triangular and is rather top heavy. Some have a narrow base; others, with uneven numbers of men and women, may be rather lop-sided.

Three factors control the shape of any population pyramid: births, deaths and migrations operating over a period of 70 or 80 years. At a national scale, births and deaths are usually most important. However, when we look at population pyramids for smaller areas, such as cities and regions, migration becomes much more important. We are going to look more closely at the age–sex pyramids for Sweden and Nigeria to see how differences in birth and death rates have influenced the pyramid shape.

CASE STUDY

4.7 Nigeria

Nigeria, with an estimated population of 127 million in 2001, is Africa's most populous country. It is also one of the world's poorest countries. Its population pyramid (Fig. 4.19) has two features that are typical of LEDCs:

◆ A very broad base.
◆ Steeply tapering sides and a narrow top.

The broad base of Nigeria's pyramid reflects the large proportion of children (44 per cent under 15 years) in the population. The crude birth rate is high (45 per 1000) and each woman will give birth to an average of 5.5 children. Two factors explain the large number of births:

◆ Women have large families because they expect to lose some in infancy (one in five children do not reach their fifth birthday).
◆ Only one in every ten women practise any form of birth control.

The steeply sloping sides of the pyramid tell us that death rates are high. Life expectancy is about 50 years and is falling as a result of AIDS. In 1999, 5.5 per cent of Nigeria's adult population was HIV positive. Nigeria can expect at least 5 million deaths from AIDS between 2000 and 2010.

Figure 4.18 A youth group near Ibadan, Nigeria, talk to a person with AIDS about the illness

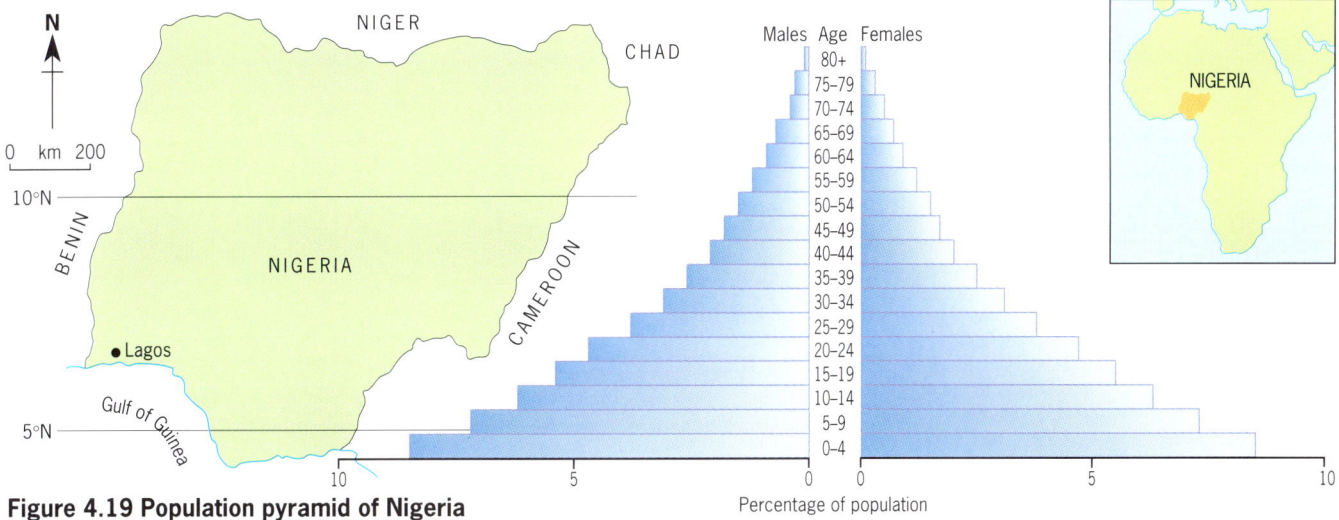

Figure 4.19 Population pyramid of Nigeria

4 Population: distribution and change

Most AIDS deaths will occur among young adults. This could have a devastating impact on the country because:
- Young adults are economically the most productive part of the population, supporting dependants such as children and old people.
- Young adults may have to give up work to care for relatives stricken by AIDS.
- Many millions of children, whose parents die of AIDS, will become orphaned.

The AIDS epidemic is not confined to Nigeria. It is the leading cause of death in sub-Saharan Africa. Indeed, of the 34 most AIDS-affected countries in the world, 29 are in Africa.

CASE STUDY

4.8 Sweden

Sweden, as one of the world's richest countries, is at the opposite end of the development scale to Nigeria (Fig. 4.20). This shows in its population pyramid (Fig. 4.21). Its narrow base and straight sides indicate an old population, with low birth rates (9.9) and low death rates (10.6). In fact, only 18 per cent of Sweden's population are aged under 15 years. On average, Swedish women have only two children and the average age of mothers when they have their first child is 27 years. Unlike Nigeria, contraception is universal in Sweden. In addition, excellent medical care and high standards of living mean that infant deaths are low. Indeed, death rates are high only in extreme old age. Swedish women can expect to live for 81

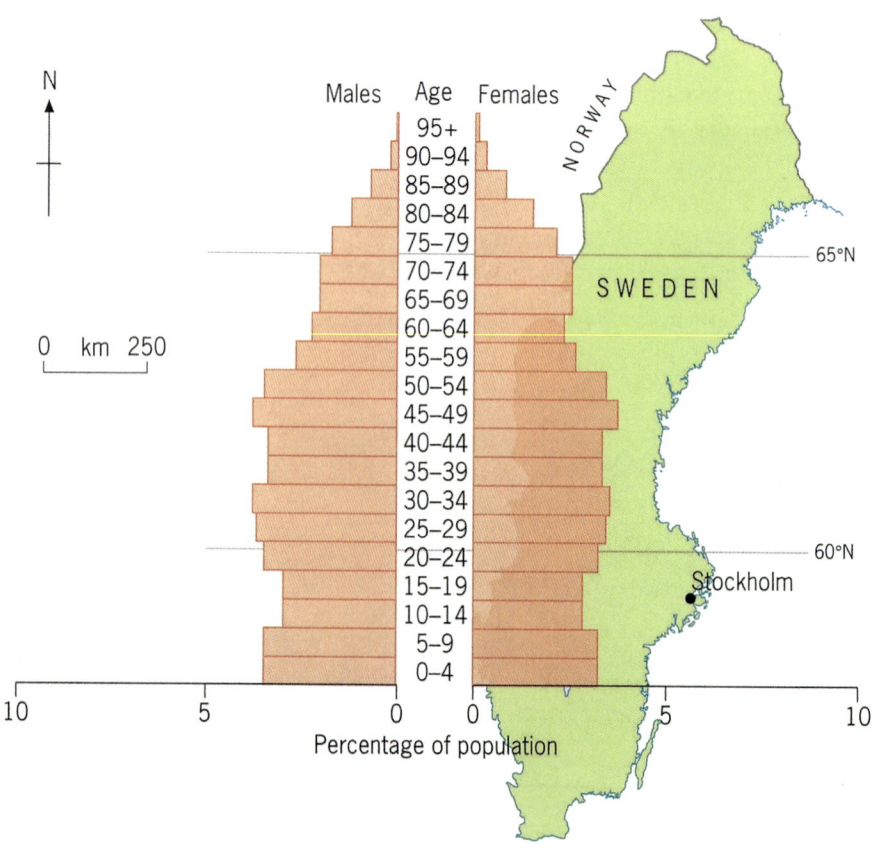

Figure 4.20 Stockholm, Sweden

Figure 4.21 Population pyramid of Sweden

4 Population: distribution and change

years, and Swedish men for 76 years. Because Sweden's death rate currently exceeds its birth rate, without substantial immigration, the country's total population will slowly decline in future.

Table 4.6 Country A: age–sex structure

Age group	% males	% females
0–4	6.9	6.6
5–9	6.5	6.2
10–14	6.0	5.8
15–19	6.1	6.0
20–24	5.1	5.1
25–29	4.2	4.2
30–34	3.3	3.4
35–39	2.7	2.8
40–44	2.2	2.3
45–49	1.8	1.8
50–54	1.4	1.5
55–59	1.2	1.4
60–64	0.9	1.0
65–69	0.7	0.8
70–74	0.4	0.5
75–79	0.3	0.4
80 and over	0.3	0.4

EXERCISES

12a Read through the case studies of Nigeria and Sweden. Then draw a table to compare the main statistics for births, deaths and population change for the two countries.
b Comment on the contrasts shown in your table.
13a Use the figures in Table 4.6 to draw a population pyramid for country A.
b* Describe and explain the probable pattern of births, deaths and natural increase in country A.
c Is country A likely to be a MEDC or LEDC? Give reasons for your answer.

4.9 Problems of population change

Economic problems

Population change often brings with it considerable problems. In most LEDCs, the scale of population growth is a problem in itself. Overwhelmed by rising numbers, many governments cannot hope to provide even basic services or meet people's demands for more housing and jobs. Today, there is a growing awareness that the most successful countries in the economically developing world are those with the lowest rates of population growth (Fig. 4.22).

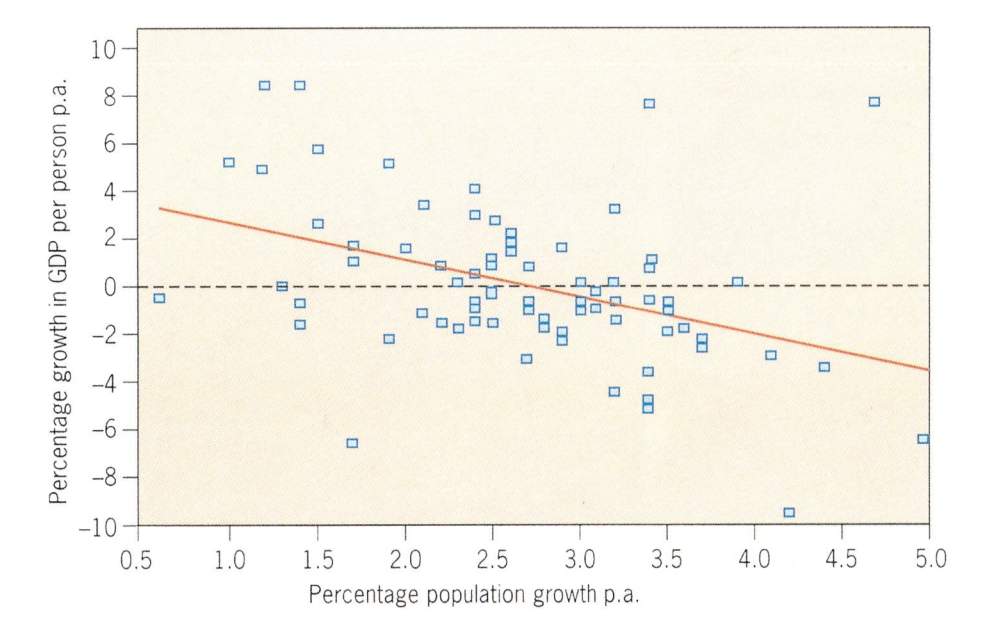

Demographic problems

Population growth also changes age structure. In LEDCs, the proportion of children has increased. Meanwhile, in MEDCs, there has been an increase in the proportion of old people. Both cause problems because both young and old people depend for their support on the adult working population (Fig. 4.23). Any increase in the proportion of children or old people puts greater pressure on already limited resources. This is the problem of age dependency; for different reasons it hits rich and poor countries alike.

Figure 4.22 Economic growth and population growth rate

4 Population: distribution and change

EXERCISES

14 Dependency can be measured as a ratio of the percentage of children and the percentage of aged people in a population, to the percentage of adults. We call this the dependency ratio:

$$\frac{\% \text{ children} + \% \text{ aged}}{\% \text{ adults}}$$

(children: 0 – 14 years
adults: 15 – 64 years
aged: 65 and over)

a Calculate the dependency ratios for Nigeria and Sweden (Figs 4.19 and 4.21).
b Both have high rates of dependency but for different reasons. Try to explain the differences.
c* Explain how today's youthful population in the economically developing world is almost certain to cause massive population growth in future.

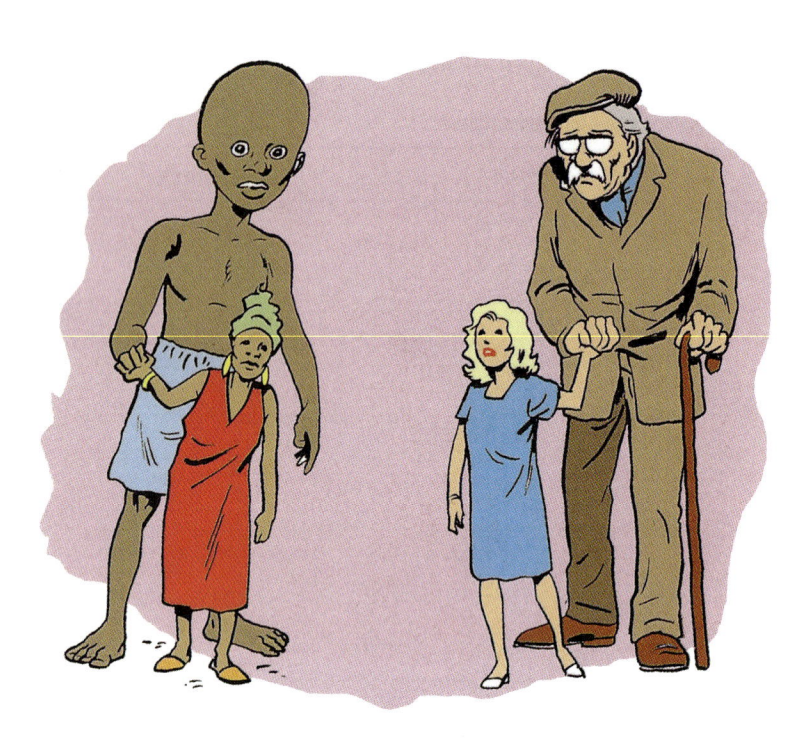

Figure 4.23 The problem of age dependency in rich and poor countries (Source: United Nations Population Fund)

CASE STUDY

4.10 Family planning in Bangladesh

Figure 4.24 Bangladesh

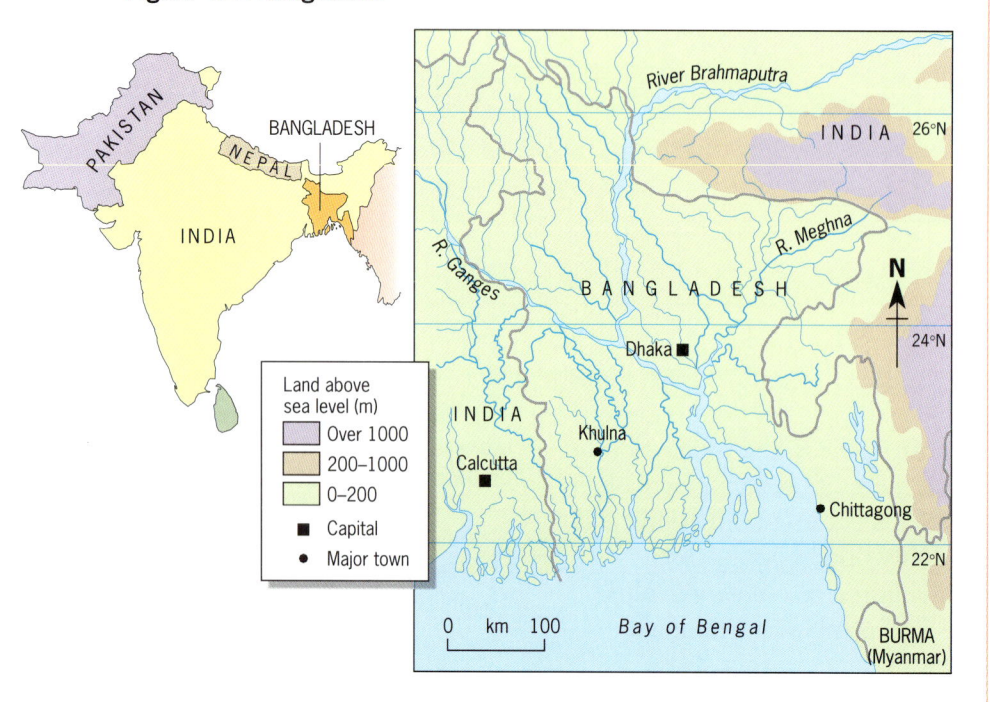

FACTFILE

- Bangladesh has a land area similar to England and Wales.
- The country's total population in 2001 was 131 million – more than twice the population of England and Wales.
- Bangladesh has the highest population density of any country in the world – 980 per sq km in 2001.
- 90 per cent of the country's population are rural dwellers, and most of them work in agriculture.
- Most rural dwellers are poor: one-third are landless and the average size of farm holdings is less than 1 hectare.
- Bangladesh is badly affected by natural disasters. Every year, floods, which last for several months, cover two-thirds of the country.
- The coastal areas are extremely vulnerable to floods caused by typhoons and tidal surges.

4 Population: distribution and change

Population growth and control

Despite its poverty, Bangladesh's population has grown rapidly since 1950 (Fig. 4.26). Alarmed at this growth, the government has, since 1975, adopted a policy of family planning. This policy includes laws that have raised the age of marriage to 18 for women and 21 for men; and that support full-time field workers who provide a contraception service.

The government also works to improve the health of mothers and their babies and women's education. This is because when women are sure that their children will survive they will decide to limit their families to just two or three children. Compared to neighbouring Pakistan and Nepal, population policies in Bangladesh have been relatively successful. Whereas in 1981 only 18 per cent of the population practised family planning, by 2001 the figure was nearly 50 per cent. As a result, the average number of children born to each woman fell from 6 in 1981 to 2.8 in 2001. However, in spite of the success of family planning, Bangladesh's youthful population will ensure continuing growth in future. By 2050, Bangladesh's population should reach 200 million.

Problems

Of course there are still many obstacles to overcome. Because Islam is the dominant religion in Bangladesh, many women are governed by purdah. This means, among other things, that they cannot leave home without permission. Thus, it is often difficult for government agencies and field workers to make contact with women and give them advice about health, nutrition and family planning. Also, despite government legislation, early marriage is still common. Girls often marry by the age of 13 and have their first baby within a year.

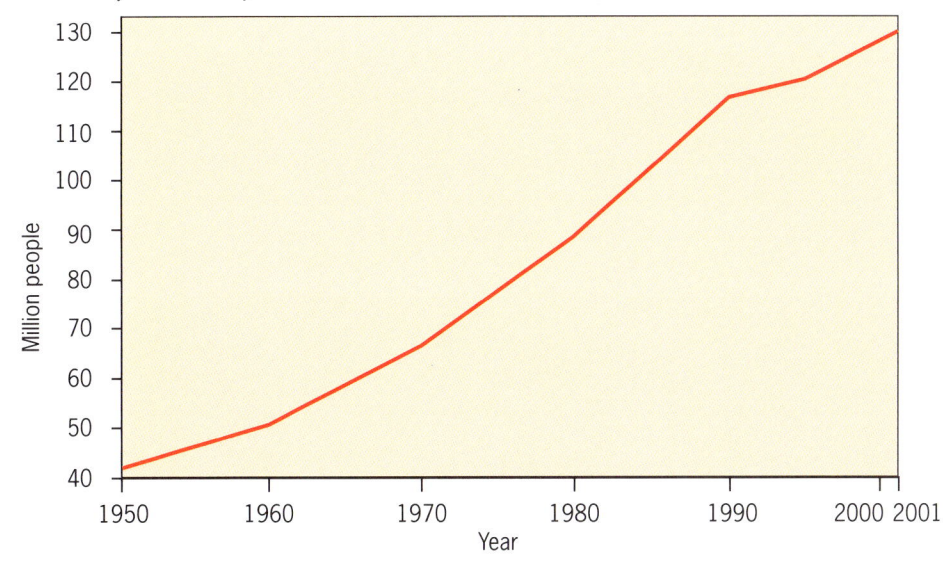

Figure 4.25 Irrigating a rice field, Bangladesh

EXERCISES

15 A family planning officer visits a young Bangladeshi farmer and his wife in the village where they live. They already have three children.
a As the family planning officer, set out the arguments you would use to persuade the couple to accept family planning.
b As the peasant farmer (or his wife), say why you reject the idea of family planning.

REMEMBER
The age structure of MEDCs and LEDCs both cause problems of dependency, although they are very different. MEDCs face the problem of ageing and increasing proportions of old people. LEDCs have youthful populations and struggle to find resources for education, healthcare etc.

Figure 4.26 Population growth in Bangladesh, 1950–2001

4 Population: distribution and change

EXERCISES

16 Study Figure 4.27.
a What is this cartoon saying about • world population growth • the effectiveness of family planning?
b* Do you agree with the message of the cartoon? Justify your response.

Figure 4.27 Family planning? (Source: *The Independent*, 20 May 1992)

TOO LATE!

The greying of the world's population

Problems of ageing populations

The proportion of old people in the world is increasing steadily. Today, average life expectancy exceeds 73 years in North America, Europe and Oceania, and is rising fast throughout the economically developing world (Fig. 4.28). By 2025, nearly one in five of the world's population will be aged 65 and over. This so-called greying of the world's population is most advanced in the rich countries of the developed world.

Ageing causes problems because old people are consumers and not producers – in MEDCs, they are given state pensions. They make heavy demands on medical services, and very old people (aged 80 and over) often require expensive nursing care. Unfortunately, while the number of old people is expanding rapidly, the number of people in work will stay roughly the same.

EXERCISES

17 Log onto www.census.gov/ipc/www/idbagg.html.
a Select two countries (one MEDC and one LEDC).
b For each country, search for information on the following:
(1) CBR and CDR; (2) natural increase/decrease rate;
(3) current population total and forecast total in 2020;
(4) population density; (5) age structure (i.e % population aged 0–14, 15–39, 40–64, 65+).
c Present the information as a table, then describe and explain the differences in population growth and population structure between the two countries in 2001.

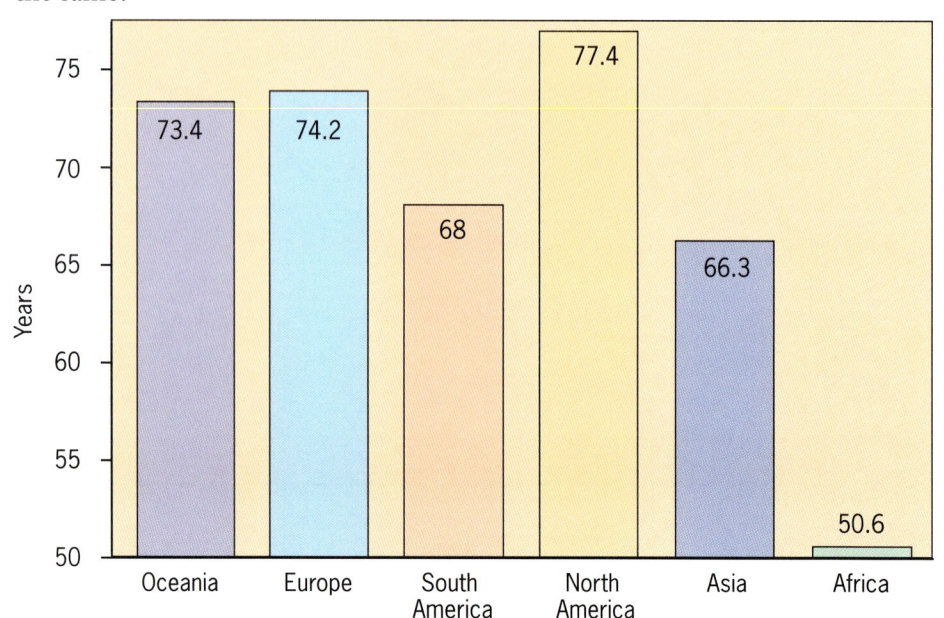

Figure 4.28 Life expectancy at birth in 2001

Region	Years
Oceania	73.4
Europe	74.2
South America	68
North America	77.4
Asia	66.3
Africa	50.6

4 Population: distribution and change

Solving the ageing crisis

Governments throughout the developed world face the same problem: how to pay for the rising numbers of old people. Various solutions have been proposed.

- Encourage immigration of young adults from poorer countries.
- Give couples financial incentives (e.g. tax concessions, family allowances) to have children.
- Raise the age of retirement so that old people remain in work (and therefore do not receive pensions) for longer.
- Raise taxes on the working population to fund the growing demand for pensions.

Ageing is not confined to MEDCs. In 2001, there were twice as many old people in poor countries as in the rich world. With fewer resources available in LEDCs, the greying of the economically developing world will create huge problems in future.

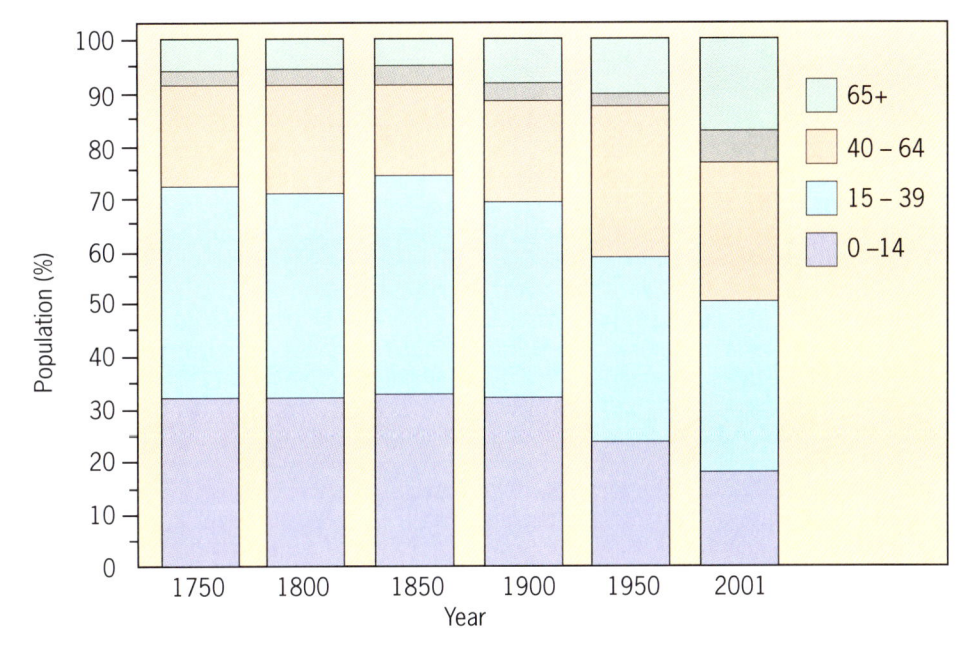

> ### EXERCISES
>
> **18** Sweden is the 'oldest' country in the world. Show how Sweden's population has aged since 1950 by calculating the old-age index from the data in Figure 4.29.
>
> Old-age index = $\dfrac{\% \text{ aged } (65+)}{\% \text{ adults } (15\text{–}64)}$
>
> **19** What are your views on the UK's ageing population? Write out your solution, explaining your reasoning in full. Be prepared to state and justify your viewpoint in a class discussion on the issue.

Figure 4.29 (left) Sweden's changing age structure, 1750–2001

Figure 4.30 (below) China's children face a heavy burden supporting older generations

4 Population: distribution and change

4.11 Summary: Population: distribution and change

KEY SKILLS OPPORTUNITIES
C1.2: Ex. 5a, 11a; C1/2.3: Ex. 4, 9, 15a, 15b, 16; C2.1: Ex. 19; C2.2: Ex. 2, 5b, 5c, 5d, 10b, 12a; N1/2.1: Ex. 8c, 10c, 11c, 14a; N1/2.2: Ex. 1, 3a, 6a, 6b, 7a, 8a, 13a, 14a, 18; N1/2.3: Ex. 3b, 3c, 6c, 6d, 7b, 8b, 13b, 14b; IT1/2.1: Ex. 17a, 17b; IT1.2: Ex. 17c.

Key ideas	Generalisations and detail
The global distribution of population is very uneven.	• The average density of population (excluding Antarctica) is 47 per sq km. However, there are large geographical variations. • Large population concentrations are found in Europe, North-east USA, East and South-east Asia. • Very low densities exist in most of Oceania, Africa north of the equator, Siberia and Amazonia. The southern hemisphere is more sparsely populated than the northern hemisphere.
Physical and human factors influence the global distribution of population.	• Physical factors determine the broad pattern of population distribution, providing both opportunities for, and limits to, settlement. Climate, relief and altitude (mountains and plains) are important factors. • Human factors include the types of economic activity in a place (i.e. farming, industry, services); levels of technology and development; length of occupancy, etc.
The world's population is growing rapidly.	• Growth in the second half of the 20th century has been especially rapid. By 2000, the world's population exceeded 6 billion. Growth will eventually level out in the mid 21st century, but not before the population has reached 10 or 11 billion.
At the global scale, the difference in numbers of births and deaths causes population change.	• Births are measured by the crude birth rate (CBR); deaths by the crude death rate (CDR). The difference between the CBR and the CDR is the natural increase rate. • The current rapid growth of world population is due to an excess of births. This results from a fall in the CDR but a continuing high CBR.
LEDCs are mainly responsible for the current rapid growth of the world's population.	• Population growth is mostly in LEDCs. Here CBRs remain high, but CDRs have fallen because of improvements in health and hygiene. • MEDCs contribute only 5 per cent of current world population growth.
Population change is often related to advances in economic development.	• The demographic transition describes the population changes that occurred in MEDCs from the early 19th century. These changes accompanied industrial development and urbanisation. • Population changes in today's LEDCs appear to be more rapid and are occurring for different reasons. • AIDS will have a big impact on population growth in LEDCs in the next 20 years. Some countries in Africa may suffer a decline in population.
Social, religious, political and environmental factors also affect birth rates.	• Apart from levels of economic development, factors that influence birth rates include: the status of women; attitude of religion towards contraception; health/hygiene and the survival of children; diet; housing; and government policies aimed at either encouraging or discouraging births.
Age–sex structure is summarised in population pyramids.	• LEDCs, such as Nigeria, typically have broad-based, triangular-shaped pyramids. They result from high CBRs and CDRs, and indicate youthful populations. • In MEDCs, population pyramids are narrow and straight-sided. They represent older populations, with low CBRs and low CDRs, e.g. Sweden.
Changes in age structure give rise to difficult issues.	• In LEDCs, age dependency is high because of the large proportion of children. Children have to be supported by the working population and make heavy demands on education and health care services. Also, large numbers of children mean further rapid growth 10–20 years later. The burden of dependency will increase significantly in LEDCs most badly affected by AIDS. • MEDCs have a problem of increasingly large numbers of old people. This group relies on the working population for pensions and health care. At present, the proportion of working adults is decreasing, thus making the dependency problem worse.

5 Population: migration and resources

5.1 Introduction

In the last chapter we saw how the difference between births and deaths causes population change at the global scale. However, when we look at population change at a smaller scale (within a country or region) a third factor – **migration** – begins to have an influence. Migration is the focus of the first part of this chapter. In the second part we deal with the effect of population growth on resources such as food, forests, soils and water.

5.2 What is migration?

Migration is the permanent relocation of an individual or group. The term usually describes movement over some distance, at least from one region to another. If such movement occurs between countries (i.e. international migration) we usually describe it as either **immigration** (migrants coming into a country) or **emigration** (migrants leaving a country).

Migrations vary not only in their length, but also in their direction. For instance, in LEDCs migration from rural to urban areas is most common. This contrasts with MEDCs where most migration is in the opposite direction, from urban to rural areas.

5.3 Why do people migrate?

People migrate for one of the following reasons: economic, social, political and environmental. For a majority of migrants, economic reasons are probably most important. In particular, people move to improve their job prospects, income and standard of living. Sometimes social reasons may also be crucial. Migrants may be attracted by better educational

EXERCISES

1 Consider the following population movements:
- nomadic herders moving with their livestock;
- people moving from the countryside to live in towns and cities;
- your journey to school;
- people moving from one country to settle in another;
- people moving house within a town or city;
- a shopping trip to the local supermarket.

For each of these movements, say whether or not it is migration. Give reasons for your decision.

Figure 5.1 Rural-urban migration in the developing world: push and pull factors

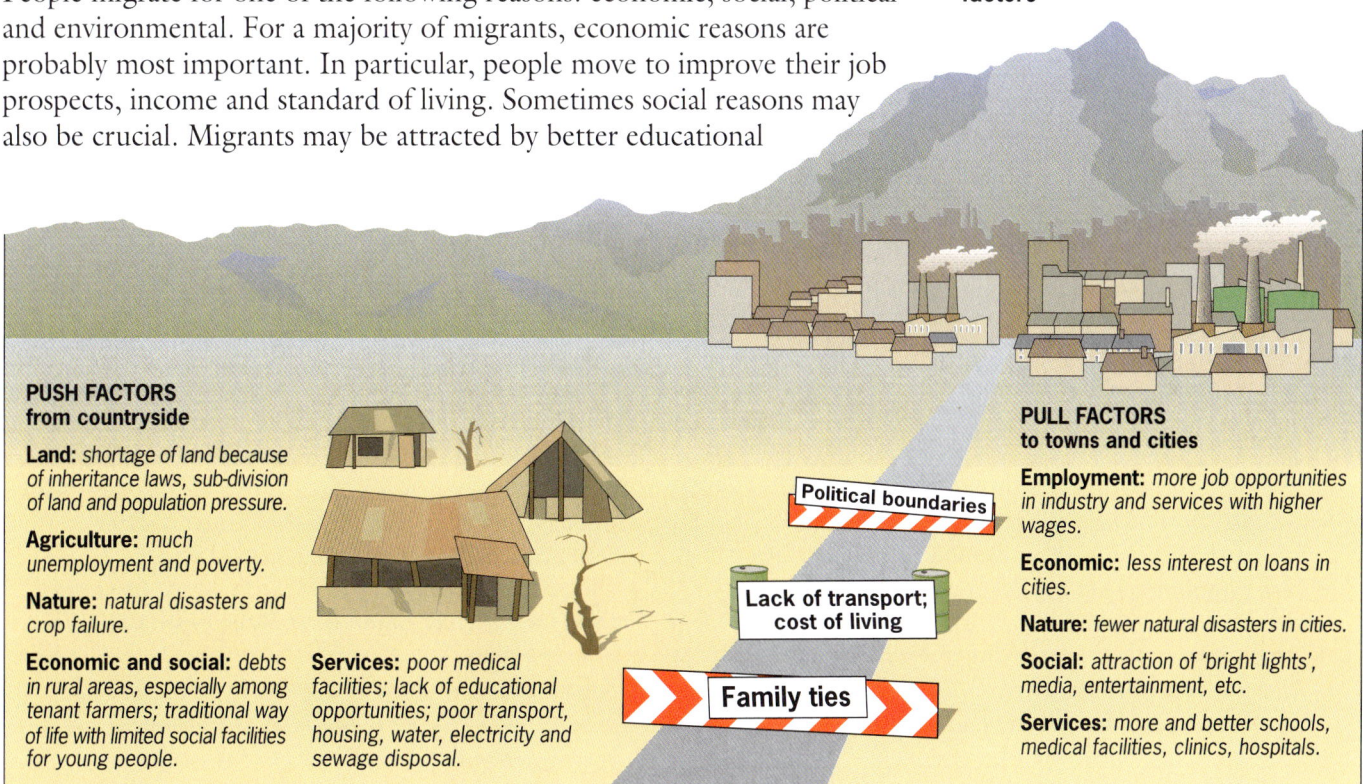

PUSH FACTORS from countryside

Land: shortage of land because of inheritance laws, sub-division of land and population pressure.

Agriculture: much unemployment and poverty.

Nature: natural disasters and crop failure.

Economic and social: debts in rural areas, especially among tenant farmers; traditional way of life with limited social facilities for young people.

Services: poor medical facilities; lack of educational opportunities; poor transport, housing, water, electricity and sewage disposal.

Political boundaries

Lack of transport; cost of living

Family ties

PULL FACTORS to towns and cities

Employment: more job opportunities in industry and services with higher wages.

Economic: less interest on loans in cities.

Nature: fewer natural disasters in cities.

Social: attraction of 'bright lights', media, entertainment, etc.

Services: more and better schools, medical facilities, clinics, hospitals.

5 Population: migration and resources

Figure 5.2 (above left) Nanjing Road, Shanghai, China

Figure 5.3 (above right) Living in poor conditions, Poona, India

opportunities, better medical services or the desire to join family or friends who have already moved.

All of these reasons have one thing in common: they are voluntary. However, many other migrants have to move. For example, people suffering religious or political persecution, or caught up in wars or environmental disasters, may be forced to flee for their lives.

Push and pull factors

Push and **pull factors** provide an explanation for migration.
- Push factors are the disadvantages found at a migrant's place of origin. They might include a lack of jobs, poor services, religious persecution and so on.
- Pull factors are the advantages of a migrant's place of destination. These positive factors are most often the opposite of push factors (e.g. availability of jobs, educational opportunities etc.).

Many people ignore both push and pull factors and decide not to migrate. Often, family ties are just too strong for people to uproot themselves. Also there may be obstacles that block the movement of potential migrants. These include the cost of moving, isolation and remoteness, the difficulty of crossing national frontiers and so on.

EXERCISES

2 Study Figures 5.2 and 5.3.
a You are a peasant farmer in an LEDC. Your image of the city is like that in Figure 5.2. What features of the city might tempt you to migrate there?
b* You are a migrant to the city and live in the area shown in Figure 5.3. How does your experience of the city differ from that of the peasant farmer's image?

CASE STUDY

5.4 Rural-urban migration in Peru

FACTFILE

- In 2001, Peru had a population of 27.5 million, making it the fourth-largest country in South America.
- Peru's average population density is low: just 21.5 per sq km.
- Peru comprises three regions (Fig. 5.4): the desert that follows the Pacific coast; the Andes mountains; the hills and forest-covered lowlands of the remote east.
- Just over half of the country's population lives in the desert region. Most are in the capital, Lima.
- One in three Peruvians live in the Lima metropolitan area. Overall, three-quarters of Peru's population live in urban areas.
- **Rural-urban migration** dominates population movements within Peru and is the main cause of rapid urban growth and **urbanisation**.

5 Population: migration and resources

Causes of rural-urban migration

Compared with standards in the developed world, living conditions in towns and cities in the developing world are poor. Jobs are scarce; decent housing is in short supply; and essential services like clean water, electricity and sewerage systems are often non-existent (Fig. 5.5). Despite this, migrants continue to pour into cities like Lima. Most are farmers and farm labourers from poor regions like the Andes (Fig. 5.6). They know that wages are higher in the cities and that there are more jobs available (Table 5.1). Of course, not all migrants succeed in finding a job. Many have to rely on self help, and survive by washing cars, running errands, selling newspapers, recycling waste, and so on.

Although most migrants remain poor, few want to return to the countryside. According to the UN, rural households in Peru are three times more likely to be poor than urban households; infant mortality is twice as high, and malnutrition more common. Although one-third of Lima's inhabitants live in squalid camps, housing conditions in the countryside are even worse.

Selectivity of rural-urban migration

Some types of people are more likely to migrate than others. We can say, therefore, that migration is a selective process. In Peru, most migrants are:
◆ Young adults between the ages of 15 and 40.
◆ Better educated and have more skills than non-migrants.
◆ Predominantly female. Women are more likely to migrate than men because towns and cities offer women a wider range of jobs, particularly in office cleaning, domestic service, shops and street selling.

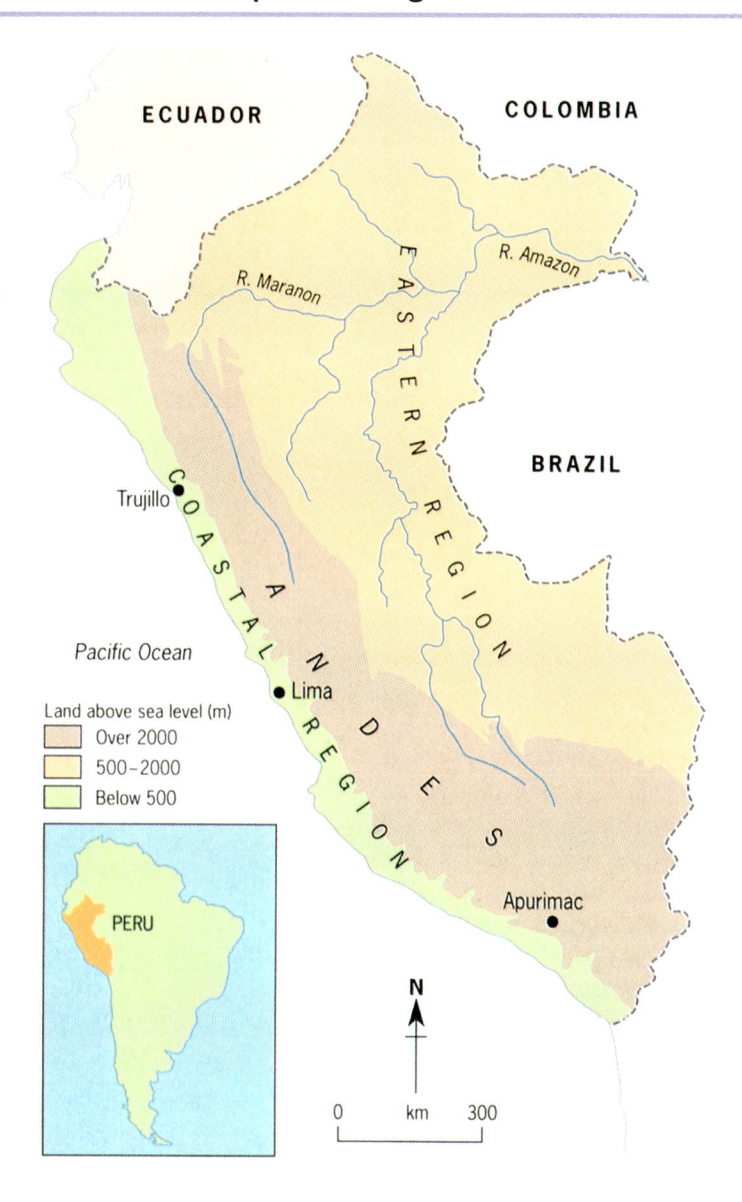

Figure 5.4 (above) Peru: principal cities and regions

Figure 5.5 Slum dwellings on the banks of the River Rimac, Lima, Peru

Figure 5.6 Peasant farm on eastern slopes of the Andes

83

5 Population: migration and resources

Table 5.1 Reasons for migrating from the countryside to urban areas in Peru

Reason	Percentage citing reason
To earn more money	39
To join family	25
No work in villages	12
Work available in towns	11
Dislike of village life	11
Poverty	9
To pay for education	7

EXERCISES

3a Draw a bar chart to show the information in Table 5.1.
b Describe the importance of economic and social factors as causes of rural-urban migration in Peru.
4 The population pyramids (Fig. 5.7) show the age-sex structure for Lima, the rural region of Apurimac in the Andes, and all of Peru. The Lima and Apurimac pyramids both show the effects of rural-urban migration.
a Describe the main features of each population pyramid.
b* Identify the pyramids for Lima and for Apurimac.
c* Give reasons for your choices.

The impact of rural-urban migration

Rural-urban migration has both positive and negative effects for cities and for the countryside.

Impact on urban areas

In urban areas, the migrants provide industry and commerce with a young and cheap workforce. However, the scale of migration is sometimes so great that cities are overwhelmed. We can see the effects in lack of housing; the absence of basic services, such as clean water, sewerage, electricity and schools; and insufficient jobs. This results in millions of urban dwellers living in poverty (see Book 1, Chapter 7).

Impact on the countryside

In the countryside, rural-urban migration creates different problems. Because migration is selective, there is often a shortage of young adults. This has several effects. It may cause rural birth rates to fall. Food production, which supports urban as well as rural populations, may also decline. Finally, the loss of young people is made worse because the migrants are usually better educated and more 'go-ahead' than non-migrants.

Even so, rural areas can also gain from migration. One important benefit is money that is sent back to rural areas by the migrants who work in cities. In Peru, rural-urban migration has also helped to reduce the pressure of population on land in the Andes and coastal plain.

5.5 Migration within the UK

In the UK, regional differences in population growth and decline are largely due to migration (Fig. 5.8). Counties like Cambridgeshire and Northamptonshire grew rapidly during the 1980s and 1990s. The main reason for growth was that more people moved into, rather than moved out

Figure 5.7 Population pyramids for Peru, Lima and rural Apurimac

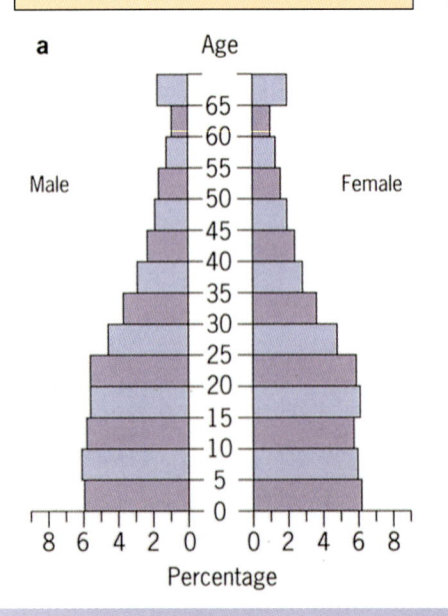

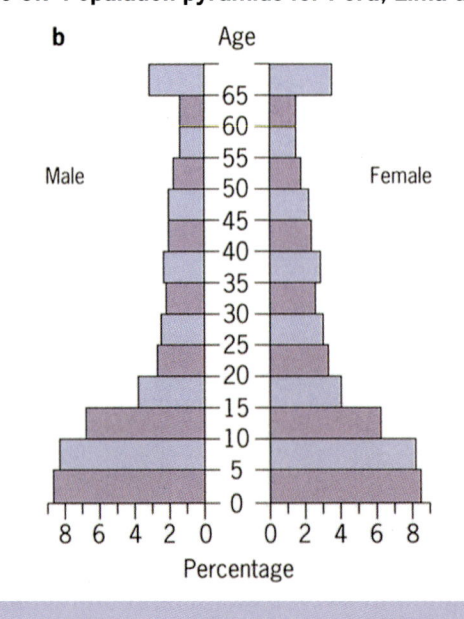

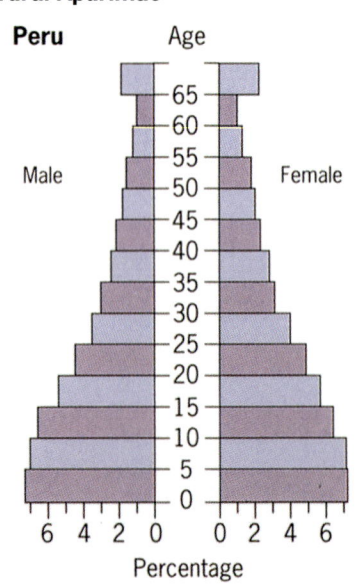

5 Population: migration and resources

of these counties. When this happens, we say that a county has a **net migration gain**. Meanwhile, the population of counties such as Merseyside and Tyne and Wear declined. This too was the result of migration, only here more people moved out than moved in. In other words, these counties suffered a **net migration loss**.

Push from the North: pull to the South

Two separate migration streams account for the pattern of population change in Figure 5.8. First, there has been a steady shift of population from North to South (Fig. 5.9). This is not new. For most of the past 100 years, the South has been more prosperous than the North. As a result, people from northern England and Scotland have moved south to find jobs.

Figure 5.8 Population change in the UK, 1981–1998

Counter-urbanisation

The second migration stream is the urban-rural shift of population, also known as **counter-urbanisation**. Counter-urbanisation describes the net movement of people out of conurbations and large cities to smaller urban centres and rural areas. Thus, while all the metropolitan counties (i.e. conurbations) declined in population between 1981 and 2001, the more rural counties of southern Britain, stretching in a broad belt from Cornwall to Lincolnshire (Fig. 5.9), grew rapidly. There are three main factors responsible for this urban-rural shift:

- Push and pull factors.
- Increased mobility, especially private car ownership.
- Retirement migration.

Figure 5.9 (below) Population change by region, 1981–1998

Push and pull factors

Many people have been attracted to smaller towns and rural areas by the promise of a better quality of life. Equally, large cities often encourage people to migrate because of pollution, traffic congestion, high rates of crime, poorer schooling and so on.

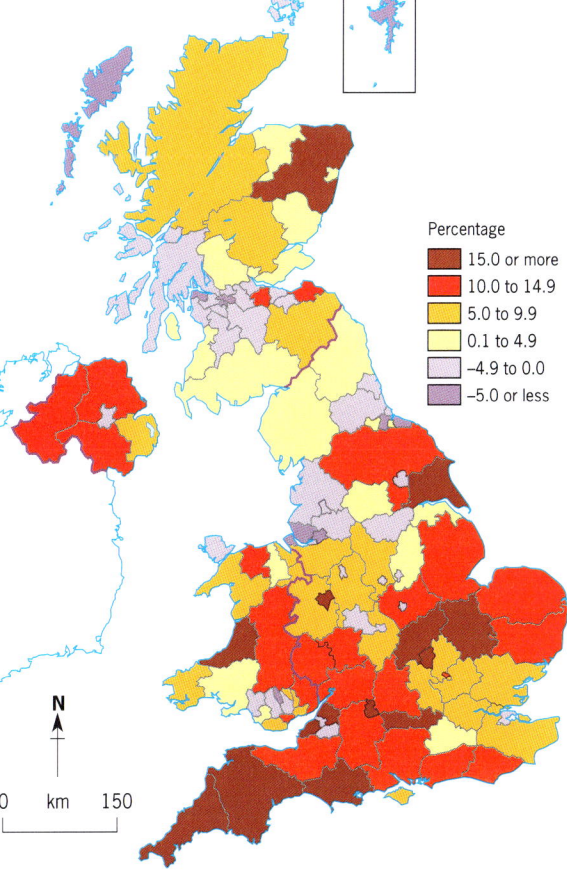

Figure 5.10 (left) Decaying inner city Bradford, West Yorkshire

85

5 Population: migration and resources

EXERCISES

5 Study an atlas map of the UK, and Figure 5.8.
a Make a list of those counties that had a growth rate of more than 10 per cent and those that lost population, between 1981 and 1998.
b* Describe the distribution of the counties that grew most rapidly and those that suffered population decline.
c* Explain the pattern of population change you described in **b**.

Mobility

Counter-urbanisation would not have been possible without private car ownership, improved transport links, and rising incomes, which have made people better off. Most migrants have moved only short distances, often to commuter settlements within an hour's journey time of a large city. However, some have moved further afield to remote rural areas, such as the Highlands of Scotland. These people often set up their own businesses or use modern telecoms (e.g. E-mail, fax etc.) to work from home.

REMEMBER
Migration in the UK operates at the national and regional scale. At the national scale, there is the drift form North to South. At the regional scale, population has decentralised from large cities to surrounding rural hinterlands.

Retirement

Thanks to occupational pensions and home ownership, many people are relatively well off when they retire. They can afford to move to environmentally attractive areas, such as the south coast of England, East Anglia and the Lake District. As a result, these areas have recorded significant population growth in the past 30 years.

CASE STUDY

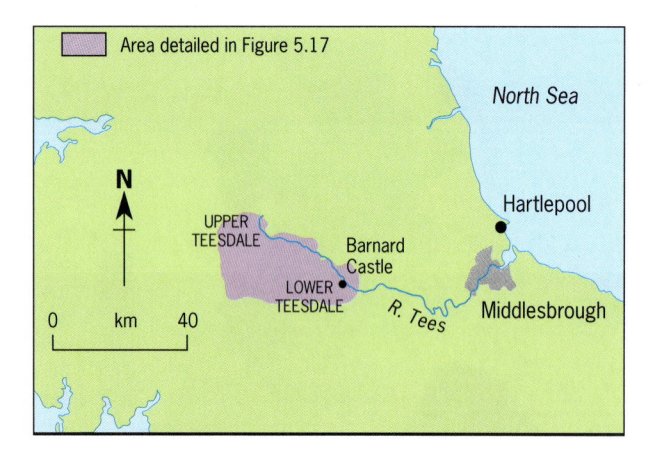

Figure 5.11 (above) North-east England

Figure 5.12 Sparse settlement, Upper Teesdale, County Durham

5.6 Population change in Teesdale

Upper Teesdale

The Tees valley above the market town of Barnard Castle in North-east England is a remote rural area (Fig. 5.11). Bleak moorlands rising to nearly 800 m dominate the upper parts of the dale. Although undeniably beautiful, Upper Teesdale is a wild and isolated place (Fig. 5.12). Its main economic activity – hill farming – supports only sparse population densities.

Like many other remote parts of highland Britain, Upper Teesdale is a region in crisis. Its population declined by almost one-third between 1951 and 1991 (Fig. 5.13).

Figure 5.13 Population change: Upper Teesdale, 1951–91

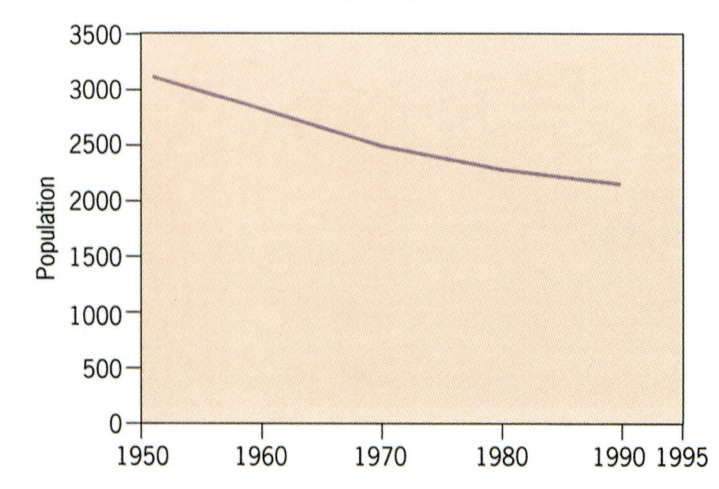

5 Population: migration and resources

The reason for this **depopulation** is out-migration. Young people are most likely to leave. They see little future in hill farming, and few other jobs are available locally. Add to this the poor level of services and the remoteness of Upper Teesdale, and migration seems the only option. This migration has two effects. First, it leaves an ageing population; and second, falling numbers lead to a further decline in services.

Lower Teesdale

Twenty kilometres down valley, the landscape is softer, farming is more prosperous and the dale is less isolated. Here, the experience of population change in the last 25 years has been very different. Since 1971, depopulation has been replaced by a population revival (Fig. 5.14). This revival has been so strong that it produced a 10 per cent growth between 1971 and 1991. How has this happened? Once again, the key is migration: the number of people moving into this part of Teesdale has exceeded the number moving out.

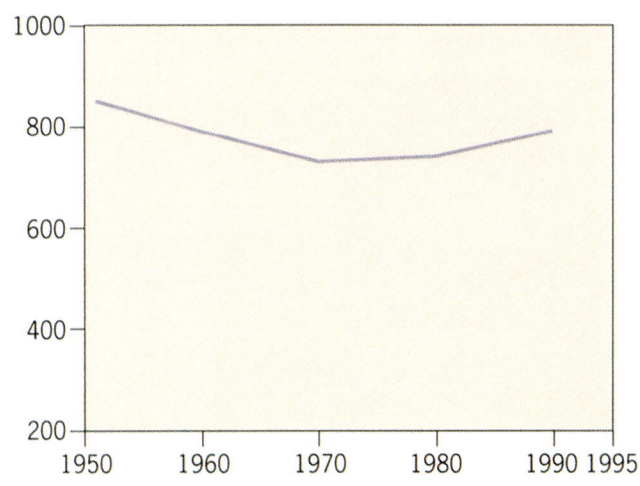

Figure 5.14 (above) Population change: Cotherstone, Hunderthwaite and Lartington, 1951–91

Figure 5.15 (left) Renovated cottages and new housing, Eggleston, Teesdale

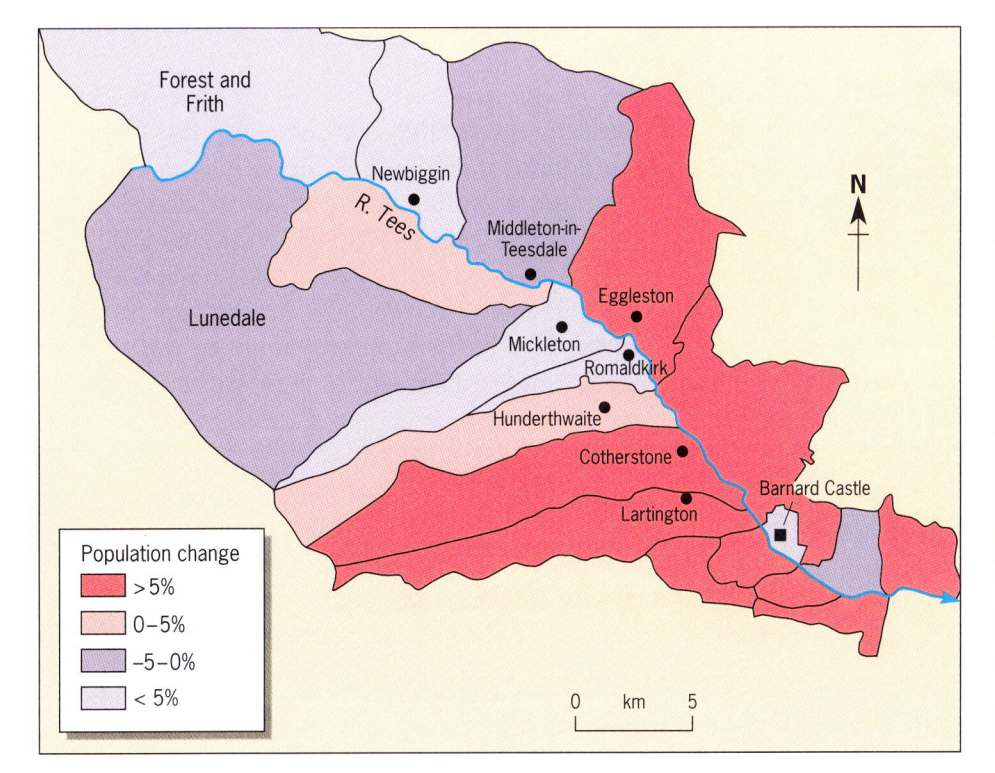

Figure 5.16 (above) Derelict village shop, Teesdale

Figure 5.17 (left) Population change in Teesdale, 1981–91

5 Population: migration and resources

> **EXERCISES**
> 6 Study Figure 5.17.
> a Describe the general pattern of population change in Teesdale between 1981 and 1991.
> b What is meant by the term 'net migration change'?
> c* Explain how net migration change influences depopulation and counter-urbanisation in Teesdale.

Population change and services

These population movements in the lower dale have led to changes in its population structure. Whereas the people moving out are mainly young and single, the in-comers are often retired couples and commuters. They are part of the process of counter-urbanisation. Consequently, many of the villages where they have settled have a new prosperity. As well as building new homes, people have converted barns and renovated farm cottages and outbuildings (Fig. 5.15). Even so, this has not stopped the decline of village services such as shops, primary schools and public transport (Fig. 5.16). This is because most in-comers have cars, are relatively well off and have no children. Because the main service centre for Teesdale – Barnard Castle – is less than 20 minutes away by car, there is little demand for village services.

> **EXERCISES**
> 7 Study Fig. 5.18.
> a Which country has (1) the largest foreign-born population; (2) the largest number of immigrants?
> b Suggest possible reasons for (1) the large proportion of foreign-born people in Australia, Canada and Switzerland (2) the small proportion of foreign-born people in Japan.

5.7 International migration

About 70 million people migrate between countries each year. We refer to these population movements as international migration. This type of migration is less common than internal population movements, such as rural-urban and urban-rural migration (see sections 5.4 and 5.5 respectively). There are two reasons why international migration is less common. First, the longer distances involved, and second, political controls make it difficult for migrants to move freely between countries.

Economic migration from Mexico to the USA

The border between the USA and Mexico (Fig. 5.19) is more than an international frontier. It is one of the few places in the world where the rich, economically developed world meets the poor, economically developing world.

Given the economic contrasts that exist on either side of the border, it is little wonder that the USA has proved an irresistible attraction for millions of Mexicans (Fig. 5.19). Today there are more than 30 million people of Latin American origin (Latinos) living in the USA, of whom 20 million are from Mexico. Every year about one million Mexicans cross the border into the USA. Seventy per cent are illegal immigrants (Fig. 5.20). Most are caught by the immigration authorities and are sent back to Mexico.

Causes of migration

The causes of migration from Mexico and other Central American countries to the USA are mainly economic. Poverty is the main push factor. Higher wages in the USA (five times those in Mexico for similar work) and the prospect of a better service in education and health care are major pull factors.

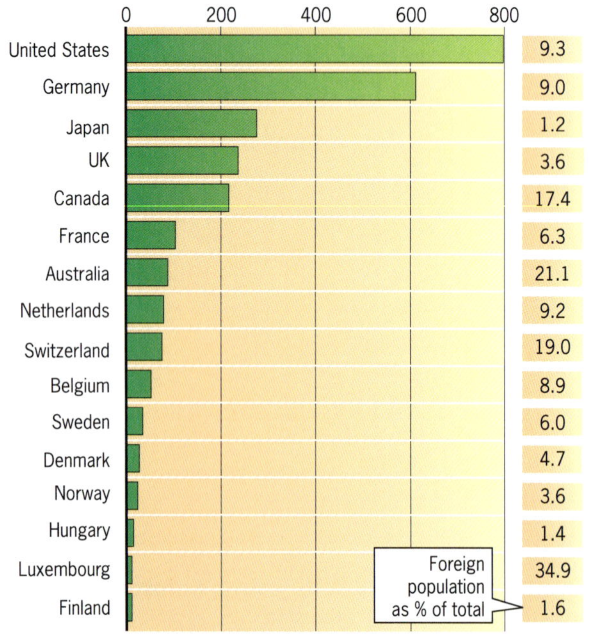

Country	Immigrants (thousands)	Foreign population as % of total
United States	~750	9.3
Germany	~600	9.0
Japan	~270	1.2
UK	~230	3.6
Canada	~210	17.4
France	~120	6.3
Australia	~100	21.1
Netherlands	~100	9.2
Switzerland	~90	19.0
Belgium	~55	8.9
Sweden	~45	6.0
Denmark	~25	4.7
Norway	~20	3.6
Hungary	~15	1.4
Luxembourg	~10	34.9
Finland	~10	1.6

Figure 5.18 Immigration into MEDCs, 1997 (thousands)

5 Population: migration and resources

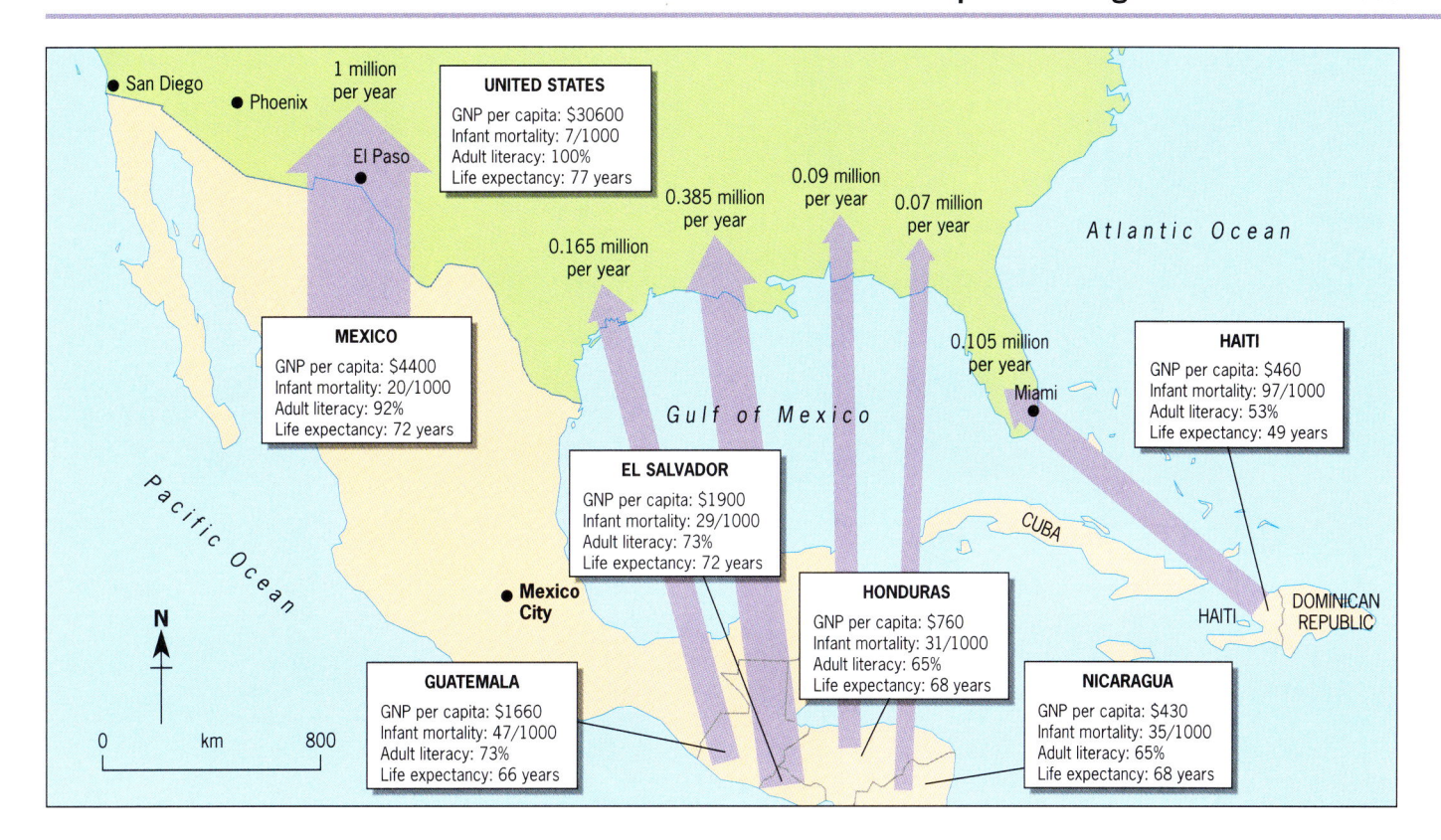

Figure 5.19 Illegal immigration from Central America to the USA

Map data:

- **UNITED STATES**: GNP per capita: $30600; Infant mortality: 7/1000; Adult literacy: 100%; Life expectancy: 77 years
- **MEXICO**: GNP per capita: $4400; Infant mortality: 20/1000; Adult literacy: 92%; Life expectancy: 72 years
- **GUATEMALA**: GNP per capita: $1660; Infant mortality: 47/1000; Adult literacy: 73%; Life expectancy: 66 years
- **EL SALVADOR**: GNP per capita: $1900; Infant mortality: 29/1000; Adult literacy: 73%; Life expectancy: 72 years
- **HONDURAS**: GNP per capita: $760; Infant mortality: 31/1000; Adult literacy: 65%; Life expectancy: 68 years
- **NICARAGUA**: GNP per capita: $430; Infant mortality: 35/1000; Adult literacy: 65%; Life expectancy: 68 years
- **HAITI**: GNP per capita: $460; Infant mortality: 97/1000; Adult literacy: 53%; Life expectancy: 49 years

Migration flows per year: Mexico to USA 1 million; 0.165 million; 0.385 million; 0.09 million; 0.07 million; Haiti to Miami 0.105 million.

The impact of immigration in the USA

Rates of immigration to the USA were extremely high throughout the 1990s. In 2000, immigrants accounted for:

- 11 per cent of the US population.
- Two-thirds of all births – immigrant birth rates were three times greater than the native-born population.

More than half of all immigrants were Latinos, and two-thirds of these were from Mexico. By 2050, there will be an estimated 100 million Latinos in the USA. They will comprise one-quarter of the country's total population.

Table 5.2 The economic advantages and disadvantages of immigration to the USA

Advantages	Disadvantages
Immigrants are youthful. They contribute to the workforce and to taxes. They lower the average age of the population and help to support the US's 'greying' population.	Immigrants have larger families than native-born Americans and make greater demands on education and welfare services.
Immigrants are consumers. Their demand for goods and services helps generate employment and wealth.	Many immigrants are poorly educated and unskilled. There is limited demand for this type of labour in a modern economy.
Immigrants often take low-paid jobs that native-born Americans are not prepared to do (e.g. domestic servants).	Immigrants may depress wages and take jobs that would otherwise be filled by the native population.
Immigrants help to keep some industries going, which otherwise might disappear e.g. clothing industry in Los Angeles, which survived in the 1980s by employing Latinos prepared to work for low wages.	Older immigrants reach retirement age and draw welfare benefits, which they have not paid for.

EXERCISES

8a Study Fig. 5.19 and suggest reasons for the scale of illegal immigration to the USA from Central America.

b Immigration into the USA, especially into states like California, is controversial. Study the economic arguments for and against immigration in Table 5.2. State your view on the issue and explain in class discussion your reasons in detail.

Figure 5.20 An illegal immigrant passes under the border fence from Mexico into the USA.

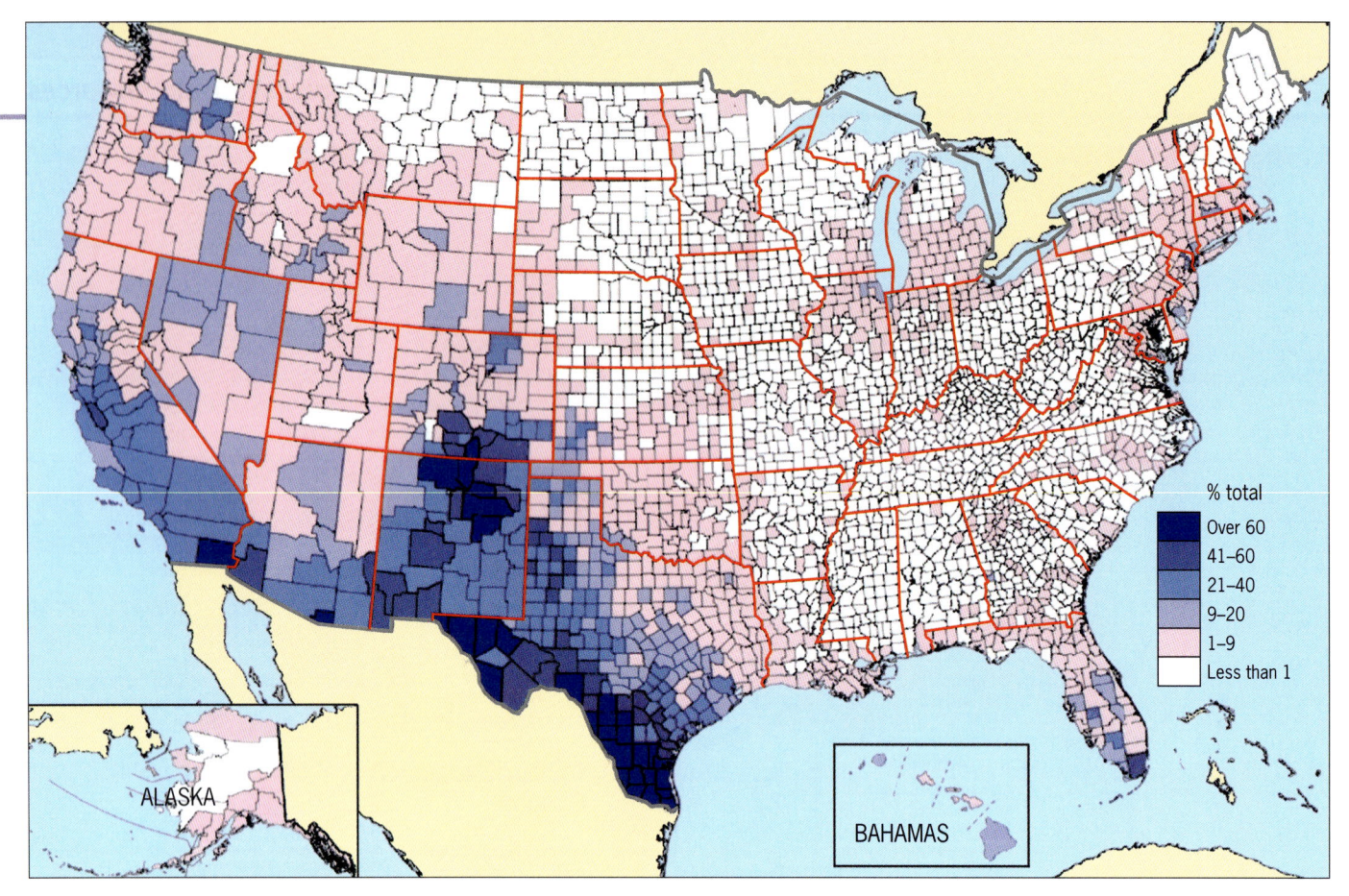

Figure 5.21 Distribution of Latinos in USA

Distribution of the Latino population in the USA

Most immigrant Latinos have settled in California and Texas (Fig. 5.21). Here they concentrate in major cities, such as Los Angeles, San Diego and San Francisco. Forty per cent of Los Angeles's population was born outside the USA and immigrants were responsible for half of all population growth in California in the 1990s.

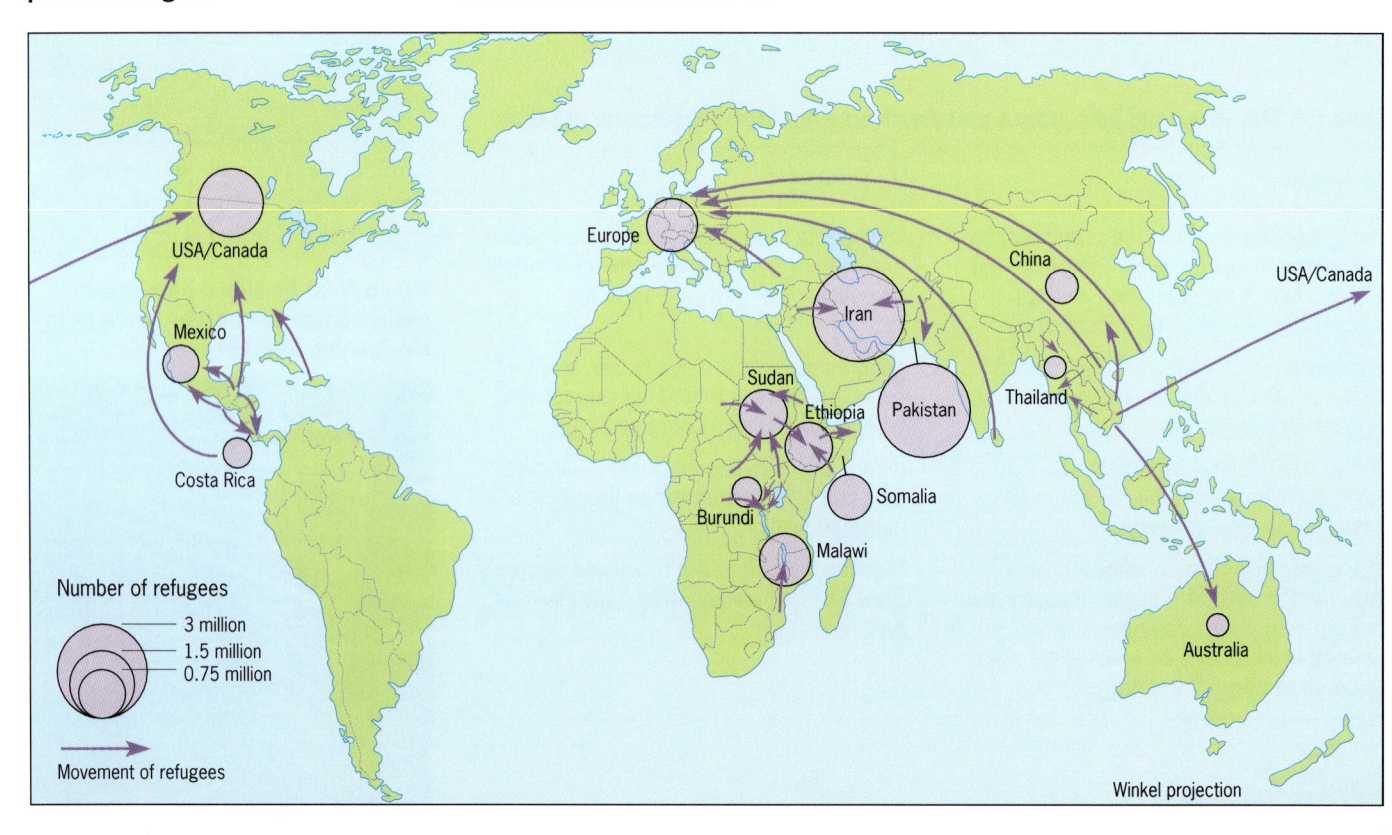

EXERCISES

9a Study Fig. 5.21 and name five states that have large concentrations of Latinos.
b Explain the distribution of Latinos in the USA.

Figure 5.22 Global movement of political refugees

5 Population: migration and resources

Refugees: unwilling migrants

People who are forced to migrate because of persecution, wars or environmental disasters are known as **refugees**. In 1999, about 7 million people were uprooted and formed part of an estimated 35 million refugees worldwide.

Internal and international refugees

Refugees may be displaced internally or internationally. For example, in 2000, 4 million Sudanese were refugees in their own country, trying to escape civil war. Often, refugees must flee their own country to escape persecution. In 1999, there were more than 100 000 international refugees in the UK, mostly from Yugoslavia, Somalia, Iraq and Sri Lanka. However, refugees are not confined to the world's richer countries. LEDCs accounted for nearly 15 million refugees in 1999, of whom 6 million were in the Middle East.

Refugees in the UK

In 1999, European countries received 363 000 asylum seekers from overseas. (An asylum seeker is someone who is applying for refugee status and permission to stay permanently in a country.) Germany was the most popular destination for asylum seekers, followed by the UK.

The number of people applying for asylum in the UK in 1999 – 112 000 – was the highest ever (Fig. 5.23). The largest group was Kosovar Albanians, displaced by the war between Kosovo and Serbia (Fig. 5.24). Most asylum seekers were young adults. Their average age was just 26 years.

The issue of asylum seekers

- Asylum seekers are granted permission to remain permanently in the UK if they prove that deportation will put their lives at risk.
- Asylum seekers are controversial and often arouse hostility from the native population.
- Many people believe that a large proportion of asylum seekers are not fleeing persecution but are economic migrants in search of a better life.
- People resent paying extra taxes to look after asylum seekers. The cost often falls most heavily on the people who live close to points of entry to the UK, such as the port of Dover and Heathrow airport. The local authorities in these places have the initial responsibility of looking after the asylum seekers. This means raising extra taxes to pay for them.

EXERCISES

10 In 2000, the world's largest groups of refugees were from Kosovo, Sudan, East Timor, Congo, Chechnya, Angola, Afghanistan, Iraq, Sierra Leone and Colombia. Use the internet to find out why each of these countries has, in recent times, produced so many refugees.

11a Study Figure 5.22. Which two countries have the largest number of refugees?
b Which are the likely countries of origin of the refugees in Iran and Pakistan?

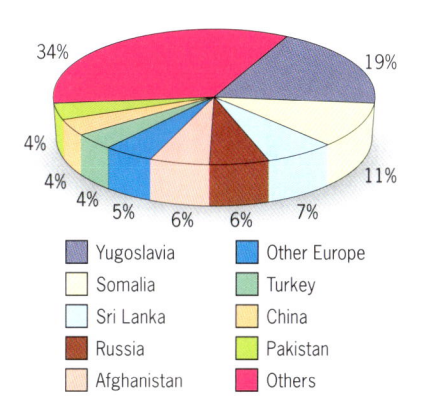

Figure 5.23 Pie chart of UK asylum seekers, 1999

- Yugoslavia 19%
- Other Europe 11%
- Somalia 7%
- Turkey 6%
- Sri Lanka 6%
- China 5%
- Russia 4%
- Pakistan 4%
- Afghanistan 4%
- Others 34%

EXERCISES

12a How do refugees differ from other immigrants?
b Outline the economic and humanitarian arguments for granting asylum seekers refugee status. (See Table 5.2)
c Why is the issue of asylum seekers in the UK controversial?

Figure 5.24 Refugees in Northern Albania, picked up after walking over the mountains from Kosovo

REMEMBER

International migration may be voluntary or forced. Voluntary migration is driven mainly by migrants seeking a better quality of life. Forced movements result from war, persecution and environmental disasters. These migrants are classed as refugees.

5 Population: migration and resources

5.8 Population growth: opportunities and problems

By 2050, the Earth's population will have grown to more than 9 billion. Such growth is bound to have important consequences for food supply. Will it be possible to feed so many people? And if so, can it be done without damaging the soils, water and forests on which we all ultimately depend?

Malthusian theory

The relationship between population growth and food supply has been controversial for more than 200 years. The debate began in 1798 when Thomas Malthus published his *Essay on Population*. He argued that:
- Food supply could never grow as quickly as population.
- Food supply grows arithmetically (1, 2, 3, 4, 5, etc.) but population increases geometrically (1, 2, 4, 8, 16, etc.).
- Food supply would always set a limit to population growth.

Malthus warned of disastrous consequences: unless people had fewer children, famine, disease and war would 'check' population growth. Eventually, the ratio between population and food supplies would balance again, but only after terrible suffering.

Since Malthus first put forward his ideas, the world's population has increased from one billion to more than six billion. However, the catastrophe he predicted failed to happen. This was because the growth in food supply *has* matched the increase in population. In the 19th century, the opening-up of new agricultural lands in the Americas and Australia boosted food production. By the 20th century, advances in farming technology (such as those in India described in Section 5.9) kept food supplies one step ahead of population growth.

CASE STUDY

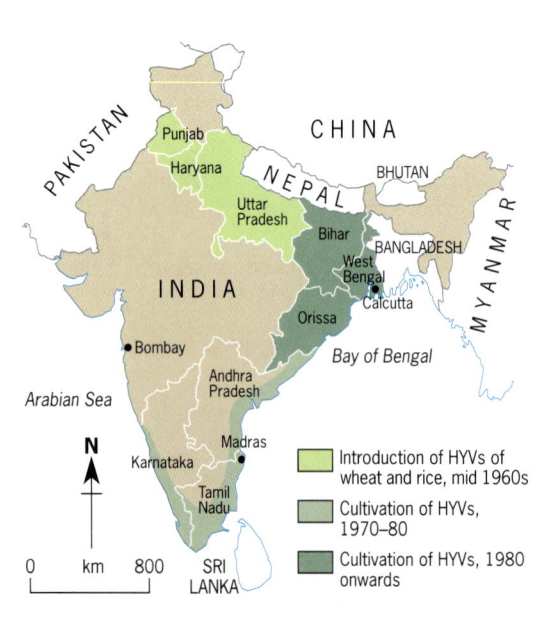

5.9 India's agricultural miracle

India has the world's second largest population (Table 5.3), and one which grew by nearly 500 million between 1960 and 1995. Despite such growth in numbers, during this period India became self-sufficient in wheat and rice; famines disappeared; and the people were better fed than ever before. What is behind this achievement?

The green revolution

In the 1950s, the Indian government aimed to provide greater **food security** (availability) for the country. First, it introduced land reform. This meant giving more farmers their own land – although often at the expense of large estates. Today, the majority of Indian farmers are owner-occupiers. This gives them greater incentive to farm efficiently, increase output and conserve the land.

Figure 5.25 India: the spread of the green revolution

5 Population: migration and resources

Then, from the mid 1960s, farmers began to grow new high-yielding varieties (HYVs) of wheat and rice (Fig. 5.25). These HYVs were part of the **green revolution** (see Book 1, Chapter 9). In addition to new seeds, the green revolution required farmers to use chemical fertilisers and irrigation. The government therefore encouraged the spread of the green revolution by providing farmers with cheap loans to buy the necessary seeds and fertilisers.

The effects of the green revolution have been dramatic (Fig. 5.27). Since the late 1970s, India has had a surplus of grain, giving the country much greater food security and reducing food imports. Moreover, this grain surplus has led to a fall in the price of wheat and rice which, above all, has helped the poorest people in India.

> *EXERCISES*
>
> 13 Study Table 5.3.
> a Draw bar charts to illustrate this information.
> b Describe the growth of population and food production in India between 1980 and 2000.
> c* With reference to Table 5.3, comment on the validity of Malthus's ideas.

Table 5.3 Population growth and food production in India, 1980-2000

	1980	1990	2000
Population (millions)	690	851	1014
Food production per person (1991=100)	82.2	99.6	105.2

Figure 5.26 (below) Yields of cereals per hectare: 1980–2000

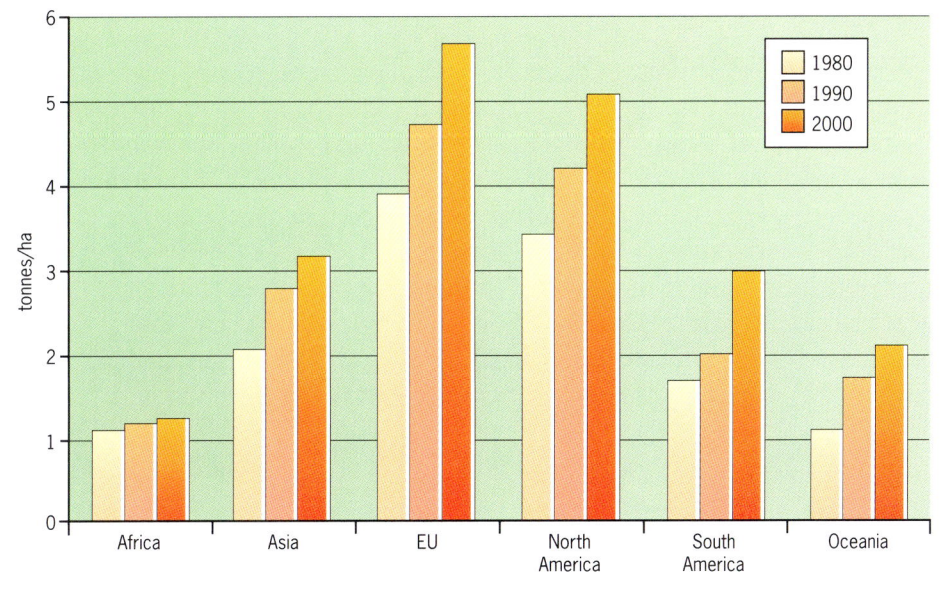

Figure 5.27 (above) Planting high-yielding rice, Karnataka, India

> *EXERCISES*
>
> 14 Study Figure 5.26.
> a Which continent had the lowest cereal yields per hectare in: 1980, 1990 and 2000?
> b Which region had the biggest percentage improvement in cereal production per hectare between 1980 and 2000?
> c* With reference to chapter 4, comment on the relationship between population change and the growth in cereal yields for the regions in Fig. 5.26.

5.10 Environmental impact of population growth

Future population growth will put added pressure on environmental resources. These resources include water, forests and land. By 2025, two-thirds of Africa's population is likely to face water shortages (Fig. 5.28). In the 1980s, deforestation caused annual losses of 15 500 sq km of tropical

5 Population: migration and resources

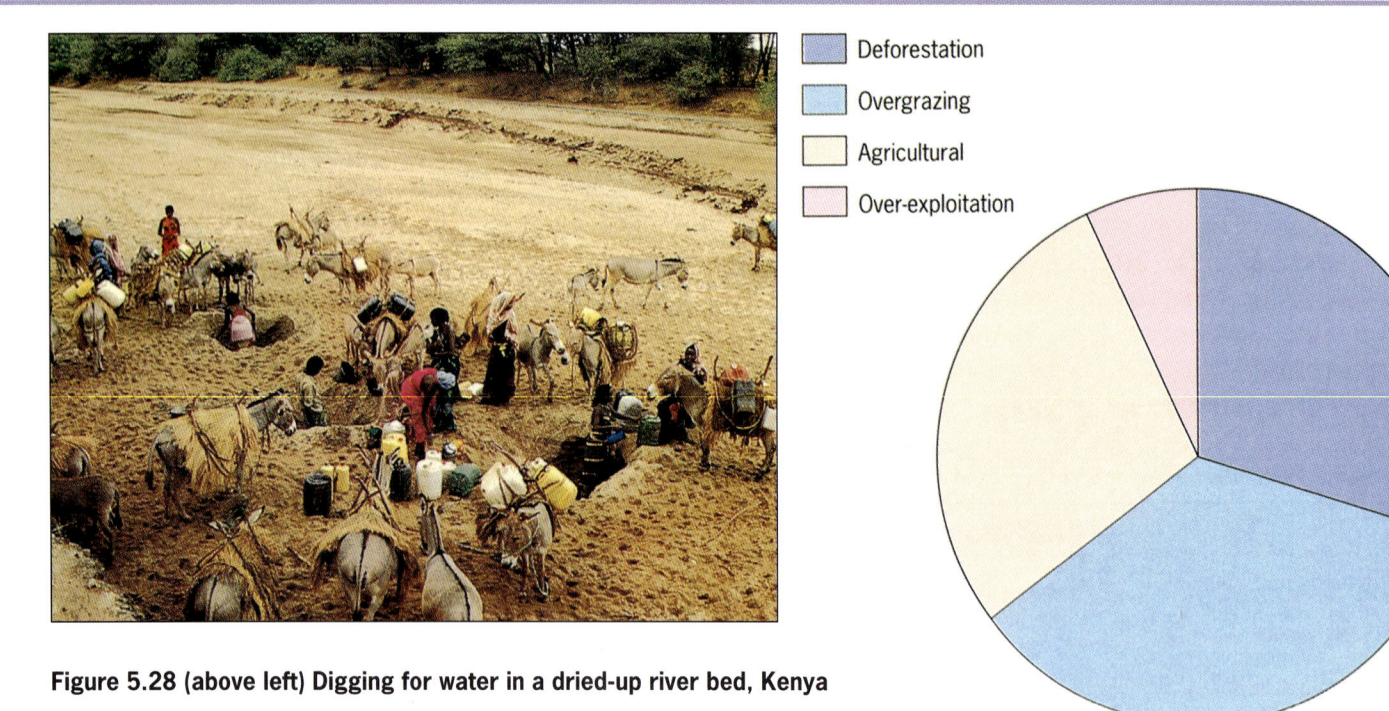

Figure 5.28 (above left) Digging for water in a dried-up river bed, Kenya

Figure 5.29 (above right) Causes of land degradation

> **REMEMBER**
> Land degradation is found in both MEDCs and LEDCs. Often, land is not degraded to the point of abandonment. Degradation may result in declining yields and resources that will not sustain long-term production.

forest. Much of this cleared land was then used for unsustainable agriculture. In many countries, the outcome was soil erosion, land degradation and abandoned farms.

Land degradation

Land degradation describes the decline in the quality and productivity of land as a result of human action. It is a worldwide problem but is most serious in dryland areas in the economically developing world. The most common types of land degradation are soil erosion, salt accumulation in soils, and deforestation (Fig. 5.29).

Population growth and poverty are important causes of land degradation. They force desperate farmers to over-exploit land. Such practices result in permanent soil damage. They are also unsustainable because they ultimately destroy the resources on which production depends.

The UN estimates that one-quarter of the Earth's agricultural land is at risk of becoming desert through degradation (Figs. 5.30 – 5.33).

Figures 5.30–5.33 Causes of land degradation

> **Deforestation**
> Removal of natural vegetation cover of forest and woodland. Clearance may be for new farm land, urban development, fuelwood and forestry. Grazing by domestic animals may then prevent regeneration.

Figure 5.30 Clearing the Amazon rainforest, Brazil

5 Population: migration and resources

Figure 5.31 (above) Comparison of heavily grazed and lightly grazed moorland, Borders, Scotland: heavy grazing on the right has left degenerated heather.

Figure 5.32 (above) Eroded land, Tumbes, Peru

Overgrazing
Destruction of vegetation cover by grazing, and by trampling of soils by domestic animals. Removal of vegetation results in soils being eroded by wind and water.

Over-cultivation
Cultivation of land without putting back sufficient nutrient fertilisers. Fallow periods may be too short, not allowing soils time to recover fertility. Natural supplies of silt to flood plains may be cut off by dams. Burning vegetation also depletes nutrient supplies. Exhausted soils are easily eroded by wind and water.

Figure 5.33 (left) Salt forming in a drainage ditch, which will eventually make the land useless, Elkargah, Egypt

Salinisation
Surface accumulation of salts (and alkalines e.g. sodium). Caused by forest clearance and over-irrigation which lead to rising water tables. High temperatures and high rates of evaporation draw salts to the surface of soils.

Figure 5.34 (below) Population problems: people or resource use? (Adapted from *The Guardian*, 6.9.94)

Environmental problems: who is to blame?

It would be wrong to assume that the only cause of land degradation is population growth and poverty in the developing world. Two other factors – wealth and technology – enter the equation. Rich countries consume more resources per person than poor countries (Fig. 5.34), and they have the technology to exploit these resources on a large scale, resulting in more pollution. As a result, many rich countries have more potential to degrade the environment than poorer countries.

What counts in the population crisis?

WHERE is the population problem worst? It depends on one's point of view. If the problem is defined simply as numbers of people, growth is largest in India.

Many more people are added to the population there than in any other country: 18 million a year. China, at 13 million a year, comes second. But India and China are already huge countries. In percentage increases, these countries are not growing nearly as fast as many others. India is growing 1.9 per cent a year and China just 1.1 per cent.

Almost all African countries are growing half as fast again: 2.7 per cent a year on average.

If, however, the population problem is defined as one of harm to the environment and using up natural resources, the worst problems are posed by the richest countries.

By that standard, the United States is way out in front. Americans use 43 times as much petroleum a head as the citizens of India.

According to one estimate, the average American has 30 times the environmental impact of an average person in a developing country.

Moreover, the US population is growing by about the same as China's.

Close behind, in terms of the amount of resources they use, are Japan and Western Europe, but both use less per person than do Americans, and their population growth rates are quite small.

5 Population: migration and resources

EXERCISES

15 Environmental impact results from population, wealth and technology (Fig. 5.35).
a Use this formula to estimate the potential environmental impact of the countries in Table 5.4.
b Which countries (rich or poor) have the greatest potential impact on the environment?
c Why does Japan have a potentially greater impact than Bangladesh?
d* Suggest other factors that might affect environmental impact. Explain their influence.

EXERCISES

16 Read Figure 5.34.
a State briefly the main points made in the article.
b Argue a case for a global solution to the population-environment crisis.

$$I = P \times W \times T$$

IMPACT Represents the environmental impact

POPULATION Represents population (absolute size, growth, distribution etc.)

WEALTH Represents per capita consumption of that population and is determined by income and lifestyle

TECHNOLOGY Represents the polluting influence of the specific technology the consumption involves

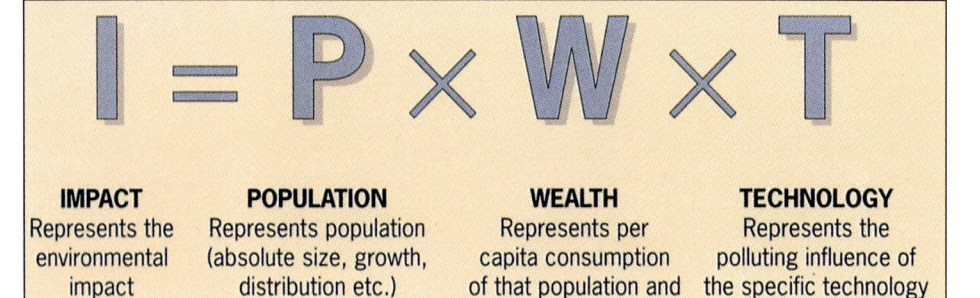

Figure 5.35 Relationship between population, wealth and technology and impact on the environment

We can blame some environmental damage in the economically developing world on rich countries. One reason is because rich countries influence the economically developing world through trade (see Chapter 9). For example, the destruction of the tropical rainforest in South-east Asia (see Book 1, Chapter 4) is partly due to the demand for timber in Japan and other MEDCs of the Pacific Rim. Similarly, in the 1970s and 1980s, Brazil rapidly developed its energy and mineral resources (see Book 1, Chapter 4) in order to pay the huge debt it owed to rich countries. The consequences for the environment – soil erosion, deforestation, flooding of land for HEP – not to mention the lives of thousands of people, were catastrophic.

Table 5.4 Environmental impact

	Population (millions) (P)	GNP per capita ($) (W)	CO_2 emissions (tonnes per capita) (T)	Impact P×W×T
Bangladesh	131	370	0.2	
India	1014	450	1.0	
Japan	127	32 230	9.1	
UK	59	22 640	9.1	
USA	283	30 600	19.6	

CASE STUDY

5.11 Egypt: expanding the resource base

FACTFILE

- Egypt's population in 2001 was 63 million, half of which depended on agriculture.
- 97 per cent of Egypt is desert.
- Almost the entire population of Egypt crowds into the Nile Valley and Delta, an area not much bigger than Wales.
- In the Nile Valley and Delta, fertile alluvial soils and irrigation water support some of the highest rural population densities in the world.
- Ancient writers called Egypt 'the gift of the Nile' – a description that is just as apt today.
- Rapid population growth in the past 50 years has put acute pressure on Egypt's limited resources of farmland and water.

5 Population: migration and resources

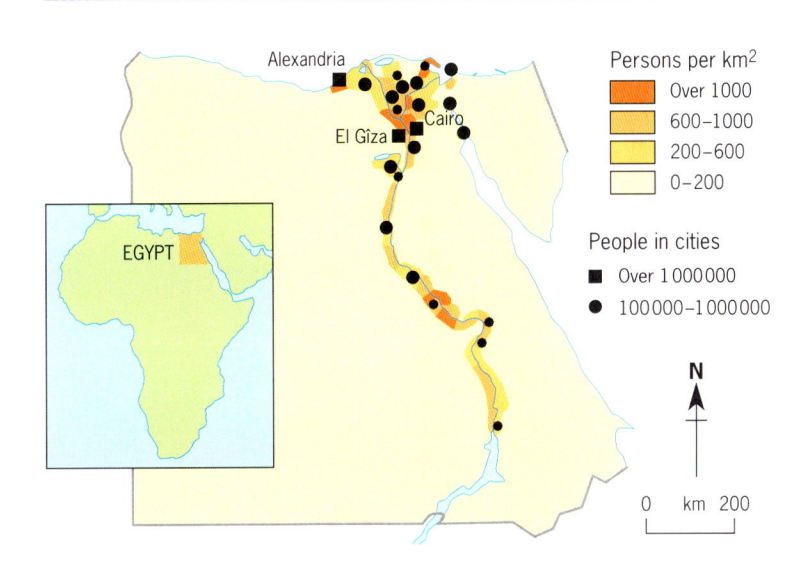

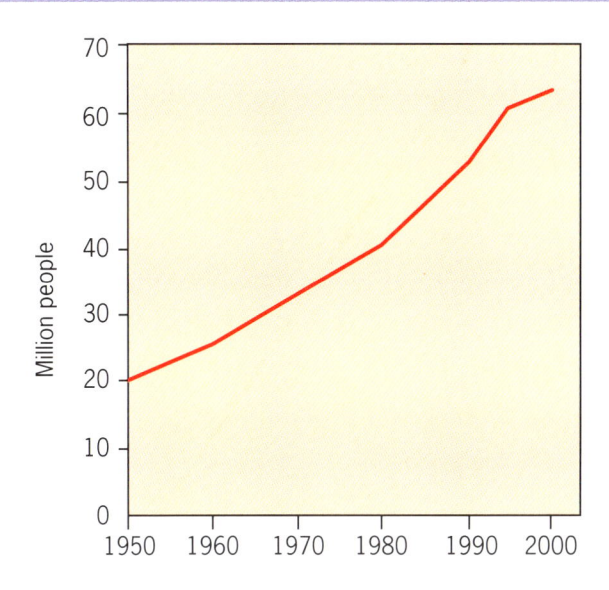

Figure 5.36 (above left) Egypt: population density

Figure 5.37 (above right) Egypt's population growth, 1950–2000

Expanding the rural resource base

Egypt's response to population growth was to increase its water resources and expand the area of farmland. It did this through a single huge project: the construction of the Aswan High Dam on the River Nile (Fig. 5.38). Completed in 1970, the dam created a huge reservoir (Lake Nasser), which is large enough to store the Nile's entire annual flood. This had two advantages. First, it provided sufficient water for farmers to grow two or three crops each year on the same land. And second, it smoothed out variations in the Nile's annual flow, ensuring enough water for farmers even

Erosion of delta and decline of fishing
The loss of silt from the Nile floods is leading to erosion of the delta (and loss of farmland) by the sea. The decline of the fishing industry is related to the loss of the fertilising effect of Nile silt.

Spread of disease
All-year-round irrigation has encouraged the spread of bilharzia, transmitted to people through snails which thrive in irrigation canals.

Salinisation
Over-irrigation has led to water-logging and salt accumulation. 2 million hectares are affected. Yields have fallen by 10 per cent in the delta. In the new lands large areas of farmland have been abandoned.

Destruction of archaeological sites
The waters of Lake Nasser flooded many important archaeological sites, including the ancient Egyptian temples at Abu Simbel.

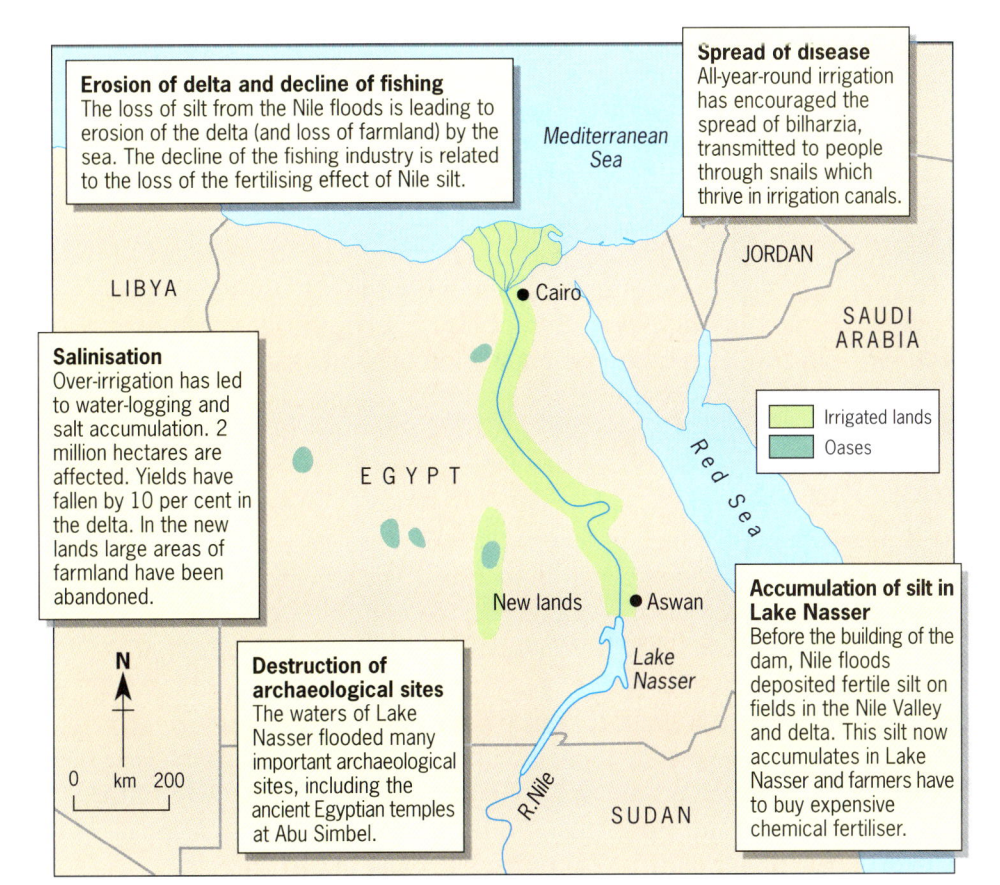

Accumulation of silt in Lake Nasser
Before the building of the dam, Nile floods deposited fertile silt on fields in the Nile Valley and delta. This silt now accumulates in Lake Nasser and farmers have to buy expensive chemical fertiliser.

Figure 5.38 (above) Aswan High Dam, Egypt

Figure 5.39 Environmental problems caused by the Aswan High Dam

97

5 Population: migration and resources

> **EXERCISES**
>
> **17** The Aswan High Dam has brought both advantages and disadvantages to Egypt. Assess the advantages and disadvantages and say whether you think that the dam has been a success or a failure.

in dry years. All of this raised crop yields and produced more food for Egypt's growing population. Meanwhile, the development of new irrigated farmlands in the desert also increased food output. The success of these schemes was such that cereal production per person actually grew by 70 per cent between 1980 and 2000.

While the Aswan High Dam has clearly benefited Egypt, there have been many problems too. These include salinisation of soils (see Fig. 5.33), the spread of water-borne diseases and the destruction of important archaeological sites (Fig. 5.39).

CASE STUDY

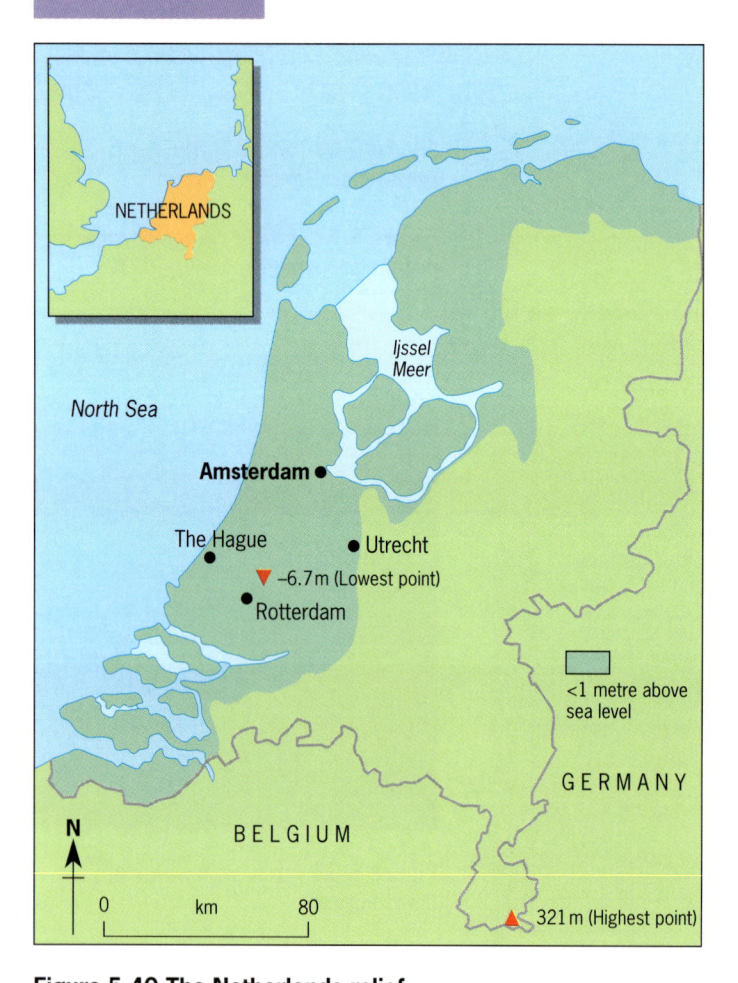

Figure 5.40 The Netherlands relief

5.12 Land reclamation in the Netherlands

Some MEDCs have responded to population growth and shortage of space by draining marshes and reclaiming land from the sea. For example, in the 17th century, a massive drainage scheme around the Wash, in eastern England, reclaimed the Fens for agriculture. More recently, large areas of land in Tokyo Bay and Osaka Bay in Japan have been reclaimed for heavy industry. And the new international airports at Osaka and Hong Kong are built on land reclaimed from the sea.

The world's experts on land reclamation are the Dutch. About one-fifth of the Netherlands is reclaimed land and nearly one-third lies below sea level (Fig. 5.40). Although reclamation has been on-going since the 13th century, the largest areas have been reclaimed over the last 80 years.

The Netherlands is a small country (about half the size of Ireland) and it has the highest population density in Europe. Land reclamation is partly a response to population pressure, and partly a need to strengthen the country's defences against flooding from the sea.

Zuider Zee

The Zuider Zee is a the biggest land reclamation project (Fig. 5.41). Started in the 1920s, it was not completed until 1968. The scheme transformed a shallow inlet of the North Sea – the Zuider Zee – into four huge reclaimed areas known as **polders**. They have added 5 per cent to the Netherlands' total land area (Fig. 5.42).

Reclamation was step-by-step, one polder at a time. The early Zuider Zee polders were used mainly for agriculture. But the decision to reclaim the fourth polder, South Flevoland, was influenced by the rapid growth of the Dutch population in the 1950s and 1960s. At this time, there was great pressure on the overcrowded western Netherlands. In particular, the

5 Population: migration and resources

Randstad conurbation, which includes the main cities of Amsterdam and Rotterdam, needed space to expand. In East and South Flevoland, therefore, the planners gave more space to housing, industry, recreation and conservation. This included two new towns, Almere and Lelystad. Almere has a target population of 250 000, mainly overspill from Amsterdam. Because it is so close to Amsterdam, most of its working population are commuters.

Land reclamation is often controversial. The original Zuider Zee scheme included plans for a fifth polder – the Markerwaard, but a decision on this has not yet been made. Today, people are more aware of the environmental disadvantages of land reclamation than in the past (Table 5.5).

Figure 5.41 The Zuider Zee project

Table 5.5 Advantages and disadvantages of land reclamation in the Zuider Zee

Advantages	Disadvantages
Increases the land area for food production.	Destroys valuable wetland habitats for birds, plants, etc.
Creates high-quality farmland (fertile soils, flat relief).	Destroys the local fishing industry. Creates flat, monotonous landscapes of low visual quality.
New land relieves overcrowding in Randstad.	Very expensive.
Improved communications (across the main enclosing dam) between Randstad and the NE Netherlands.	
Creates freshwater lakes providing recreation and leisure opportunities and a water supply.	
Shortens the coastline and reduces the risk of flooding from the sea.	

Table 5.6 Population in the Netherlands, 1950–2000 (millions)

1950	10.11
1960	11.48
1970	13.03
1980	14.14
1990	14.95
1995	15.50
2000	15.89

EXERCISES

18a Calculate the percentage population increase in the Netherlands between 1950 and 1970, and 1970 and 2000 (Table 5.6).
b* From your calculations suggest one reason why the government may decide to postpone reclaiming the fifth polder – the Markerwaard.
19 With reference to either a land reclamation scheme in the economically developed world or an irrigation scheme in the economically developing world, explain how the scheme provides for population increases.

Figure 5.42 The Ijssel Meer and the polders of the Zuider Zee

5 Population: migration and resources

5.13 Summary: Population: migration and resources

KEY SKILLS OPPORTUNITIES
C1.2: Ex. 5a, 7a, 9a, 11a, 15b, 16a; **C1/2.3:** Ex. 2, 12b; **C2.1:** Ex. 8b; **C2.2:** Ex. 4, 5b, 6a, 17; **N1/2.1:** Ex. 5c, 8a, 9b, 15c, 15d; **N1/2.2:** Ex. 3a, 13a, 15a, 18a; **N1/2.3:** Ex. 3b, 13b, 18b; **IT1/2.1:** Ex. 10.

Key ideas	Generalisations and detail
Migration is the permanent relocation of an individual or group of people.	• Migrations may be classed according to distance (e.g. international, regional, etc.) and direction of movement (e.g. rural-urban, urban-rural). Such migrations may be either voluntary or forced movements.
Migration results from a combination of 'push' and 'pull' factors.	• 'Push' factors are the disadvantages of a migrant's place of origin. They might include lack of job opportunities, poverty, political persecution, etc. • 'Pull' factors are the attractions (advantages) of the migrant's place of destination and are usually the opposite of 'push' factors. • Both 'push' and 'pull' factors may be economic (e.g. standard of living), social (e.g. educational opportunities), political (e.g. wars) or environmental (e.g. floods).
Rural-urban migration is most common in LEDCs.	• Rural-urban migration is responsible for rapid urbanisation in LEDCs today. • Rural poverty and the hope of better economic prospects in towns and cities are the driving forces of current rural-urban migration.
Rural-urban migration creates problems in both rural and urban areas.	• Because migration is selective, rural areas often lose young adults and the better educated, most go-ahead members of a community. It may also lead to an imbalance between males and females in rural (and urban) areas. • Rural-urban migration puts enormous pressure on urban services, which often cannot cope. The result is acute shortages of housing and the growth of squatter camps, and the lack of essential services such as clean water, electricity, sewerage systems, schools etc.
Urban-rural migration (or counter-urbanisation) is most common in MEDCs.	• In the UK, counter-urbanisation has led to population decline in most large cities. Meanwhile, there has been a population revival in many small towns and rural areas. Population growth has been most rapid in rural commuter hinterlands of large urban areas. Some urban-rural migration (e.g. for retirement) has been to more remote areas.
International migration may be legal or illegal.	• National frontiers are major obstacles to international migration. • The USA has the largest volume of international immigration in the world. Much of this migration is from LEDCs in Central America, especially Mexico. • Two-thirds of all Mexican immigration comprises illegal immigrants who are sent home.
Many migrants are refugees.	• Refugees are people who migrate to escape political persecution, wars or environmental disasters. Most refugee movements are between countries in the economically developing world. More than half of all refugees migrate because of environmental problems such as famine, soil erosion, land degradation, flooding, etc.
The effect of population growth and poverty on food supplies gives rise to conflicting views.	• Some writers adopt Malthus's viewpoint: that population growth eventually outstrips the growth in food supplies and results in disastrous famines, wars and disease. They argue that the recent famines and wars in Africa prove that Malthus was right. • Others believe that technology and invention allow food production to keep pace with population growth.
Population growth creates both problems and opportunities for people and societies.	• Rapid population growth and poverty are partly responsible for environmental degradation. They encourage unsustainable agriculture, which causes land degradation, i.e. soil erosion, deforestation, overcultivation and salinisation. • Population growth may encourage countries to expand food production by using new technology (e.g. the green revolution in India), expanding the irrigated farm area (e.g. Aswan High Dam in Egypt), or by expanding the cultivated area (e.g. Egypt's new lands). • In MEDCs, population growth and shortages of land may lead to land reclamation. The reclaimed land may be used for agriculture, housing, industry, conservation, etc. (e.g. Zuider Zee polders).

6 Managing natural resources

6.1 Introduction

Natural resources are things that are useful to us. They include fuels, such as oil and natural gas, and materials, such as iron ore, water and timber. Natural resources are so essential to us that we often forget the problems involved in developing and using them. These problems are the focus of this chapter. As we shall see, some are global and affect all countries, some are national, and others are very local and affect only small areas.

Figure 6.1 (below) Modern, detached houses, Cheshire, England

Figure 6.2 (below right) Women building a traditional rural dwelling, Niger, West Africa

EXERCISES

1 Study Figures 6.1 and 6.2.
a Make a list of as many natural resources that you can think of that might be used to build and heat the houses in Figure 6.1. Then make a similar list for Figure 6.2.
b From the evidence of your two lists, suggest how people's wealth might affect their use of natural resources.

6.2 Renewable and non-renewable resources

Non-renewables

Some natural resources, such as coal, oil and gas, are **non-renewable**. These are **fossil fuels**, which take millions of years to form. Once people have used them they cannot be replaced. Consequently, every year the world's stock of non-renewable resources gets smaller, and sooner or later they will run out (Fig. 6.3). At the same time, the demand for energy is increasing and will double between 2000 and 2025. Because we cannot continue to use these resources indefinitely, we say that their use is **unsustainable**.

The outlook is not quite so gloomy for other non-renewable resources. Given the right economic conditions, metals like iron, copper and aluminium, can be recycled and used again (see Section 6.9).

6 Managing natural resources

> **EXERCISES**
>
> **2** Study Figure 6.3.
> **a** On a world map, use proportional squares to show the distribution of oil reserves by country. (NB: the square sides are equal to the square root of the total oil reserves for each country.)
> **b** Which region has the world's largest oil reserves?
> **c** Which country has the world's largest oil reserves?
> **d*** Suggest one possible reason why oil reserves in Iraq and Kuwait will last longer than those in Saudi Arabia.
> **3a** Sort the following natural resources into renewable and non-renewable: coal, wave power, uranium, timber, peat, water, soil, oil, zinc, fish, wind power, geothermal power, natural gas, iron, rubber.
> **b*** Draw a labelled diagram to show how water is renewable, although it is a finite resource like oil (see Chapter 1).

Renewables

Renewable resources include plants and animals, which follow biological cycles of growth and reproduction; water, which is constantly cycled between the land, atmosphere and seas; and supplies of energy, such as solar, wind and geothermal power. These resources never run out; they are literally inexhaustible.

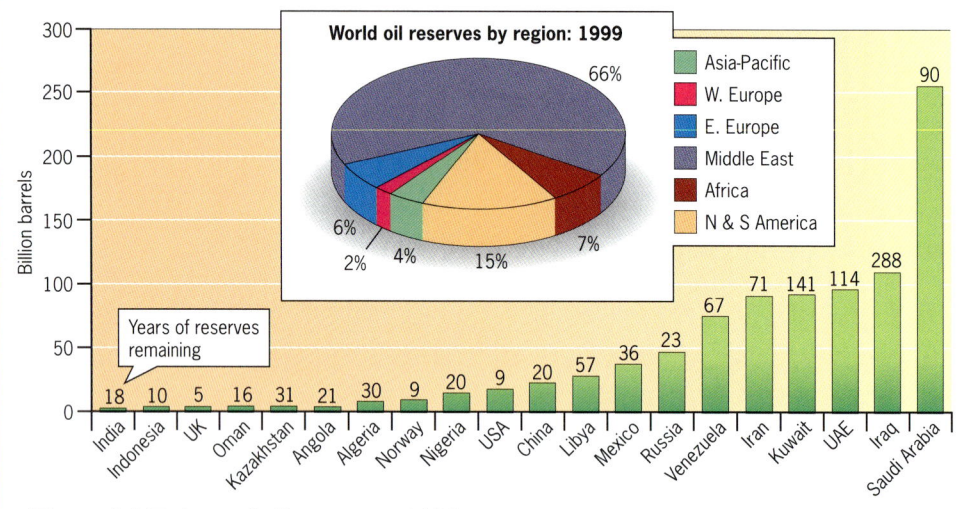

Figure 6.3 Estimated oil reserves, 1999

Figure 6.4 Loading pine logs, Miyako Port, Japan: Japan's dwindling forest reserves cannot meet its demand for soft-wood products. Japan is also the world's largest importer of tropical hardwoods.

> **REMEMBER**
> Renewables are sustainable resources – they are both non-polluting and inexhaustible.

6.3 Natural resources and levels of development

Given the importance of natural resources for producing energy, we might assume that rich countries have lots of them. In some cases this is true. Take the examples of the USA, Canada, Australia and Sweden. All are rich countries and all owe their prosperity, at least in part, to an abundance of energy and other natural resources. Even so, many countries with vast natural resources remain poor. Sierra Leone in West Africa has rich reserves of diamonds, bauxite, rutile (titanium ore), iron ore and fish, and yet is one of the poorest countries in the world (see Section 8.6). Even Brazil, with its fabulous mineral wealth in Amazonia, is a relatively poor country. Why is this?

Human resources and capital

Often, what are lacking in poor countries are human and economic resources. Generally, this means shortages of skilled and educated workers, and capital (money). Both are needed to unlock a country's natural resources.

To underline the importance of skills and capital, we only need to look at Japan. Japan is near the top of any list of the world's most prosperous nations, and yet it has few natural resources of its own. It has to import most of its timber (Fig. 6.4) and 99 per cent of its energy and mineral needs (Table 6.1). Denmark, the Netherlands, Belgium and Singapore are also examples of countries that have become prosperous in spite of having few natural resources of their own.

6 Managing natural resources

This suggests that when it comes to wealth, it is human and capital resources, rather than natural resources, that are crucial. In fact, no amount of natural resources will ensure prosperity without the human skills and capital to develop them.

Wealth and the consumption of natural resources

While the possession of natural resources has no clear link with prosperity, consumption of them does (see Fig. 6.5). The two richest countries, the USA and Canada, are also the biggest consumers of energy. Poor countries, such as Bolivia and El Salvador, consume only small amounts of energy.

Most people in the economically developed world have high standards of living, supported by their use of huge amounts of natural resources. Just think of how much energy we use. We use energy to heat our homes and schools, for transport, lighting and cooking, and to make all the things we regard as essential to our way of life. In other words, it is only by using large amounts of energy that we can sustain our high standard of living. We can say that the volume of natural resources a country consumes is a better measure of its prosperity than the size of its natural resource base.

Table 6.1 Japan and Brazil: output of natural resources and manufactured products (million tonnes), 1998

	Japan	Brazil
Bauxite	0.00	9.70
Aluminium	1.63	1.19
Copper ore	0.00	0.05
Copper	1.97	0.21
Iron ore	1.70	177.00
Steel	93.50	36.00
Oil	0.00	41.00
Petroleum production	186.00	69.00

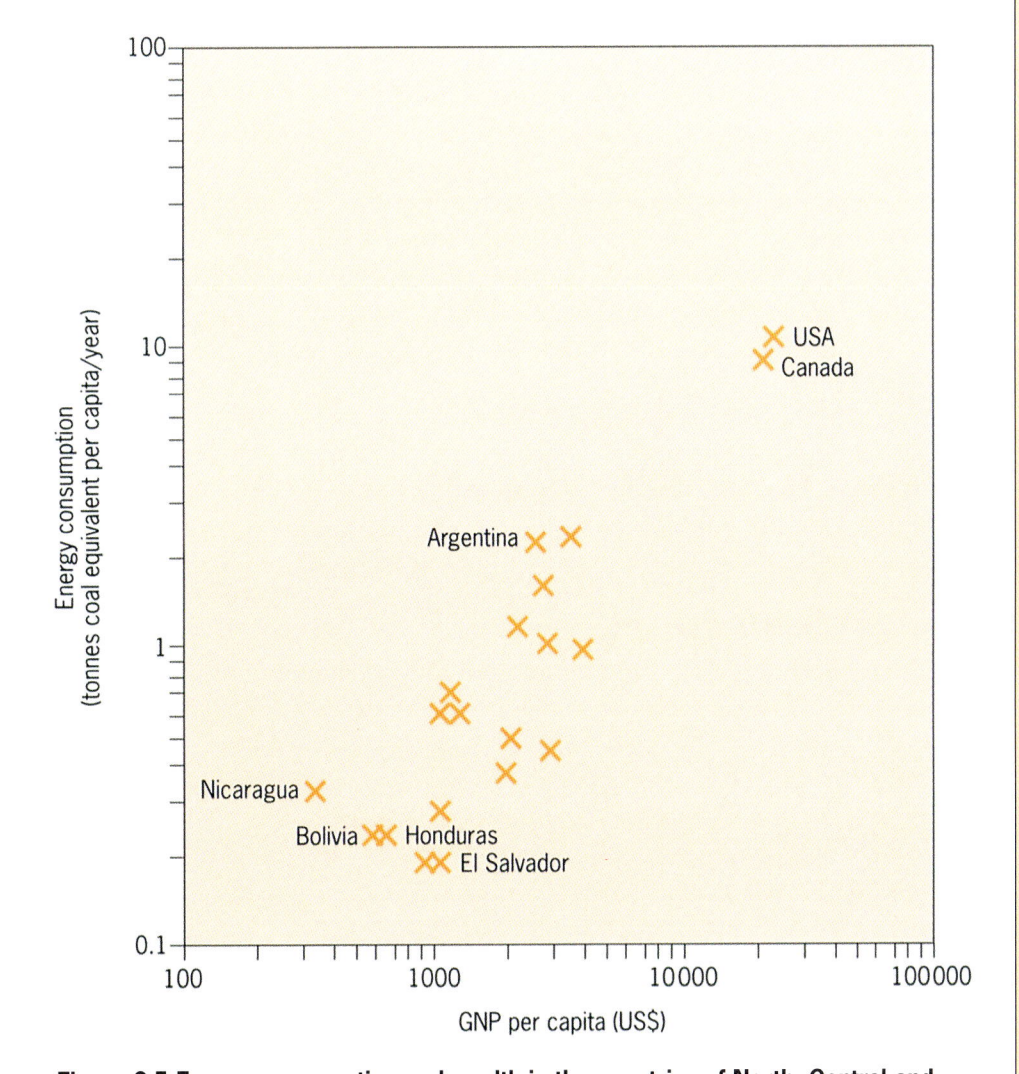

Figure 6.5 Energy consumption and wealth in the countries of North, Central and South America

EXERCISES

4a Draw a bar chart to show the production of natural resources and manufactured goods in Japan and Brazil (Table 6.1).
b* In 2000, Japan's GNP per person was US$32 230 – seven times greater than Brazil's. Using the information from your bar chart, describe the importance of natural and human resources in generating wealth.
5 Study Figure 6.5.
a Describe the relationship between energy consumption per capita and GNP per capita.
b* Give a brief explanation of the relationship you have described.
6a Compile a table to show population growth rates and energy consumption for the countries in Table 6.2. (Energy consumption rates are shown in Figure 6.5.)
b Describe what happens to energy (resource) consumption as population growth increases.
c* Discuss in class the implication for the well-being of people in poor countries in the economically developing world.

6 Managing natural resources

Table 6.2 Annual population growth, 2000 (percentage)

Argentina	1.1
Bolivia	2.0
Canada	0.4
El Salvador	2.3
Honduras	2.7
Nicaragua	2.3
USA	0.6

6.4 Fossil fuels and global warming

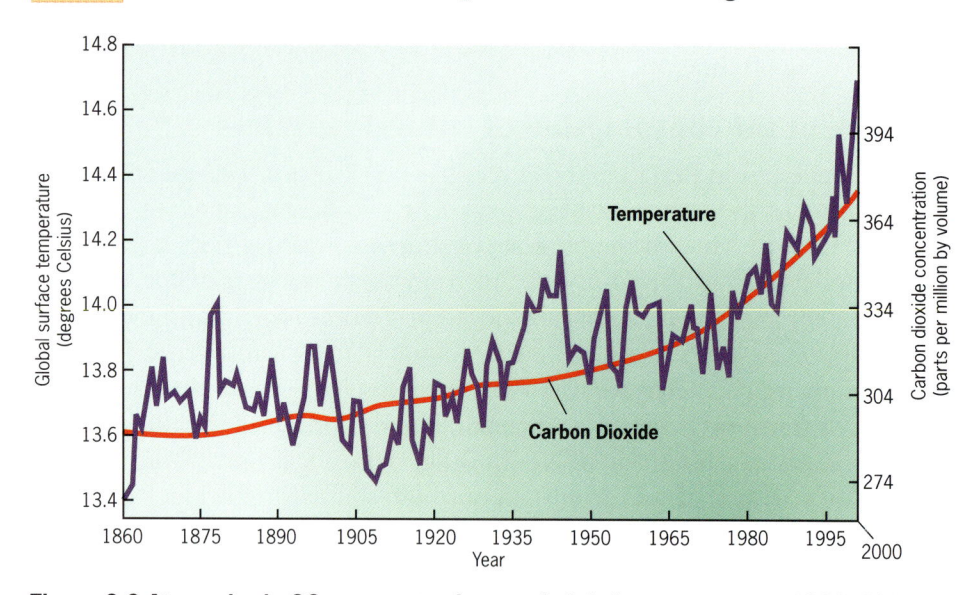

Figure 6.6 Atmospheric CO_2 concentrations and global temperatures, 1860–2000

Figure 6.7 Coal-fired power station, Brandenburg, Germany

Climate change

In the 20th century, average global temperatures rose by 0.6°C (Fig. 6.6). Most of this increase occurred after 1950. Indeed, six of the ten warmest years on record were in the 1990s. A temperature rise of just 0.6°C in a hundred years may not seem very much, but it is the most rapid increase since the end of the ice age, 10 000 years ago (see Section 2.2). And most scientists believe that the Earth's climate is going to get even warmer. By 2100, the average global temperature could be between 1.6°C and 4°C higher than today's. Over landmasses like Europe, the temperature could rise by up to 8°C.

The evidence of a warmer global climate is easy to find. In the Alps, glaciers have shrunk to half their size since 1850 (see Figs. 2.7 and 2.8). In Antarctica, huge areas of the ice shelf, the size of Spain, have melted. Rapidly melting ice, combined with the expansion of ocean water because of higher temperatures, resulted in a 50-cm rise in sea level in the 20th century.

Causes of global warming

The theory behind global warming is called the 'greenhouse effect' (Fig. 6.8). This is a natural process that keeps the Earth warm and makes the planet habitable.

- Carbon dioxide (CO_2) and methane (CH_4) allow sunlight to pass through the atmosphere and heat the Earth's surface.
- Heat radiated from the Earth is absorbed by CO_2 and CH_4 and warms the atmosphere.

In the past 200 years, industrial development, population growth and rising prosperity have greatly increased the demand for energy (Fig. 6.7). Most of this energy has come from burning fossil fuels, such as coal and oil. This has greatly increased the amount of carbon dioxide and other greenhouse gases in the atmosphere. As a result, more of the Earth's heat is trapped, causing temperatures to rise (Fig. 6.8).

6 Managing natural resources

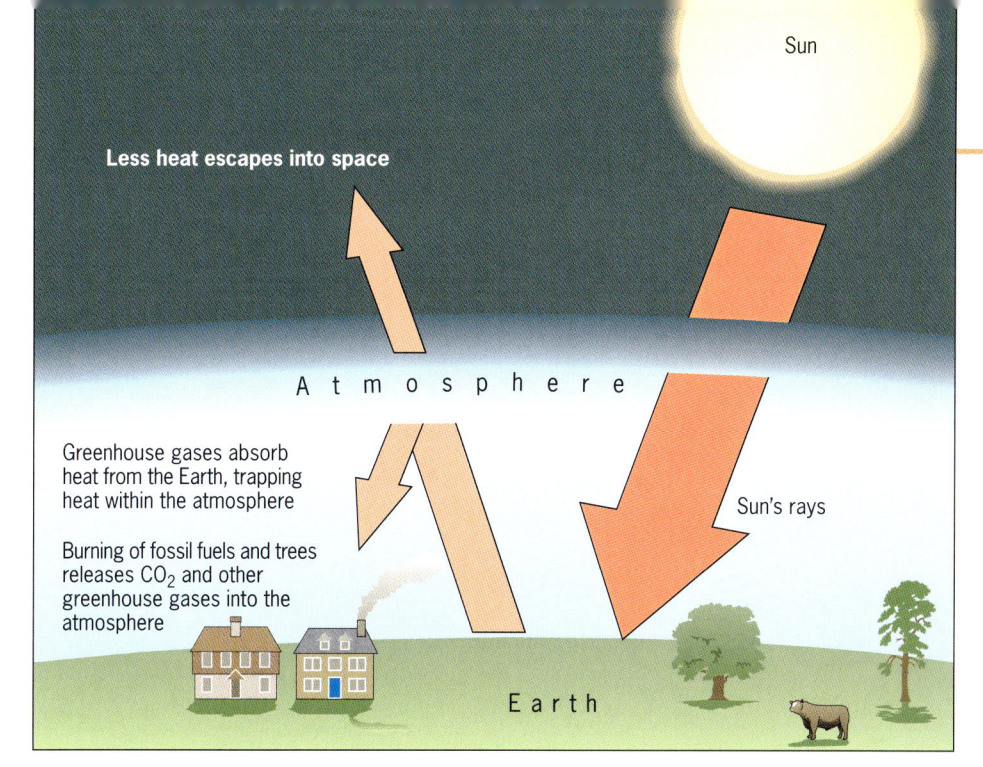

Figure 6.8 The greenhouse effect

Figure 6.9 (below) UK consumption of primary fuels, 1970–1997

Who is responsible for global warming?

The main responsibility for global warming rests with MEDCs, such as the USA, Germany and the UK. Because the UK has large reserves of coal, oil and gas it has always relied heavily on fossil fuels (Fig. 6.9). Power stations burn these **primary fuels** and convert them to **secondary energy**, i.e. electricity. Industries also use some primary fuels directly. For example, oil refineries make oil into petroleum products for transport vehicles, and steelworks use coal for smelting iron. Meanwhile, gas is used widely for domestic heating and cooking.

UK CO$_2$ emissions

About 2 per cent of the CO$_2$ emissions caused by humans each year come from the UK, mainly from coal-fired power stations. Compared to the USA, which produces one-quarter of the world's CO$_2$, the UK's emissions are relatively small. Moreover, thanks to the switch from coal to gas as the leading primary fuel (Fig. 6.9), CO$_2$ emissions in the UK declined by 15 per cent between 1990 and 2000.

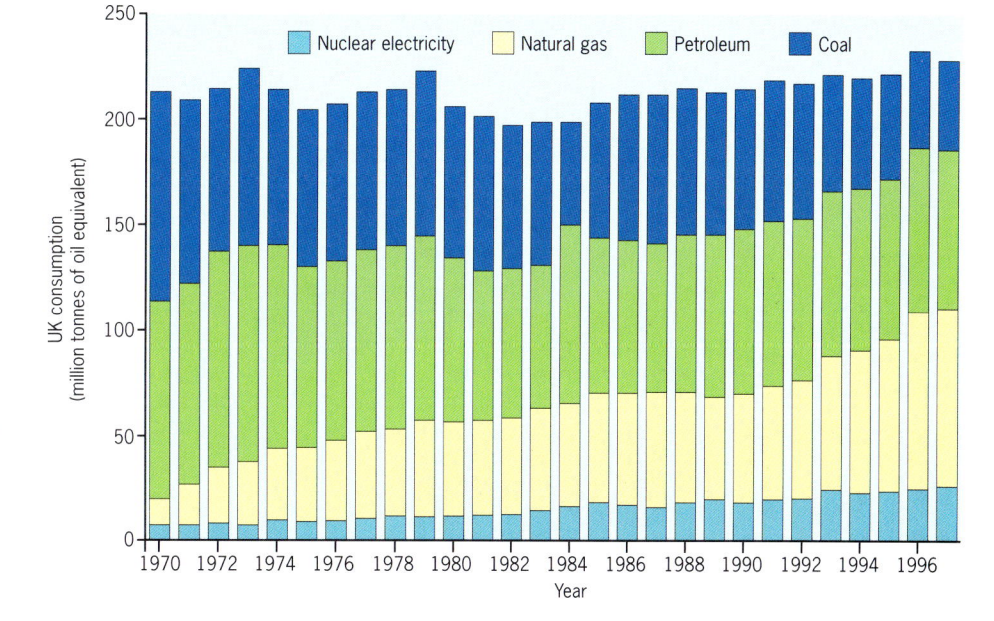

Table 6.3 Carbon emissions and GNP per person, 2000

Country	GNP per person (US$)	Carbon emissions (tonnes/person/year)
USA	30 600	19.6
Canada	19 320	14.8
Australia	20 050	15.8
Germany	25 350	10.2
UK	22 640	9.0
Japan	32 230	9.1
Italy	19 710	7.1
France	23 480	5.8
Sweden	25 040	5.5
South Korea	8490	8.4
Mexico	4400	3.5
China	780	2.7
India	450	1.0

105

6 Managing natural resources

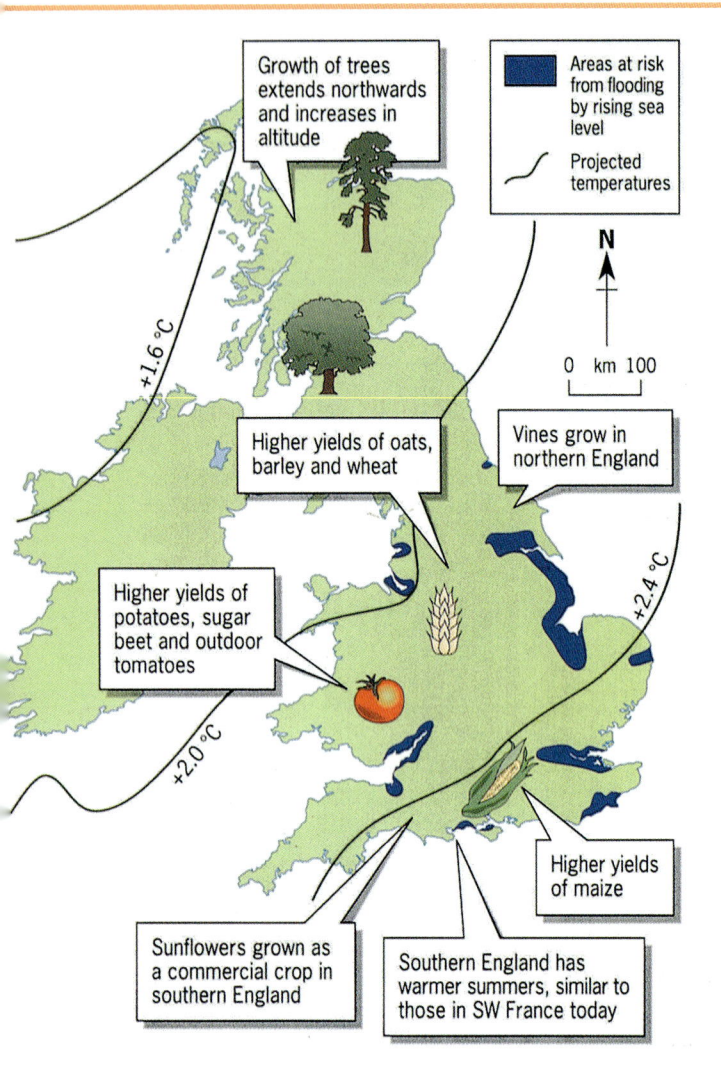

The consequences of global warming

Global warming has an effect at local and national scales, as well as at a world scale (Table 6.4). Although we hear most about the harmful effects of global warming – disruption of the world's climate, rising sea levels, spread of tropical diseases, etc. – it is not all bad news. In the UK for example, climatic change might bring a number of benefits (Fig. 6.10).

Table 6.4 The impact of global warming

World

Rising sea levels	• Melting glaciers and ice sheets and thermal expansion of water in the oceans could raise sea levels by 1.5 m by 2100. • Several small island states, such as the Maldives and the Marshall Islands, could disappear altogether (Fig. 6.17). • Large parts of Egypt, Bangladesh and South China could be flooded, displacing up to 150 million people in these countries.
Climate change	There could be major disruption of the global climate: • Drought, storms and floods could become more frequent (e.g. Mozambique floods and Hurricane Mitch in 2000). • Harvests could drop by one-fifth in Africa, and in South and South-east Asia: famine would increase. • Siberia could become warmer, allowing greater crop yields. • Ocean currents like the North Atlantic Drift could break down, giving much colder winters in North-west Europe. • Deserts are already advancing from North Africa into southern Europe.

National

Pests and diseases	There could be an increase in insect pests such as aphids, mites, cockroaches and fleas. As a result, some crop yields could fall. Tropical diseases could spread to temperate regions. The malarial mosquito could re-establish itself in southern England.
Wildlife	Flooding of estuaries and salt marshes could destroy the habitats of millions of birds. Natural ecosystems would not be able to adjust to rapid climate change. Sub-arctic species of plants could disappear, unable to compete with the spread of trees and more vigorous plants. The Scottish Highlands could lose animals such as mountain hares, snow buntings and ptarmigans, which are adapted to snowy conditions in winter (see Figs 2.36 and 2.37).
Climate	Southern Britain could become drier, causing severe water shortages. Wetter conditions could prevail in northern Britain.
Sea level change	Low-lying areas of England, e.g. Fens, Somerset and west Lancashire could be flooded unless sea defences are strengthened. Protecting the UK coast from flooding could cost £30 billion in the next 50 years.
Winter sports	Rising temperatures and lower winter snowfalls could mean economic disaster for ski resorts in the Alps and Scotland.

Figure 6.10 (top left) Possible effects of global warming in the UK by 2050

Figure 6.11 (left) Children queuing for food rations, Somalia

Figure 6.12 (right) Malarial mosquito

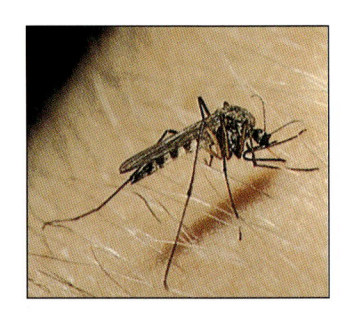

6 Managing natural resources

Figure 6.13 (left) Very low water levels, Thruscross Reservoir, North Yorkshire

Figure 6.14 (bottom left) Aerial view of the Fens, eastern England

Figure 6.15 (bottom right) At the ski lift, French Alps

EXERCISES

7a Plot the information in Table 6.3 as a scattergraph with GNP on the x-axis and carbon emissions on the y-axis.
b What effect does wealth (GNP per person) appear to have on carbon emissions?
c* Why does the scatter of points on your graph not follow a straight line? Suggest what other factors might affect the level of carbon emissions.

EXERCISES

8 Study Figures 6.11–6.15 and Table 6.4. Suggest how each photograph might be linked to global warming.

Global warming: an international environmental problem

Global warming, along with the depletion of the ozone layer, deforestation and overfishing, is an international environment problem (IEP). In other words, it is a problem that affects the entire global community. It is happening because of the misuse of the atmosphere by all countries. The atmosphere is a shared resource, but because no one country owns it, it is very hard to get governments to agree on a common plan to halt the pollution responsible for global warming.

Global action

As global citizens we have a responsibility to conserve shared resources and plan for their sustainable use. Global warming is the single most urgent IEP. Attempts have been made to get governments to come to an agreement on how to tackle global warming (for example at Kyoto in 1997 and The Hague in 2000). The Kyoto meeting set MEDCs the target of reducing emissions of carbon dioxide by 12.5 per cent by 2010. Some countries, such as the UK and Germany, have already achieved this target. Others, notably the USA, lag well behind. The most recent meeting in The Hague in 2000 broke up without any agreement on cuts in carbon emissions.

6 Managing natural resources

> **EXERCISES**
>
> **9** You are a delegate at a meeting to discuss global climate change. You represent one of the following countries or organisations: an oil exporting country; a member of AOSIS; a rapidly industrialising country in Asia; a trans-national car manufacturer; Greenpeace. There is a proposal to reduce global carbon dioxide emissions by 20 per cent by 2010.
>
> **a** Choose one of these roles and state your view on the proposal.
>
> **b*** Using the information on global warming in this chapter, explain, in the context of global citizenship, why countries have a responsibility to solve the problem of global warming.

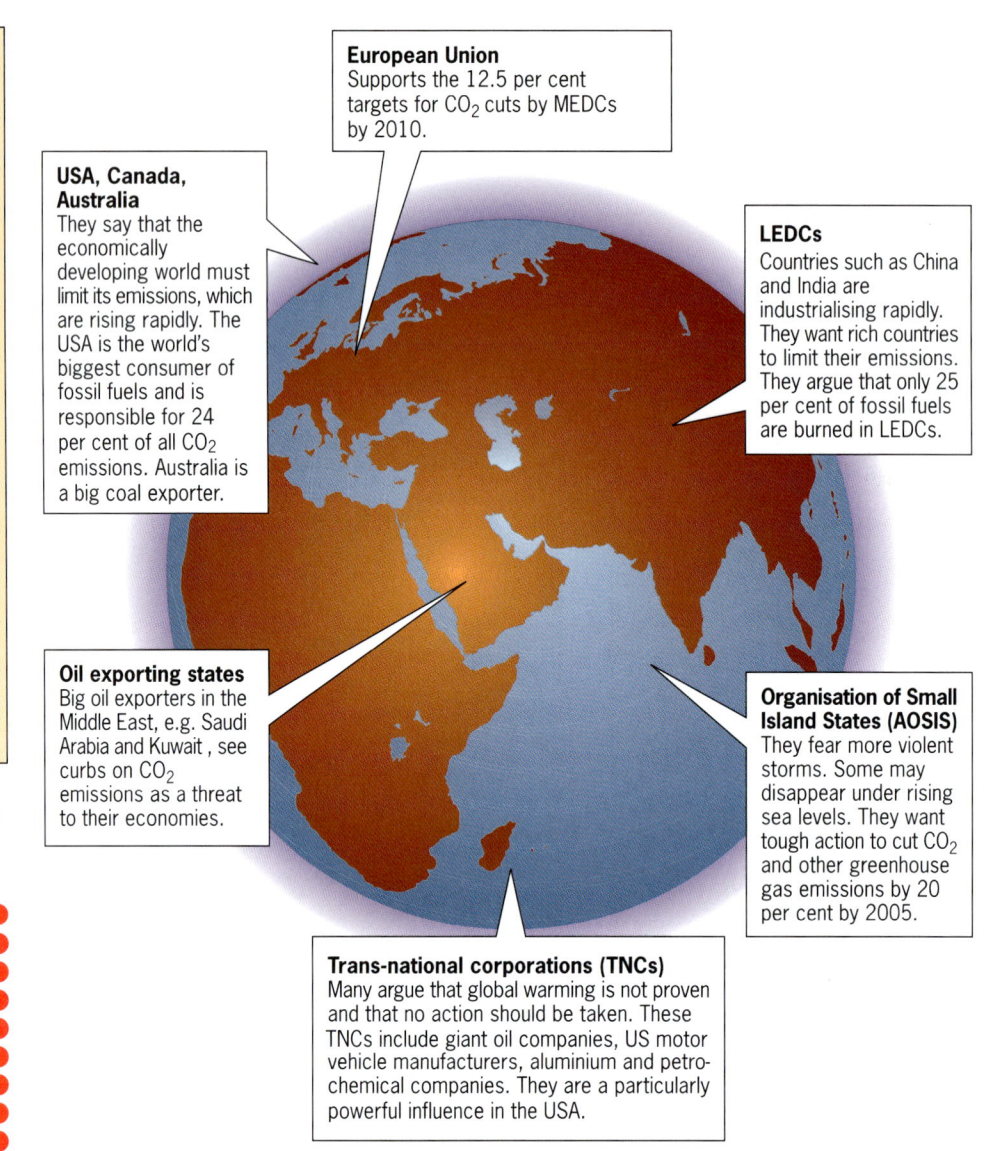

Figure 6.16 Attitudes towards global warming

> **REMEMBER**
> Because IEPs involve the use of common resources, such as the oceans and atmosphere, they are the most difficult of all environment problems to solve.

European Union
Supports the 12.5 per cent targets for CO_2 cuts by MEDCs by 2010.

USA, Canada, Australia
They say that the economically developing world must limit its emissions, which are rising rapidly. The USA is the world's biggest consumer of fossil fuels and is responsible for 24 per cent of all CO_2 emissions. Australia is a big coal exporter.

LEDCs
Countries such as China and India are industrialising rapidly. They want rich countries to limit their emissions. They argue that only 25 per cent of fossil fuels are burned in LEDCs.

Oil exporting states
Big oil exporters in the Middle East, e.g. Saudi Arabia and Kuwait, see curbs on CO_2 emissions as a threat to their economies.

Organisation of Small Island States (AOSIS)
They fear more violent storms. Some may disappear under rising sea levels. They want tough action to cut CO_2 and other greenhouse gas emissions by 20 per cent by 2005.

Trans-national corporations (TNCs)
Many argue that global warming is not proven and that no action should be taken. These TNCs include giant oil companies, US motor vehicle manufacturers, aluminium and petro-chemical companies. They are a particularly powerful influence in the USA.

Attitudes towards global warming

The attitudes of governments towards global warming are largely based on self interest.

Countries in favour of carbon cuts
- EU countries, especially the UK, have led the way in cutting CO_2 emissions. These countries believe that it is in the long term interests of the global community to halt global warming. Thus, in 2000, the UK's environmental minister stated: 'There is no doubt that climate change is the greatest threat to the future of mankind .... the measures we are putting in place are to save us from the worst effects of global warming.'
- The 36 countries that form the Association of Small Island States (AOSIS) demand urgent action on global warming. Rising sea level threatens the very existence of these countries (Fig. 6.17). Indeed, some countries, e.g. the Maldives, could disappear in the next 100 years.

Countries against carbon cuts
- Poorer countries, such as China and India, currently undergoing rapid

6 Managing natural resources

industrial growth, argue that MEDCs are to blame for global warming. They are not prepared to abandon their development plans, which might lift millions of people out of poverty.
- The USA – the world's biggest polluter – was instrumental in the failure of talks at The Hague in 2000. Americans consume huge amounts of energy. Any US government that urged cuts in energy consumption would be unpopular with most Americans. Agreement is made even more difficult by the political influence of the oil companies and other powerful industrial corporations in the USA.
- Major oil exporting countries, such as Saudi Arabia, Kuwait and the United Arab Emirates, oppose any carbon cuts that might reduce the demand for crude oil.

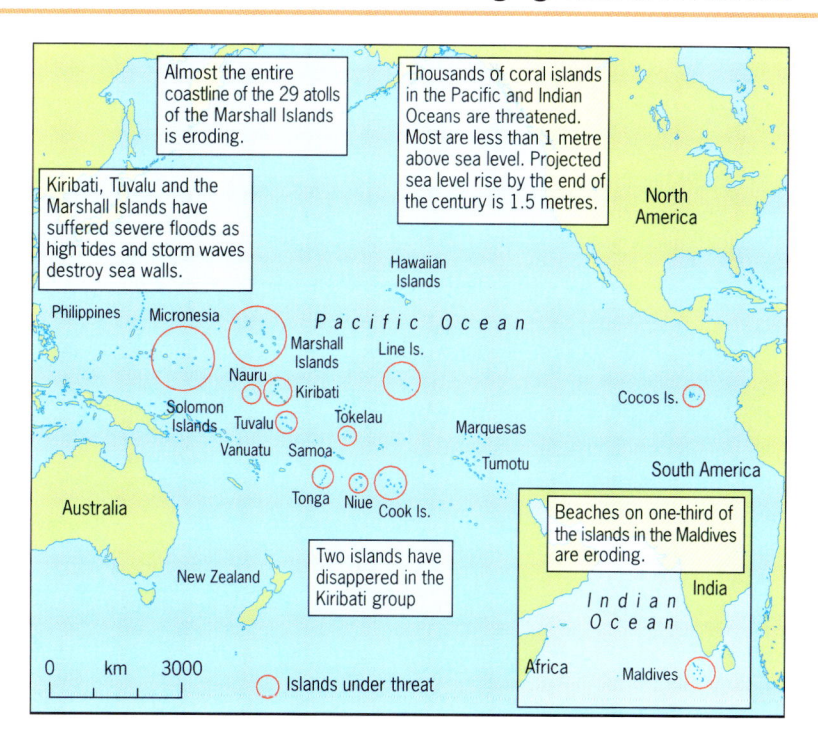

Almost the entire coastline of the 29 atolls of the Marshall Islands is eroding.

Thousands of coral islands in the Pacific and Indian Oceans are threatened. Most are less than 1 metre above sea level. Projected sea level rise by the end of the century is 1.5 metres.

Kiribati, Tuvalu and the Marshall Islands have suffered severe floods as high tides and storm waves destroy sea walls.

Two islands have disappeared in the Kiribati group

Beaches on one-third of the islands in the Maldives are eroding.

○ Islands under threat

6.5 The disappearing ozone layer

Ozone is a type of oxygen found in the Earth's atmosphere. Concentrated between 20 and 30 km above the surface, it absorbs a large part of the sun's ultraviolet (UV) radiation. UV radiation is what gives us a suntan, but in large concentrations it causes skin cancer and eye cataracts, as well as damaging crops and plankton (which is at the base of the oceans' food chain).

How the ozone layer is damaged

Today the ozone layer is under threat from artificial gases, particularly chlorofluorocarbons (CFCs) used in industry to make propellents in aerosols, refrigerants and plastic foam packaging. Each spring in the Arctic and Antarctic, massive holes appear in the ozone layer. They result from complex chemical reactions between ozone, sunlight, ice particles and chlorine from CFCs (Fig. 6.18).

Fortunately, the Arctic and Antarctic are sparsely populated so health risks are small in these areas. But there are concerns that the Antarctic ozone hole is beginning to affect parts of Australia, New Zealand and southern Chile (Fig. 6.19). Because of this, people in Australia and New Zealand are warned about the risk of skin cancer from exposure to the sun.

Ultra-violet B radiation

Chlorine accumulates in clouds during winter. This is released in spring and reacts with sunlight to destroy ozone.

Ozone Layer

25km

Ozone (O$_3$) absorbs and filters out UV-B radiation

CFCs slowly rise up through the atmosphere

Earth's surface

▲ CFCs

Figure 6.17 (top) Islands threatened by rising sea level

Figure 6.18 (right) Thinning of the ozone layer

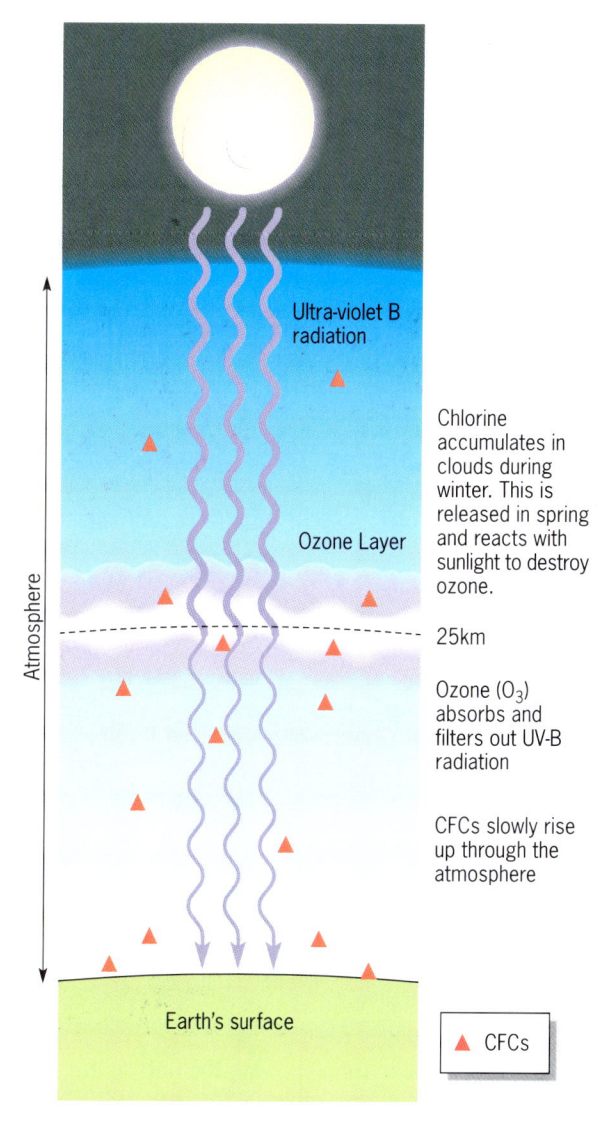

109

6 Managing natural resources

EXERCISES

10 Read through Figure 6.19.
a Find the latitude of Punta Arenas in an atlas.
b Describe the effects of ozone thinning on people and animals in Punta Arenas.
c How do people protect themselves from the dangers of UV radiation?
d What are the economic effects in Punta Arenas of thinning ozone?
e* Why do you think that it has been easier to get international agreement on the ozone problem than on global warming?

Figure 6.19 Ozone hole keeps Chilean city under wraps (Calvin Sims, *The Guardian* 7.3.95)

Protecting the ozone layer

Ozone thinning is an IEP that demands action by all countries. In the late 1980s, an international agreement banned production of CFCs. Although CFCs will remain in the atmosphere for many years, and cause further loss of ozone, the ozone layer should recover within the next 50 years.

The people of Punta Arenas, a quiet port, lying on the Strait of Magellan at the bottom of the world, do not venture outside without first rubbing sun block on their exposed skin and donning dark glasses.

For this the 113 000 residents of Punta Arenas have one man to thank, or to despise. He is Bedrich Magas, an electrical engineering professor at the city's University of Magallanes.

It was Dr Magas who first alerted his home town to the dangers of a large hole in the ozone layer which exposed the area, he says, to unsafe levels of solar radiation.

For the past eight years he has appeared on television and radio programmes and lectured to community groups, school children and agricultural associations, warning people to avoid prolonged exposure to the sun.

As he walks through the streets of Punta Arenas, Dr Magas is approached by mothers who complain that the fair-skinned children turn bright pink when playing outdoors, and farmers who say their sheep are going blind from cataracts they attribute to the sun's rays.

He tells them to shield themselves when the sun is high in the sky and to put pressure on the government to finance research into the impact the radiation may have on them.

'It is much too early in the process to say for certain that the problems these people are experiencing are due to ozone depletion,' Dr Magas said. 'But what we do know is that such high levels of radiation are dangerous and destructive. We are facing a worldwide emergency that is starting in Antarctica and spreading north and something must be done.'

But...the last mayor said Dr Magas was destroying the lucrative ecological tourist industry that Punta Arenas desperately needs as its sheep farming declines.

The city has become the main gateway to the Antarctic, and the cruise ships and flights that use it have created local jobs.

What is certain is that Punta Arenas, the world's southern most city, is the only place where large numbers of people live under the Antarctic hole in the upper atmospheric ozone layer. This is believed to be caused by chlorofluorocarbons (CFCs), once heavily used as propellants for aerosols, as refrigerants, and in the productions of plastic foams.

6.6 Acid rain

In the industrialised countries of the northern hemisphere, much of the rainfall is sour and acid. The cause of this acid rain is air pollution (see Fig. 6.7). The main culprits are power stations and oil refineries burning fossil fuels and giving off waste gases, especially sulphur dioxide (SO_2) and oxides of nitrogen (Fig. 6.21).

Results of acid rain

Acid rain has done serious harm to forests, soils, lakes, rivers and the stonework of buildings. For example, in Scandinavia, large tracts of coniferous forest are dying and thousands of lakes, acidified by pollution, are lifeless. Acid rain is an IEP. The Scandinavian states blame the UK for many of these problems. They argue that prevailing south-westerly winds dump acid rain generated in the UK on their forests and lakes. Within the UK, the problem of acid rain varies from place to place. Worst affected are upland areas with high rainfall and hard, impermeable rocks (Fig. 6.22). Not only does high rainfall deposit more acid on these areas, but the impermeable rocks cause acid water to run unaltered into rivers and lakes. In contrast, where permeable rocks like chalk and limestone are found, acid water seeping into the ground is neutralised before it reaches streams and rivers.

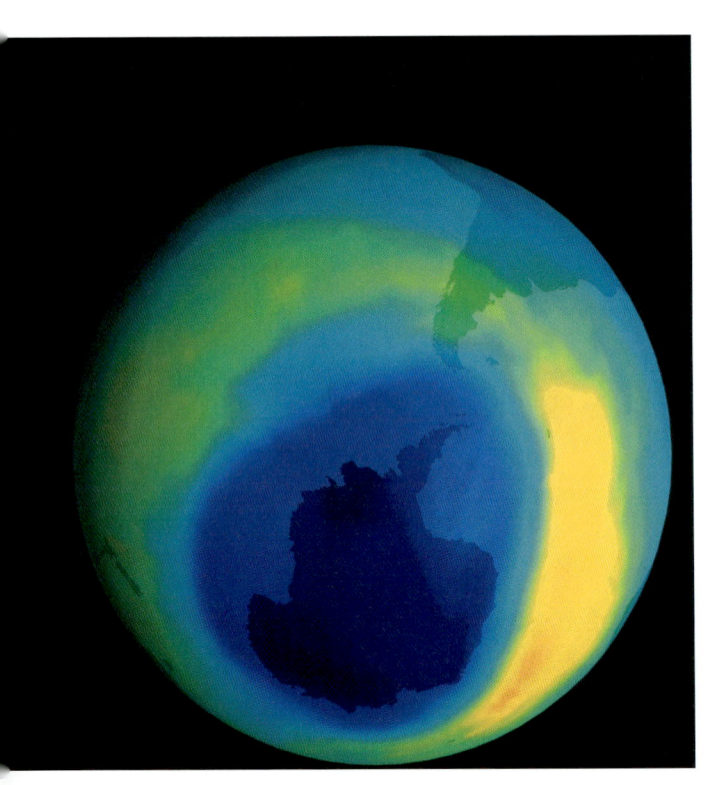

Figure 6.20 Satellite map showing the hole in the ozone layer over Antarctica on 6 September 2000.

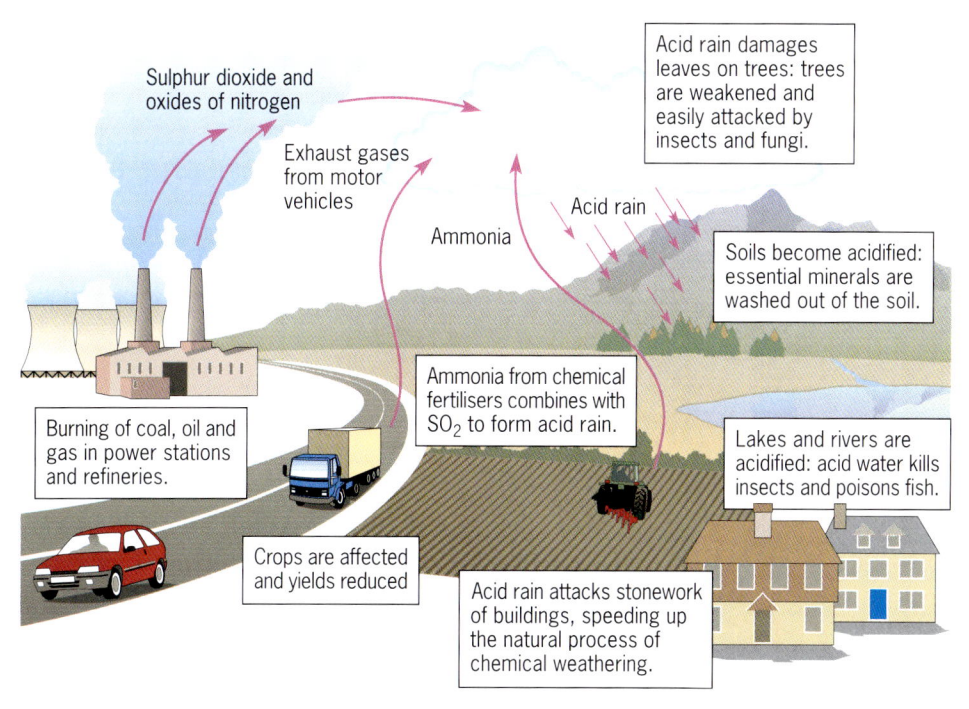

6 Managing natural resources

Figure 6.21 Acid rain: causes and effects

- Sulphur dioxide and oxides of nitrogen
- Exhaust gases from motor vehicles
- Ammonia
- Acid rain
- Burning of coal, oil and gas in power stations and refineries.
- Ammonia from chemical fertilisers combines with SO₂ to form acid rain.
- Acid rain damages leaves on trees: trees are weakened and easily attacked by insects and fungi.
- Soils become acidified: essential minerals are washed out of the soil.
- Lakes and rivers are acidified: acid water kills insects and poisons fish.
- Crops are affected and yields reduced
- Acid rain attacks stonework of buildings, speeding up the natural process of chemical weathering.

EXERCISES

11 Study Figure 6.22 and Table 6.5.
a Make a list of the places in Figure 6.22 and state the likely acidity (high, medium, low) of surface water at each place.
b* Give a general summary of how the effects of acid rain will vary across the UK.

Table 6.5 Rainfall acidity

Mean annual rainfall	Acidity on impermeable rocks	Acidity on permeable rocks
More than 1500 mm	High	Medium
1000–1500 mm	High	Medium
Less than 1000 mm	Medium	Low

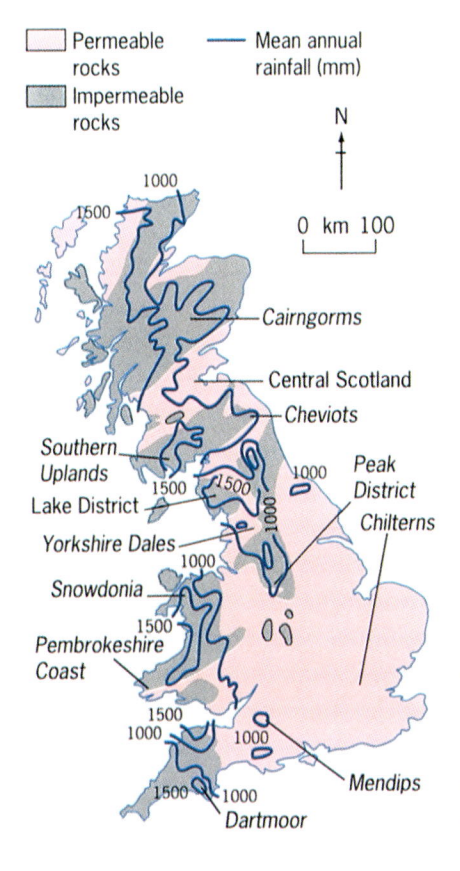

Figure 6.22 (below left) Rainfall, rock type and acid rain in the UK

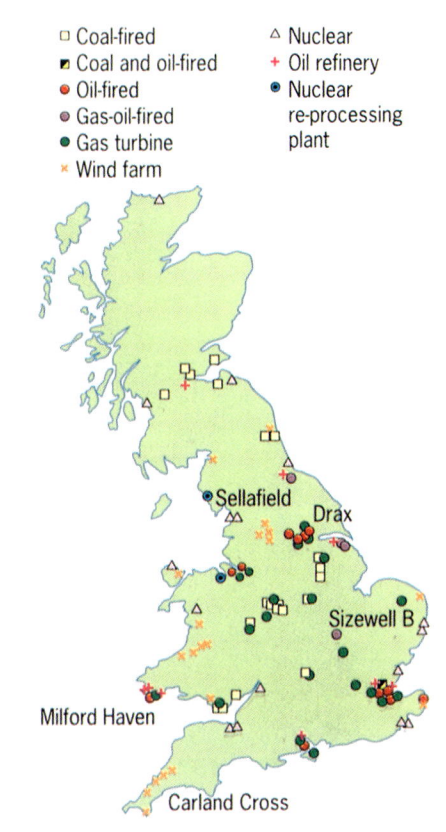

Figure 6.23 (below) Distribution of power stations in the UK

Key (Fig 6.22):
- Permeable rocks
- Impermeable rocks
- Mean annual rainfall (mm)

Places labelled: Cairngorms, Central Scotland, Cheviots, Southern Uplands, Lake District, Yorkshire Dales, Snowdonia, Pembrokeshire Coast, Peak District, Chilterns, Mendips, Dartmoor

Key (Fig 6.23):
- Coal-fired
- Coal and oil-fired
- Oil-fired
- Gas-oil-fired
- Gas turbine
- Wind farm
- Nuclear
- Oil refinery
- Nuclear re-processing plant

Places labelled: Sellafield, Drax, Sizewell B, Milford Haven, Carland Cross

Reducing sulphur dioxide

The EU has taken steps to reduce emissions of gases that cause acid rain. An EU directive in 1988 called for a 40 per cent reduction in SO₂ emissions within 10 years. The UK's emissions halved between 1970 and 1993, and further significant reductions occurred between 1993 and 2000. The cuts were mainly due to the rising popularity of gas as a primary fuel for home heating, and the switch from coal-fired to nuclear- and gas-fired power stations.

Sulphur dioxide can be removed from the chimneys of coal-burning power stations. Drax is the UK's largest coal-fired power station (Fig. 6.23) and the first to be fitted with desulphurisation equipment. It burns more than 10 million tonnes of coal a year, most of which comes from the nearby Selby coalfield (see Fig. 6.25). The desulphurisation equipment uses

111

6 Managing natural resources

Figure 6.24 Coal production in the UK, 1990–1998

limestone, and removes 90 per cent of sulphur dioxide from waste gases. But desulphurisation is expensive and adds significantly to the cost of electricity.

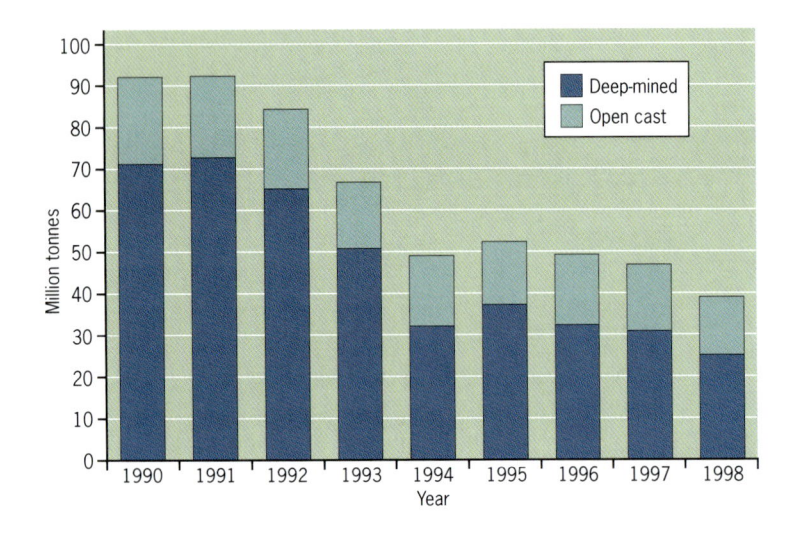

6.7 Opencast mining in the UK

We have seen how, at a world scale, burning coal and other fossil fuels causes atmospheric pollution and problems such as global warming and acid rain. At a local scale, coal extraction also leads to environmental damage.

- Mountains of waste from coal mines are dumped to form unsightly spoil heaps.
- Abandoned mine workings cause surface subsidence, which can damage buildings and disrupt drainage.
- Highly toxic water seeps from old mine workings and pollutes rivers and streams.

The environmental impact of open casting

Since the mid 1980s, the UK coal industry has undergone massive decline (Fig. 6.24). But this has not always reduced the pressure of coal mining on the environment. One reason for this has been an expansion of opencast mining (Fig. 6.25). Essentially, open casting is quarrying coal from the land's surface. It is much cheaper than traditional deep mining. However, it is one of the most environmentally destructive processes carried out in Britain. Opencast operations often occupy huge sites and create holes up to 150 m deep. Great mounds of earth act as screens around the site to reduce the noise of blasting and wind-blown dust. Within this protective screen, open casting creates a lunar landscape (Fig. 6.26). Today, more than one-third of the UK's coal production comes from opencast sites.

Figure 6.25 Coalfields and coal mines in the UK

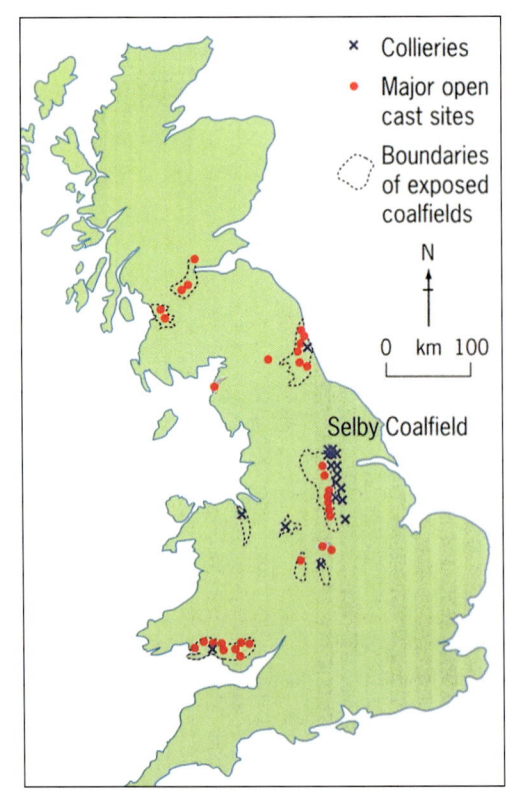

Open casting and land restoration

By law, mining companies have to manage opencast sites in order to minimise their effects on local residents and on the landscape. No-one could argue that opencast operations are anything other than ugly. Although land restoration methods are improving, it still takes many years for agricultural land to return to its former productivity, and for trees and hedgerows to become established. Planning authorities also try to control open casting by not allowing development in environmentally sensitive areas such as National Parks, Areas of Outstanding Natural Beauty and Nature Reserves – areas that are also attractive to tourists (see Chapter 7).

The case for open casting

In its defence, opencast mining is only temporary. Once mining has ended, the mining companies have to restore the site. In areas badly affected by deep mining in the past, open casting can even improve the landscape. Open casting also creates jobs, often in parts of the country where unemployment is high, and provides cheaper coal for electricity generation and for industry. Meanwhile, if the only way to maintain our standard of living is to burn large amounts of fossil fuels, many people would argue that some environmental damage is a price worth paying.

6 Managing natural resources

Figure 6.26 (left) Opencast coal mining, Hirwaun, South Wales

EXERCISES

12 What is your attitude to opencast mining? Explain your view (which you should be prepared to defend in discussion) taking account of the following: • the need for cheap coal • the need for employment in coalfield areas • the effect on agriculture, local residents, wildlife and the landscape.

6.8 Alternative energy resources

Our dependence on fossil fuels cannot continue indefinitely. There are two reasons for this:
◆ Fossil fuels will eventually run out, so that alternative fuels will have to be found.
◆ The burning of fossil fuels creates serious IEPs.

So what is the answer? We shall examine two possibilities in this section: an expansion of nuclear energy and a gradual shift away from non-renewable to renewable resources.

Nuclear power

Nuclear energy is produced by the **fission** of uranium in nuclear reactors. Although uranium is a non-renewable resource, there is enough available to supply the nuclear industry for the foreseeable future. (It is also possible to recycle uranium, making future supplies even more secure.) Nuclear energy has another advantage: it does not produce carbon dioxide and other waste gases and therefore does not contribute to global warming or acid rain.

About 20 per cent of the UK's electricity comes from 14 nuclear power stations, but with little recent investment this will fall. Some of the oldest nuclear power stations, dating from the early 1960s, have already closed down and others will follow in the next few years. Because new stations can take up to 10 years to build, old stations cannot be replaced quickly. This means that the proportion of electricity generated by nuclear power will fall in future.

Safety issues

The expansion of nuclear power is not popular. The main reason is concern over its safety. Attitudes to nuclear power changed after the world's worst nuclear accident at Chernobyl in the Ukraine in 1986. Since then, hundreds of people in eastern Europe have suffered from radiation-related cancers, and thousands more will be affected in future. Even today parts of upland Britain are still contaminated with radioactive fall-out from Chernobyl. Although the British nuclear industry has a good safety record, since Chernobyl the UK has built only one new nuclear power station – Sizewell B in Suffolk (1994). (Fig. 6.23).

Estimated percentage contribution to electricity production by renewables in 2010 (10 per cent of total production)

- HEP 0.75%
- Landfill gas 1.4%
- Wind 5%
- Biofuels 3%
- Solar 0.5%
- Tidal 1%
- Wave 0.5%

Potential contribution of renewables to UK's future electricity demands

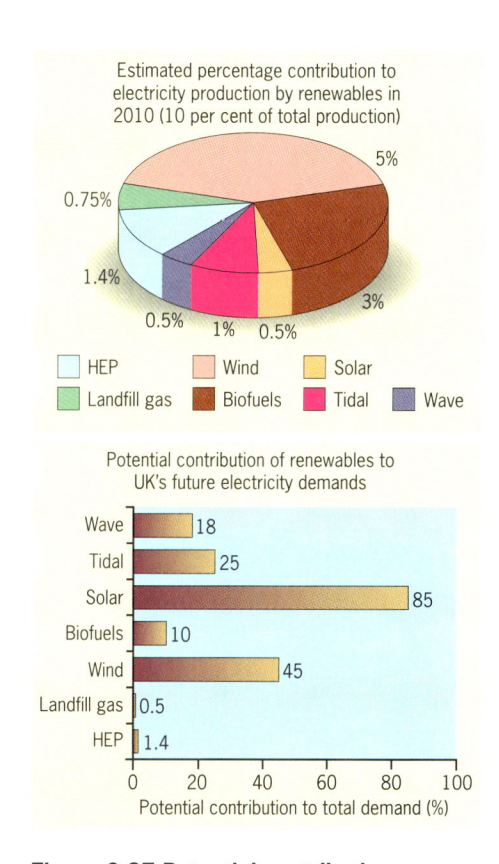

Source	Potential contribution to total demand (%)
Wave	18
Tidal	25
Solar	85
Biofuels	10
Wind	45
Landfill gas	0.5
HEP	1.4

Figure 6.27 Potential contribution and estimated contribution of renewables to the UK's electricity production in 2010

EXERCISES

13 Should nuclear power replace fossil fuels as the world's main energy source in future? State and explain your opinion on this matter in class discussion.

6 Managing natural resources

> **EXERCISES**
>
> **14** Use Figure 6.23 to describe and explain the distribution of wind farms in the UK.
>
> **15a** Describe the likely combination of renewables to UK electricity demand in 2010 (Fig. 6.27).
>
> **b** Discuss the potential of renewable energy resources in the UK Fig. 6.27).

The disposal of spent radioactive fuel

A major headache for the nuclear industry is the disposal of radioactive waste. The most toxic, high-level waste will remain radioactive for thousands of years. At the moment, all high-level radioactive waste in the UK is stored on the surface at the nuclear re-processing plant at Sellafield in Cumbria (see Fig. 6.23). Eventually it will have to be put into specially designed storage areas deep underground. But people have objected strongly about the siting of these, and no such storage area has yet been completed by any country.

Renewable resources

In future, coal-, oil- and gas-fired power stations could be replaced by wind farms and other forms of renewable energy (Fig. 6.29). In fact, the UK government aims to produce 10 per cent of the country's electricity from renewables by the year 2010.

Wind power

Advantages and disadvantages

Wind power is renewable, efficient and non-polluting. Moreover, modern wind turbines occupy only a small area, so that people can use wind farms not only for generating electricity, but also for other activities. Wind farms, unlike nuclear and thermal power stations, are also easy to build and to dismantle. However, even wind power has some disadvantages. Many people feel that wind turbines spoil the appearance of the landscape. Unfortunately, the best places to site wind farms are often the most scenic. Other people, who live close to wind farms, complain about the noise made by the blades of the turbines. A further disadvantage is that because wind is intermittent, it is not always possible to generate electricity when it is needed.

Figure 6.28 Carland Cross wind farm, Cornwall

Figure 6.29 Other forms of renewable energy

Geothermal power
Feasible where hot rocks, due to volcanic activity, lie close to the surface. Iceland gets most of its electricity from geothermal power (see Book 1, Chapter 1). Granite formations in South-west England have the potential for development. Unlike the wind, the tide is a predictable source of energy.

Wave power
There is great potential on the UK's stormy western coastlines. An experimental wave station has been set up on Islay.

Tidal power
The UK, with its large tidal range, could produce huge amounts of electricity from barrages across estuaries. Disadvantages include the high capital costs; the disruption of navigation; and the destruction of important wildlife habitats (mudflats, salt marshes) in estuaries.

Solar power
Solar power can be used to heat water directly for domestic heating; to make steam to generate electricity; to generate electricity directly using photo-voltaic cells. Places in low latitudes with cloudless climates have the greatest potential although solar power is relevant even in cloudy, high latitude climates. But it is costly and solar power plants need large areas to collect sunlight. Also, power production stops at sunset.

Hydro-power
Hydroelectric power (HEP) already provides nearly 2 per cent of the UK's electricity. HEP is important in countries with high mountains and large rivers, e.g. Norway and Sweden. There are significant environmental costs, particularly where dams are built and valleys are flooded.

6 Managing natural resources

Wind power in the UK
In 2001, the UK produced just 3 per cent of its electricity from renewables. Most of this came from wind power. The latest wind turbines have been built offshore. The UK's first offshore turbines were sited at Blyth in Northumberland in 2000. The two turbines at Blyth produce enough energy to power 1500 homes. By 2010, up to 1000 turbines like those at Blyth should generate one per cent of the UK's electricity. Siting wind turbines offshore overcomes most environmental objections to wind power. Also, offshore turbines can be bigger than those on land, and can exploit the higher wind speeds over the sea.

Fuel wood
Wood provides 14 per cent of the global demand for energy, much of it in LEDCs. One advantage of fuel wood is that it does not contribute to global warming. Although CO_2 is released when wood is burned, it is a small enough amount to be absorbed by trees during their growth. Another advantage is that fuel wood, like other forms of **biofuels**, is renewable.

Eggborough: the UK's first wood-burning power station
The UK's wood-burning power station is in Eggborough, North Yorkshire. It is fuelled by fast-growing willow saplings, which occupy 2000 hectares of land near the power plant. Wood is harvested from the trees on a three-year rotation. Annual production of dried wood is between 8 and 20 tonnes per hectare. Treated sewage sludge is used as a fertiliser.

The Eggborough power station generates 8MW – enough to supply electricity to a town of 16 000 people. The large area of willow saplings also provides a valuable habitat for wildlife.

> ### EXERCISES
> **16** Study Figure 6.28.
> **a** Suggest a possible advantage of this site for wind turbines.
> **b** What other type of land use is found within and around the wind farm?
> **c** Give your opinion of the impact of wind farms on the landscape in Figure 6.28.
> **d*** Forty per cent of Europe's usable wind power is in the British Isles. Suggest reasons for this.
> **17a** Summarise the advantages and disadvantages of renewable energy resources.
> **b** In your opinion, which type of renewable energy has the greatest potential for the future? Give reasons.
> **c** Select a renewable energy resource and write a report on its potential advantages and disadvantages etc. in the UK or another country of your choice. Use the web as a data source (e.g. www.greenenergy.org.uk).
> **d** Give a presentation to your class based on your report.

6.9 Recycling and conserving resources

In nature, there is a constant cycling of energy and materials. Nothing is wasted: anything produced by one organism as waste is used by another. Today, the challenge is to design products that can be fully recycled in an economic system without producing any waste (Fig. 6.30). Only in this way can we manage the Earth's non-renewable resources and achieve truly sustainable growth.

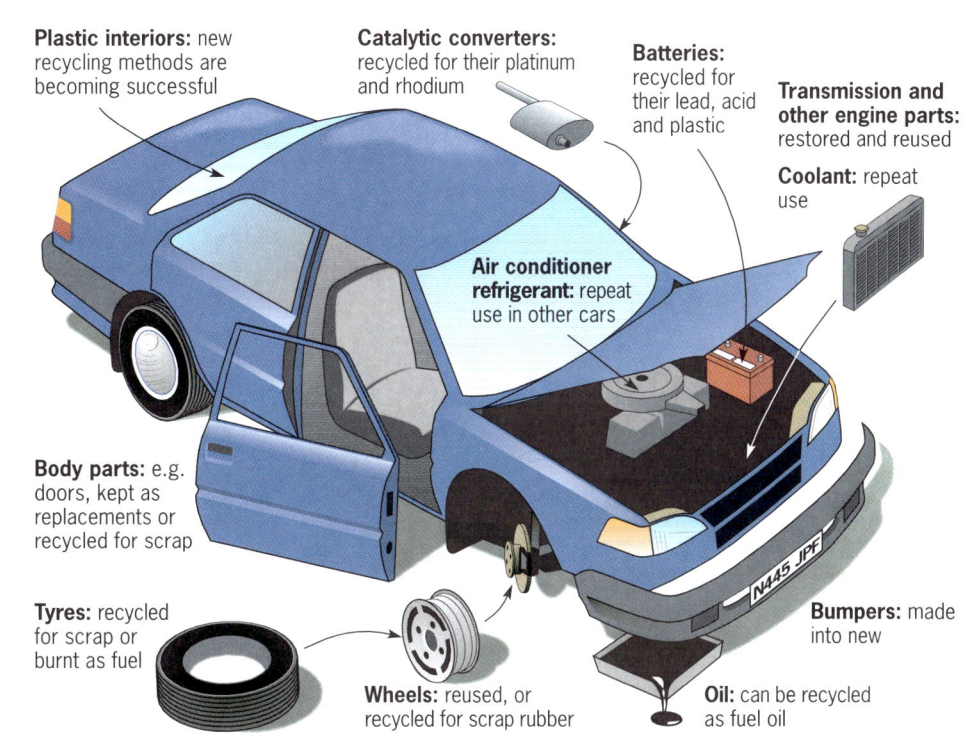

Plastic interiors: new recycling methods are becoming successful
Catalytic converters: recycled for their platinum and rhodium
Batteries: recycled for their lead, acid and plastic
Transmission and other engine parts: restored and reused
Coolant: repeat use
Air conditioner refrigerant: repeat use in other cars
Body parts: e.g. doors, kept as replacements or recycled for scrap
Tyres: recycled for scrap or burnt as fuel
Wheels: reused, or recycled for scrap rubber
Oil: can be recycled as fuel oil
Bumpers: made into new

Figure 6.30 Car components that can be recycled

115

6 Managing natural resources

> **EXERCISES**
>
> **18a** Why, as global citizens, should we encourage the recycling and conservation of resources?
> **b** Use Figure 6.32 to discuss the environmental advantages of recycling waste paper.
> **19** Make a sketch of your own design for an energy-efficient house. Add labels to explain its main energy-saving features.

Recycling resources

Metals such as steel, copper and aluminium have been recycled for many years. Recycling has a number of environmental benefits:
- It reduces the demand for new resources of non-renewables such as metals and oil (used in plastics);
- Compared to extracting and processing new resources, it reduces emissions of carbon dioxide by 10 per cent;
- It reduces the volume of waste materials, which are either dumped in landfill sites, or incinerated.

The pressure to find new landfill sites in densely populated regions, such as South-east England and the Tokyo metropolitan area, is acute. Landfill sites often result in a loss of countryside and the leakage of toxic waste into the environment. This problem of recycling has to be solved if cities are to be truly sustainable in future.

Household waste

Some local councils in the UK encourage households to recycle domestic waste, and provide separate bins for glass, plastics, paper etc. The success of such schemes depends on the price of recycled materials. In 1998, for example, the price of recycled paper collapsed to £5 a tonne and many councils closed paper banks and stopped kerbside collections of paper.

Conserving resources

More than half the energy consumed in the UK is used in buildings. The UK government gives grants for improving insulation in homes to reduce heat loss. We can conserve resources by using them more efficiently, e.g. by burning fossil fuels and wood in more energy-efficient furnaces and stoves. We could also design more energy-efficient houses. For example, because heat rises it makes more sense to have the living area on the first floor and

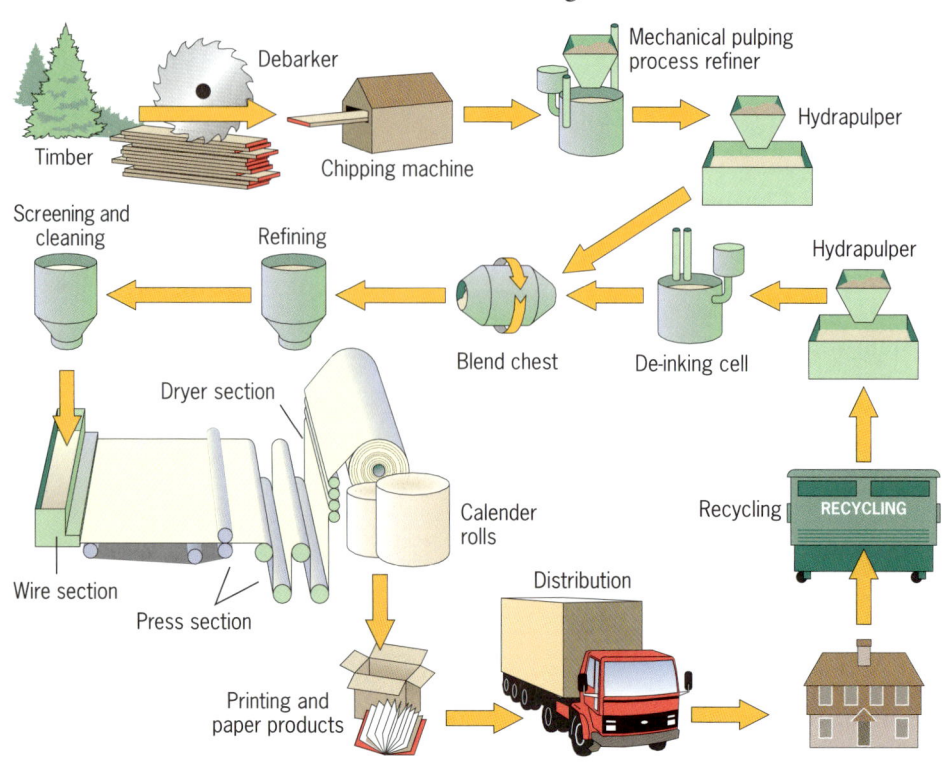

Figure 6.31 (below) Recycling: Municipal waste in the USA

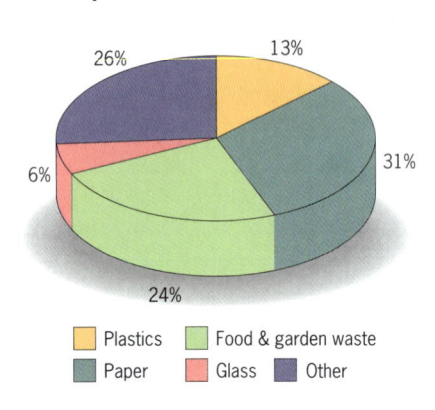

- Plastics: 13%
- Paper: 31%
- Food & garden waste: 24%
- Glass: 6%
- Other: 26%

> **REMEMBER**
> Recycling is essential if we are to create a sustainable global economy. Most materials can be recycled. However, it is often more profitable to use new materials rather than recycle old ones.

Figure 6.32 (right) Waste paper recycling

6 Managing natural resources

the bedrooms and bathroom on the ground floor. Similarly, more advantage could be taken of free solar energy. Houses aligned east-west and with plenty of windows on south facing walls (and few on north-facing walls) will maximise the solar energy gain.

6.10 Summary: Managing natural resources

> **KEY SKILLS OPPORTUNITIES**
> **C1.1:** Ex. 6c, 12, 13, 17b, 18;
> **C1.2:** Ex. 1a, 2b, 2c, 8, 10, 11a, 14; **C1/2.3:** Ex. 9b, 12, 15b; **C2.1:** Ex. 12, 13, 16c, 17b, 17d, 18; **C2.2:** Ex. 5a, 5b, 6b, 11b, 15a, 17a; **N1/2.1:** Ex. 16a; **N1/2.2:** Ex. 2a, 4a, 7a; **N1/2.3:** Ex. 4b, 4c, 7b, 7c; **IT2.2:** Ex. 17c; **IT2.3:** Ex. 17c.

Key ideas	Generalisations and detail
Natural resources are things that are valuable to people.	• Natural resources include fuels, minerals, water, timber, soil, fish, etc.
Natural resources can be divided into renewables and non-renewables.	• Non-renewable resources, such as fossil fuels, are finite and will eventually run out. • Renewable resources are either constantly recycled (e.g. water), or renewed (e.g. plants and animals), or are inexhaustible (e.g. solar energy).
No simple relationship exists between a country's natural resources and its level of development.	• Possessing a wealth of natural resources does not guarantee development. Many of the world's poorest countries have a huge natural resource potential (e.g. Sierra Leone). • Several rich countries, such as Japan and the Netherlands, have few natural resources. Their success is based on human resources (i.e. education and skills of their populations). However, the amount of natural resources consumed per person is a good indicator of development.
The world's reliance on non-renewable resources is not sustainable and is responsible for international environmental problems (IEPs).	• Atmospheric pollution through the burning of fossil fuels is responsible for global warming and acid rain. • Global warming will cause major disruption to the world's climate and rising sea levels in the 21st century. • Acid rain is a more local problem, which destroys forests, aquatic life in lakes and rivers and attacks the stonework of buildings. • Ozone thinning, caused by the use of CFCs in industry, poses a health risk through increased levels of UV radiation.
It is difficult to get agreement on tackling IEPs.	• Global warming is connected with the atmosphere, which is not the responsibility of any one country. • Countries often have conflicting interests. For poor countries, rapid industrialisation (based on fossil fuels) may bring advantages that outweigh the disadvantages of global warming. Major coal, oil and gas exporters don't want any cut-back in the use of fossil fuels. On the other hand, many small island states (e.g. Maldives) will be drowned by rising sea levels unless global warming is stopped.
The development and use of resources at a local scale can have a significant environmental impact.	• Opencast mining has a devastating effect on local environments. Minimising these effects requires careful management by mining companies and planners. • Planners must take account of other interests, including agriculture, conservation and tourism, etc.
Greater emphasis must be put on renewable/alternative resources to achieve sustainable growth in future.	• Nuclear power could be expanded, but at the moment there are doubts concerning its safety. • Renewable energy resources, such as wind, solar and hydro power, and biofuels will become more important. But in the short term they cannot fully replace fossil fuels.
Governments need to encourage greater efficiency in the use of resources and more recycling to achieve sustainable growth in future.	• Manufactured items, such as cars, should be designed so that all of their materials can be recycled. • Fuels should be burned more efficiently; buildings should conserve energy and make maximum use of solar energy.

7 Tourism

EXERCISES

1a Name the places you have visited as a tourist in the last year or two. Make a list of the natural and human resources found at these places.

b Study Figures 7.1–7.5. For each photograph describe the possible resources for tourism. Suggest what type of tourist activities might be developed around these resources.

7.1 Introduction

When we have free time, or **leisure**, we can choose how to spend it. There are many options: we might read a book, play a sport, go to the cinema or simply do nothing. Tourism is one way of spending some of our leisure time. It involves visiting places, either on a day trip or by staying away from home for at least one night. In the last 50 years, the demand for tourism has soared. As a result, tourism has become one of the world's fastest-growing industries. In this chapter we shall look at the growth and importance of tourism, and the advantages and disadvantages it brings. But first, we'll consider the resources that are the basis of tourism.

Figure 7.1 (above left) Double Arch – natural arches at the Arches National Park, Utah, USA

Figure 7.2 (above centre) Mauna Kea beach resort, Hawaii

Figure 7.3 (above right) Elephants in Amboseli National Park, Tanzania, with Mt Kilimanjaro in the distance

Figure 7.4 (far left) Tourists crossing St Mark's Square in Venice, Italy, during a flood

Figure 7.5 (left) Warriors from the Masai tribe, Kenya

7.2 Resources for tourism

Tourism, like any other economic activity, depends on both natural and human resources. Mediterranean countries, such as Greece, have warm, sunny climates and sandy beaches; the Alps offer stunning mountain scenery, snow and lakes; while East Africa is home to some of the world's most

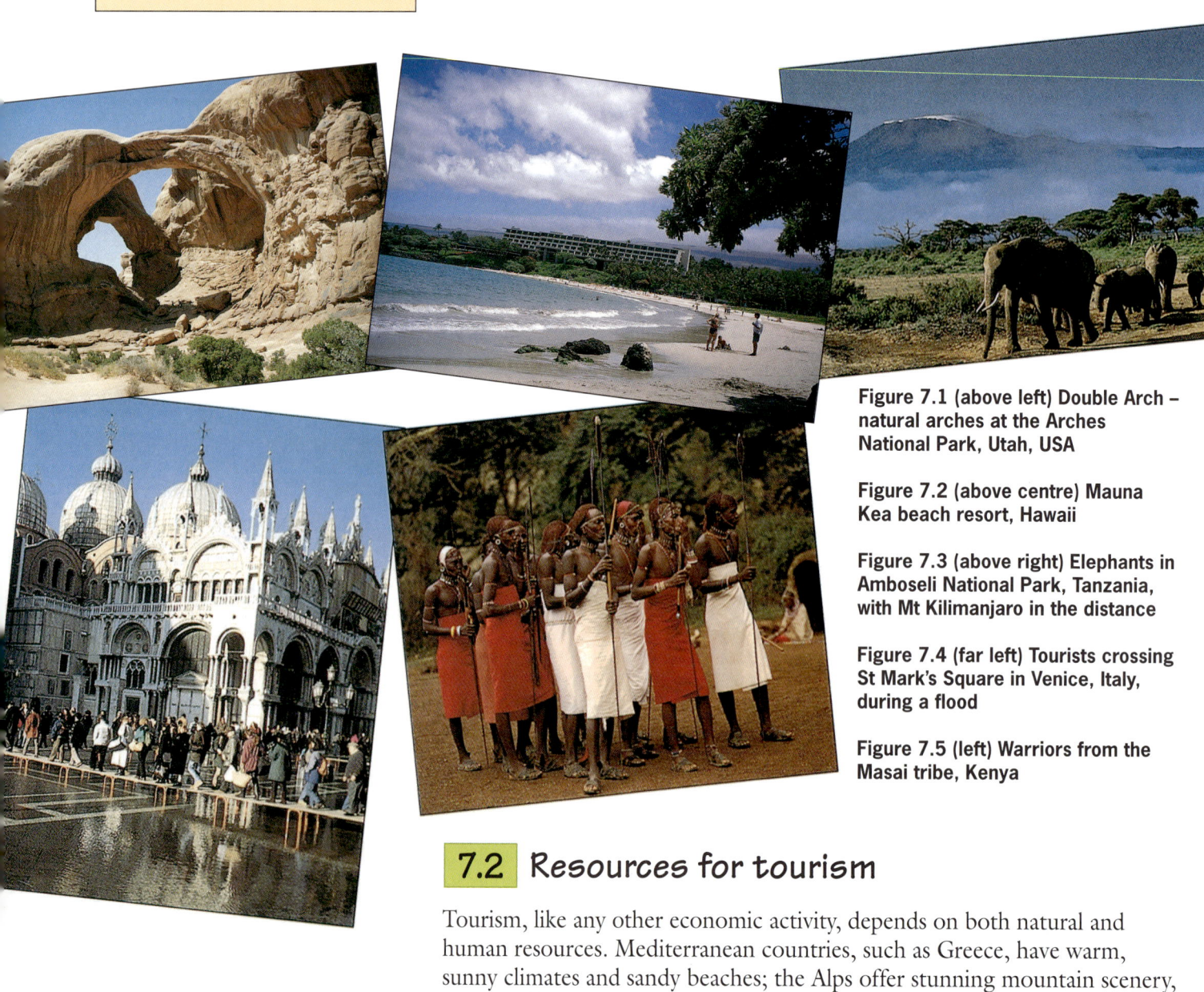

7 Tourism

spectacular wildlife. All of these attractions are natural resources that provide opportunities for recreation. However, some resources for tourism are made by humans. For example, tourists visiting Paris or Rome will be attracted by human resources. These might include the cities' buildings, culture, history and night life.

7.3 Tourism in the UK

Mass tourism in the UK began in the second half of the 19th century. Seaside resorts such as Scarborough and Southend grew rapidly to serve people visiting from the expanding industrial towns and cities. Two things made this possible:
- The development of the railways offering cheap and rapid transport for thousands of tourists and day trippers.
- The introduction of paid holidays, which made a week's holiday away from home affordable to factory workers.

EXERCISES

2a Most Victorian seaside resorts developed to serve nearby industrial towns and cities. Find the location of the following resorts in an atlas: Blackpool, Brighton, Clacton, Margate, Skegness, Southport, Scarborough, Southend, Weston-super-Mare and Whitley Bay.
b Suggest which major urban centres they would have served during the 19th century.
c* There were no large Victorian resorts in Cornwall, Norfolk or West Wales. Suggest possible reasons for this.

CASE STUDY

7.4 Blackpool: a Victorian seaside resort

FACTFILE
- *Blackpool was the first seaside resort for mass tourism.*
- *Blackpool is the largest and most popular seaside resort in the UK.*
- *17 million tourists visited Blackpool in 2000.*
- *Tourism accounts for 75 per cent of economic activity in the town and is worth £500 million a year.*

EXERCISES

3 Study Figure 7.6. State and explain two possible reasons for Blackpool's continuing popularity as a resort.

The origins of tourism

Until the mid 19th century, Blackpool was untouched by tourism. Only wealthier people visited the seaside, attracted by the supposed health-giving properties of sea water and sea air.

In 1846, the railway connected Blackpool to the fast-growing industrial towns of northern England. Almost overnight Blackpool became the first 'working class' resort, providing recreation and entertainment for thousands of factory workers from Lancashire and Yorkshire.

Blackpool's rapid growth

As Blackpool's popularity increased, rows of terrace houses were built to accommodate tourists. Soon the town sprawled along the sea front, from Fleetwood in the north to Lytham St Anne's in the south (Fig. 7.6).

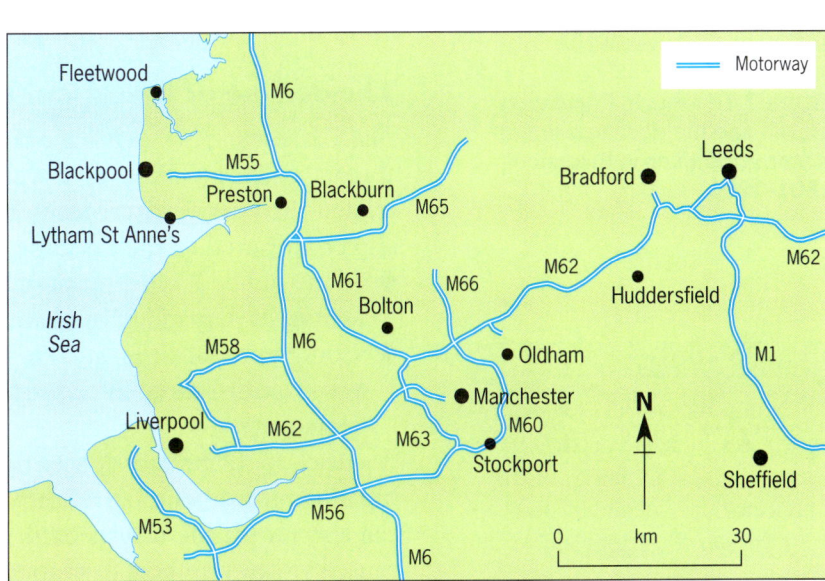

Figure 7.6 Blackpool: its relative location to motorways and major centres of population

7 Tourism

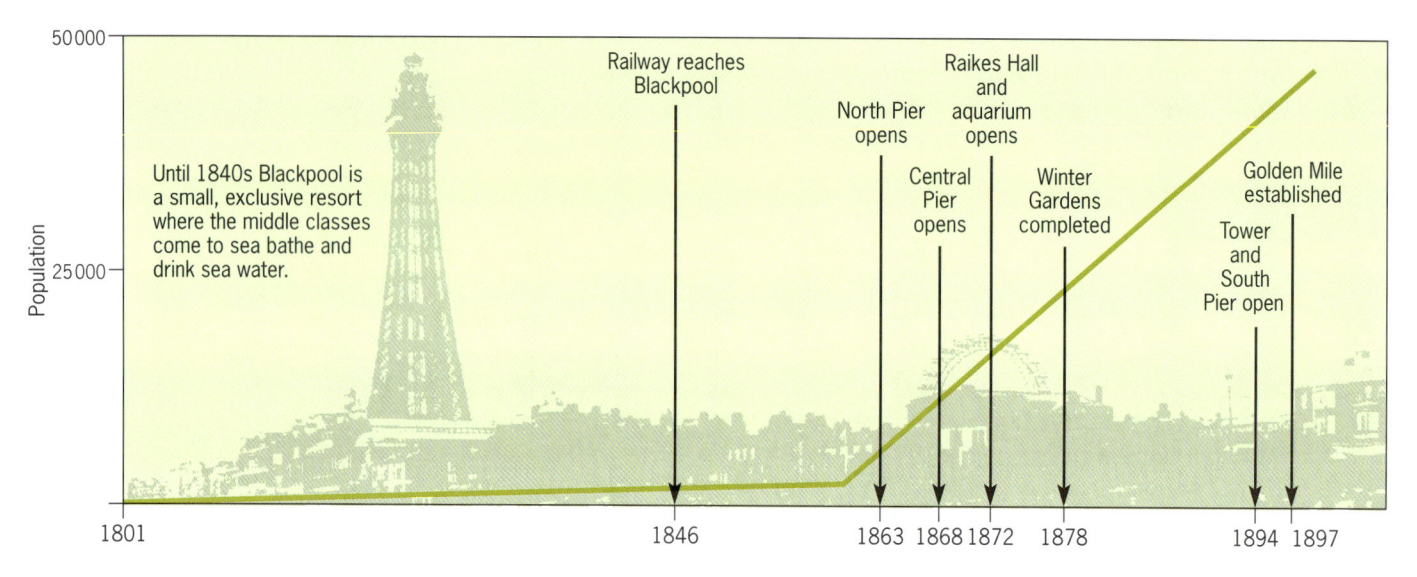

Figure 7.7 (top left) Engraving of early Victorian Blackpool, c. 1840

Figure 7.8 (top centre) Beach crowds near Blackpool Pier, c. 1890

Figure 7.9 (top right) Holidaymakers by Blackpool Tower, 1903

Figure 7.10 (above) Population growth and the development of visitor attractions in Blackpool, 1801–1901

By 1890, Blackpool was attracting more than one million visitors a year. Apart from the beaches, Blackpool offered numerous attractions created by people (Fig. 7.10). These included a promenade, three piers, the Winter Gardens, the Golden Mile and the Tower (Figs. 7.7–7.9). The town also offered all manner of entertainments, from amusement arcades and music halls, to firework displays and steamer trips to the Isle of Man.

The decline of Blackpool as a seaside resort

The UK's seaside resorts have declined in recent decades. Package holidays abroad, based on cheap air travel and increased domestic competition, have eroded their popularity. Meanwhile, lack of investment in visitor facilities has also contributed to their decline.

- Many old bed-and-breakfast establishments in Blackpool fall below the standards demanded by visitors.
- Conference facilities, such as the Winter Gardens, are out-dated compared to those in centres such as Brighton, Bournemouth and Harrogate.

Blackpool, like most other traditional seaside resorts, suffers relatively high unemployment, particularly in the winter months. Unreliable weather can also hit seaside resorts hard. For instance, the poor summer of 1999 caused visitor numbers to plunge by 20 per cent.

EXERCISES

4 Study Figure 7.11. State three ways in which the model of a seaside resort differs from the concentric ring model (see Book 1, Chapter 6) of a typical town.

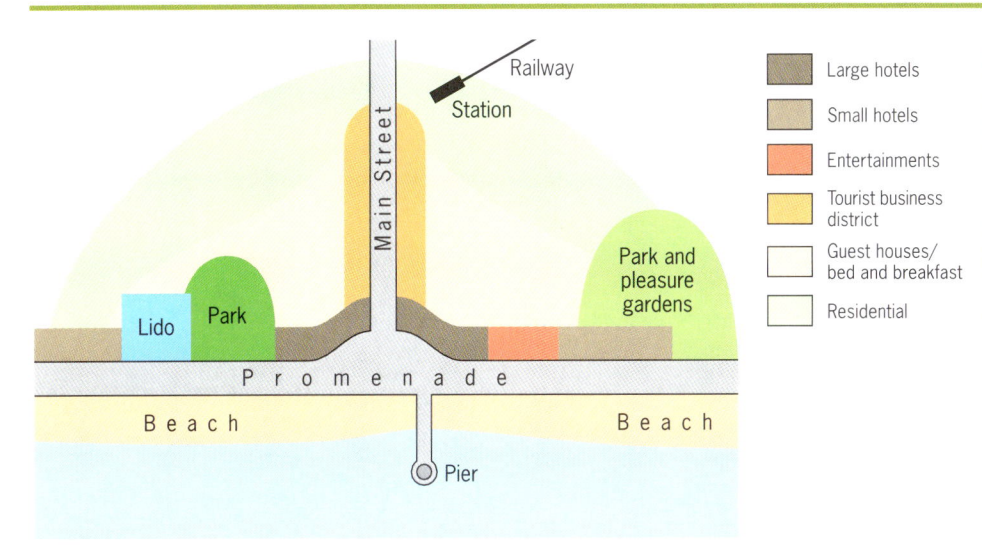

7 Tourism

EXERCISES

5 Study Figure 7.13.
a State the direction in which the camera is pointing. Give your reasoning.
b Describe Blackpool's shape.
c* Suggest a possible reason for Blackpool's shape.

Figure 7.11 (left) Model of a seaside resort

Figure 7.12 (above) Britain's leading seaside resorts

Regeneration

Blackpool now relies more on short-stay visits and package deals, rather than on family holidays. But it wants to diversify its leisure activities to make it an all-year-round resort. Investment in new conference facilities, and ambitious plans to reinvent Blackpool as the UK's version of Las Vegas have been proposed. Meanwhile, Blackpool is bidding to win UK and EU funds to assist its regeneration.

Table 7.1 Origins of UK visitors to Blackpool

	Visitors %	UK population %
North-west	24	11.7
Yorkshire & Humberside	17	9.0
Scotland	13	9.4
North	12	5.4
West Midlands	11	9.5
South-east	8	30.2
East Midlands	6	7.2
Rest of UK	9	17.6

Figure 7.13 (below) Aerial view of Blackpool, 1996

EXERCISES

6a Plot the origin of visitors in Table 7.1 as a pie chart.
b Describe Blackpool's visitor catchment area in the UK.
c* With reference to Table 7.1 and Figure 7.12, suggest a possible explanation for the differences in the proportion of visitors coming to Blackpool from the different regions.
d* Write a report suggesting strategies for making traditional seaside resorts more attractive to young visitors. Present your report in class as a document for discussion.

121

7 Tourism

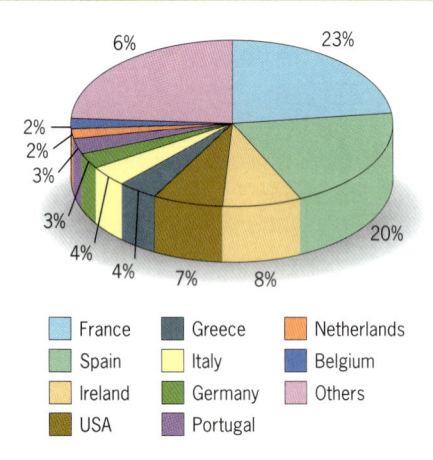

France — Greece — Netherlands
Spain — Italy — Belgium
Ireland — Germany — Others
USA — Portugal

Figure 7.14 (above) Destinations of international tourists from the UK

7.5 Travelling abroad

Today people travel much further than their nearest seaside resort. In 1999, the most popular foreign destinations were France, Spain, Ireland and the USA (Fig. 7.14). Foreign tourism to the UK has also grown dramatically in the past 30 years (Fig. 7.15). There are several reasons for this.

- People have longer paid holidays.
- They are better-off and can afford to spend more on tourism, often having several foreign holidays a year.
- Improvements in transport (in particular, cheap air travel) have made international tourism much easier.

Tourists visiting the UK

Nowadays, tourism in the UK is a huge industry. In 1999, foreigners made 26 million trips to the UK, spending more than £12 billion. By 2003, it is estimated that overseas visitors will spend £18 billion in the UK. Employment provides further evidence of tourism's importance, with more than 1.7 million people working in hotels, restaurants, cafes, pubs, travel agencies, museums, sports centres and so on (Fig. 7.16).

7.6 National parks in England and Wales

We have seen that the tourist attractions of seaside resorts like Blackpool are largely made by humans. Very different, though, are the national parks of England and Wales. The parks' principal attractions are their natural beauty and wildlife. However, these are fragile resources, easily damaged by visitors. Unlike seaside resorts, tourism is **sustainable** in national parks only through careful management of the countryside and control of visitors.

EXERCISES

7 Study Figure 7.17.
a Summarise the regional pattern of tourism for British residents and for foreign visitors.
b* Suggest possible reasons for the differences in the regional patterns.

Figure 7.15 (right) Value of foreign tourism to the UK, 1988–99

Figure 7.16 (below) Employment in tourism in the UK

Hotels, accommodation — Travel agencies, tour operators
Restaurants, cafes — Libraries, museums, etc.
Bars, night clubs, etc. — Sports, recreation, etc.

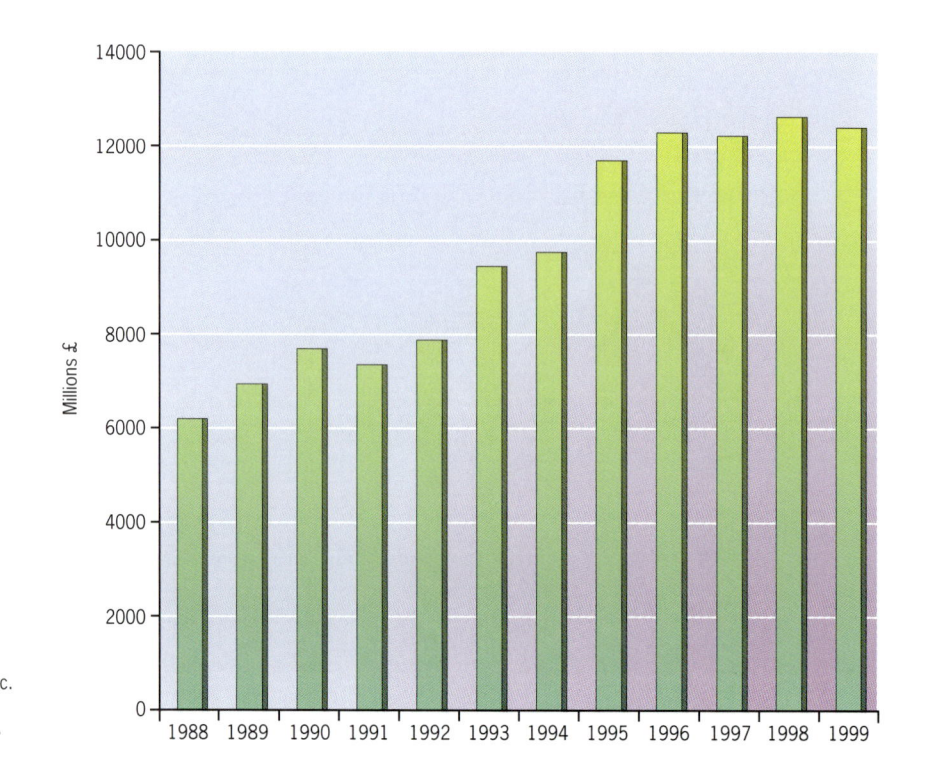

7 Tourism

Features of national parks

National parks are large tracts of relatively wild land defined by Act of Parliament. Most are areas of mountain and moorland (Fig. 7.18). Because of their outstanding natural beauty, they are protected by legislation against development. Altogether, there are ten national parks in England and Wales plus the Broads in East Anglia, which has a similar status, and the proposed New Forest National Park in Hampshire. National parks cover 7 per cent of the land area of England, and 20 per cent of Wales.

Each park is looked after by a National Park Authority (NPA). NPAs have three main purposes:

- ◆ To protect the environment.
- ◆ To promote 'quiet enjoyment and understanding of the parks'.
- ◆ To take account of the economic and social needs of local people.

Two features of national parks make these responsibilities particularly difficult to carry out. First, despite their name, most of the land in national parks belongs to private individuals. And second, national parks are not areas set aside exclusively for conservation. A wide range of economic activities use national parks, including forestry, agriculture, mining and quarrying, water supply, army training and tourism (see Section 7.7).

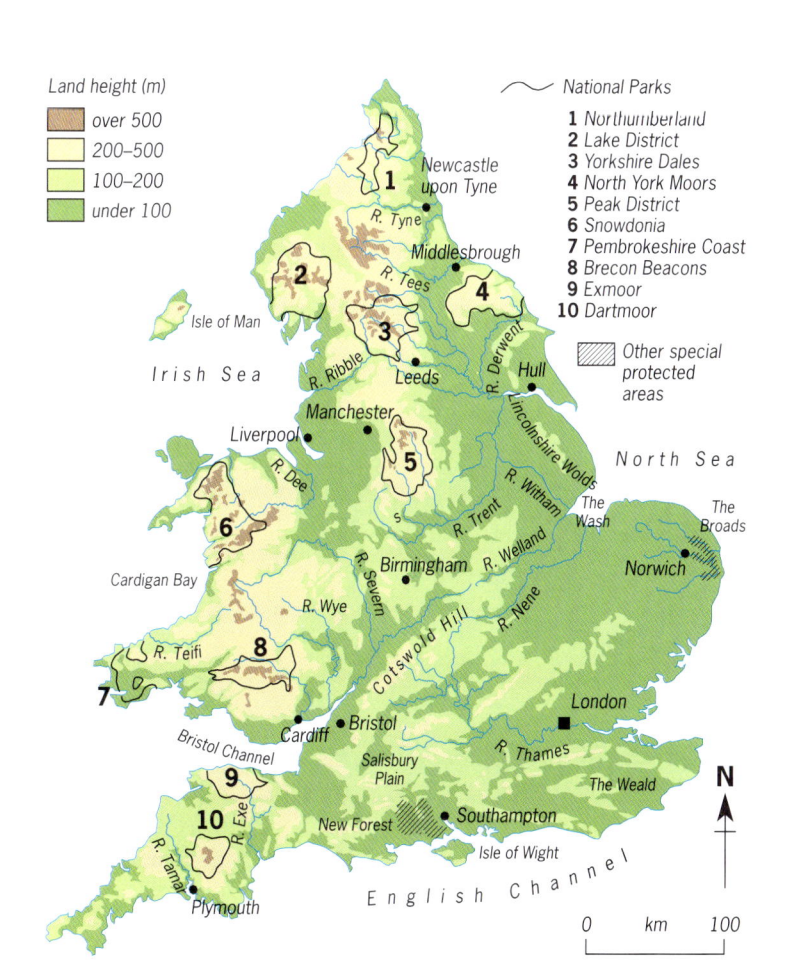

Figure 7.17 (above right) England and Wales's regional pattern of tourism, 1999

Figure 7.18 (right) Distribution of national parks in England and Wales

123

7 Tourism

EXERCISES

8a Describe and explain the distribution of national parks in England and Wales (Fig. 7.18).
b Plot the information in Table 7.2 as a series of bar charts.
c* Suggest three possible reasons for differences in the number of visitors to national parks (Table 7.2).

Table 7.2 Number of visitor days in national parks

	Visitor days (millions)
The Broads	5
Brecon Beacons	7
Dartmoor	4
Exmoor	1
Lake District	14
Northumberland	1
North York Moors	8
Peak District	12
Pembrokeshire Coast	13
Snowdonia	8
Yorkshire Dales	8

Conflicting uses of national parks

Economic activities within national parks can conflict with conservation and with each other. Quarrying, for example, is important in several national parks (see Book 1, Chapter 2). Yet its effect is to destroy the very landscape national parks are supposed to protect. In addition, army firing ranges cover large areas of national parks, including the Northumberland National Park, Brecon Beacons and Dartmoor. As well as excluding visitors, the army's activities have had a destructive effect on the environment. Such uses also conflict with the main purpose of national parks: to promote 'quiet recreation', conservation and public access. Some of the most serious conflicts concern tourism. This chapter will now examine these conflicts in detail, in the context of the Lake District National Park.

CASE STUDY

7.7 The Lake District

Figure 7.19 (right) Crinkle Crags in Great Langdale

Figure 7.20 (below) The eastern fells in winter

Figure 7.21 (below right) Haweswater

7 Tourism

Figure 7.22 (above left) Gillercomb

Figure 7.23 (above right) Bowness-on-Windermere: a tourist honeypot

EXERCISES

9 Study Figures 7.19–7.23.
a Describe the physical environment of the Lake District. (Look back at Chapter 2 to help identify some of the physical features.)
b Suggest ways in which the physical environment in Figures 7.19–7.23 provides opportunities for recreation.

REMEMBER
National parks are based on the idea of stewardship. We have a responsibility to conserve these areas and pass them on to future generations undamaged.

Figure 7.24 The Lake District National Park

125

7 Tourism

Figure 7.25 Footpath erosion caused by visitor pressure in a national park

FACTFILE

- Designated in 1951, the Lake District is the largest national park in England and Wales.
- Most of the Lake District is rugged upland, comprising ancient volcanic rocks, which form the highest mountains in England.
- The Lake District is the best example of a glaciated upland region in England, and has classic glacial landforms such as corries, U-shaped valleys and ribbon lakes.
- 14 million people visit the national park each year to sight-see, explore the fells, cruise and sail on the lakes, and tour cultural and historic sights including the homes of literary figures such as William Wordsworth, John Ruskin and Beatrix Potter.
- One-third of all employed residents are dependent on tourism.
- 90 per cent of visitors arrive and travel within the park by car.

Distribution of recreational activity

Some recreational activities, such as fell walking and rock climbing, occur widely throughout the national park. Others, such as sailing and wind surfing, focus on lakeside resorts such as Windermere, Bowness, Ambleside and Keswick. These resorts, with their hotels, guest houses, restaurants, gift shops and galleries, are the **honeypots** for mass tourism. They attract large crowds during the summer months and help to reduce pressure on the more sensitive parts of the Lake District.

Advantages of tourism

The main advantages are the jobs and money that tourism provides for the 40 000 people that live in the park. For example, half of the workforce in Windermere and Keswick depends directly on tourism (compared to 6 per cent nationally). Tourists also support local businesses such as village

Figure 7.26 Boating troubles, *The Westmorland Gazette*, 21.2.92

EXERCISES

10 Windermere attracts a wide variety of uses. Some of these uses are in conflict with each other and put a great deal of pressure on Windermere's water and shore. Read Figure 7.26.
a What problems do speedboats cause on Windermere?
b What arguments do those people opposed to the restrictions on speedboats put forward?
c* What is your view on the conflict between speedboats and other recreational uses of Windermere? Give reasons for your view.

Fight starts over 16 km/h limit on the lake

LAKE DISTRICT planners have voted overwhelmingly to impose a blanket 16 km/h speed limit on Windermere.

A costly public inquiry now seems inevitable, as opponents say they will continue to fight the Lake District Planning Board's decision to introduce the 16 km/h limit.

The speed restriction could have a disastrous effect on the economy and cause unemployment in Bowness and Windermere with affluent water-skiers and power boaters forced out of the Lake District.

But advocates of the speed limit argue that noisy and pollution-causing water-skiers will be replaced by a new breed of tourists wanting to visit the Lakes for tranquillity.

Similar speed restrictions on Ullswater, Coniston Water and Derwent Water had resulted in peaceful lakes.

During the planning board meeting, supporters of the 16 km/h speed limit argued that allowing power boats on Windermere was like allowing Formula 1 cars in a children's playground; that the immense pressure on the lake was bound to increase without the blanket speed limit; and that England's largest lake was still too small to comfortably accommodate the huge number of water-skiers and power boaters.

Canoeists, swimmers and anglers would be once more able to enjoy the lake without powerboats and proposed compromise alternatives would be difficult to police, it was argued.

But opponents argued that imposing a blanket speed limit was using a sledge hammer to crack a nut and that people seeking solitude in the Lakes already had many places to go.

7 Tourism

shops and pubs, as well as bus and rail services. In 1999, tourists spent £479 million in Cumbria, and most of this was in the Lake District National Park.

Disadvantages of tourism

Given the dependence of the Lake District on tourism, it is essential for local people that tourism continues to prosper. For this to happen, Lakeland's most valuable resource – its natural beauty – must be protected. However, getting the right balance between protecting the environment and the needs of local people is difficult and can sometimes lead to conflict.

> **REMEMBER**
> Tourism is unsustainable if the pressure of visitor numbers results in the degradation of environmental resources.

Figure 7.27 Heavy traffic in Bowness

Table 7.3 Environmental conflicts in the Lake District

Speed boats and jet skis on Windermere	After a long battle (including a public inquiry), the national park succeeded in imposing a 16 km/h speed limit on Windermere. Speed boats and jet skis are noisy and polluting, and conflict with virtually every other user of the lake. The speed limit legally comes into force in 2005.
Footpath erosion	Increased numbers of walkers in the fells has caused severe footpath erosion on the more popular routes (Fig. 7.25). The NPA and the National Trust aim to restore the landscape and create sustainable routes on the fells. In 2000, the total length of paths in urgent need of repair was 250 km. £4.2 million will be needed for footpath repair between 2000 and 2010.
Off-road vehicles and motorbikes	Over the past 10 years, there has been a large increase in off-road driving (i.e. on unsurfaced roads and tracks) by 4-wheel-drives and motorbikes. Off-roaders create noise, pollution, erosion and conflict with other users. As an experiment, a hierarchy of trail routes has been defined by the NPA. Routes are classed either as free to all users, no use by 4WDs, and no use by any vehicles. The restrictions are not legally binding, but it is hoped that off-roaders will act responsibly.
Traffic congestion	Most people travel to the Lake District by car. At weekends and on bank holidays, the major roads into the region, and the more popular valleys (e.g. Langdale, Borrowdale and Kentmere) become badly congested. A traffic management plan in 1995 recommended restricting access to more popular areas by private car, and closing some roads to traffic completely. Access for visitors would be by shuttle bus (as in US national parks such as Yosemite and Zion). The plan proved highly unpopular with local people who depend on tourism for their livelihood. They felt that restricting access in this way would greatly reduce the number of visitors to the park.

EXERCISES

11 Study the proposals for managing the problem of traffic congestion in the Lake District (Table 7.3) and the individual views shown in Fig. 7.28. Imagine that you are one of the following people: a Lakeland farmer; the owner of a popular Lakeland pub; a regular fell walker who lives outside the Lake District; or a resident who has retired to the Lake District. Write a letter to the local newspaper stating and explaining your opinion on the NPA's proposed traffic management scheme.

127

7 Tourism

Conservationist
Some routes are not essential; apart from providing local access, they could be used to provide safe routes for walkers and cyclists. This might encourage people to use their cars less and improve visitors' experience of the area.

Planner
Urgent action is needed. Many Lake District roads are narrow, steep and winding and are simply not suitable for large volumes of traffic.

Hotel owner
The valley will be cut off from traffic. People like to come here after a walk. If they can't get here by car they aren't going to walk and so they aren't going to come in.

Cumbrian Tourist Board spokesperson
If you are going to place restrictions on the lifeblood of the area, people will go away. They don't like a lot of rules and will simply find somewhere else to spend their money and leisure time.

Local councillor
Something has to be done: the motor vehicle is an increasing nuisance to resident and visitor alike.

NPA officer
We want to create a different environment, a park environment where people are encouraged to walk and cycle rather than use their car.

Figure 7.28 Issue of traffic management in the Lake District National Park. (Inset) Roadside parking by visitors, Great Langdale

7 Tourism

7.8 Tourism in the economically developing world

Although LEDCs account for only one-fifth of world tourism, this proportion is increasing rapidly. Indeed, in Africa, tourism has become the continent's fastest growing industry. For many poor countries, tourism provides one of the few avenues for development (see Chapter 8). Tourism brings many advantages (see Fig. 7.37). However, without planning and sensitive development, tourism does little to benefit local people, while the most thoughtless projects can lead to environmental disaster (see Section 8.13).

CASE STUDY

7.9 Green tourism in Zimbabwe

Since the late 1980s, Zimbabwe in southern Africa has experienced a tourist boom. Advertising itself as 'Africa's paradise', Zimbabwe saw the number of tourists visiting the country grow at an average annual rate of 18 per cent between 1989 and 1994.

Tourism has brought considerable benefits to Zimbabwe. More than one million foreign tourists visited the country in 1998. They spent more than £50 million and provided direct employment for 80 000 local people. In recent years, Zimbabwe's tourism industry has been a great success story. It has overtaken its main rival, Kenya, and now has the fourth-largest tourism industry in Africa.

EXERCISES

12a Name the two great rivers that form the northern and southern borders of Zimbabwe (Fig. 7.29).

b Despite being in the tropics, Harare's average temperature in January is just 21ºC. Using Figure 7.29, suggest a reason for this. How might this benefit tourism?

Figure 7.29 Zimbabwe

7 Tourism

Figure 7.30 (below) Major tourist attractions in Zimbabwe

Figure 7.31 (top inset) Lake Kariba, Zimbabwe

Figure 7.32 (middle inset) Victoria Falls, Zimbabwe

Figure 7.33 (bottom inset) Watching the elephants, Hwange National Park, Zimbabwe

Resources for tourism

Zimbabwe's greatest asset is its spectacular wildlife, concentrated in the country's national parks and game reserves (Fig. 7.29). These conservation areas cover 10 per cent of the country, and occupy an area roughly equal in size to Denmark. In national parks such as Hwange and Mana Pools (Fig. 7.30), tourists can see most of Africa's largest mammals, including elephants (Fig. 7.33), rhinos, lions, giraffes, hippos, buffalo, zebra and antelope, as well as hundreds of species of birds. Other major attractions include the Victoria Falls on the Zambezi River (Fig. 7.32) and Lake Kariba (Fig. 7.31).

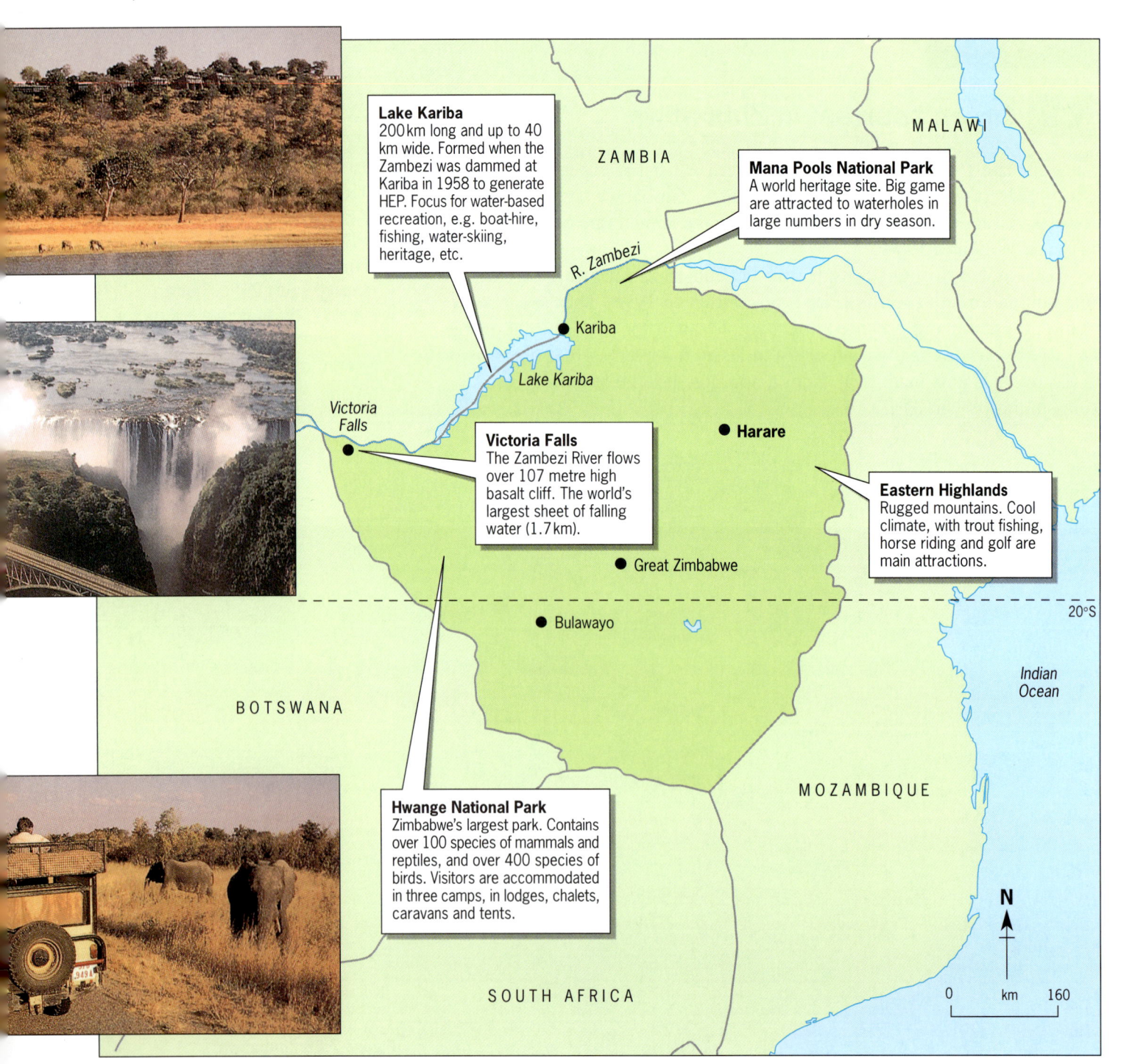

Lake Kariba
200 km long and up to 40 km wide. Formed when the Zambezi was dammed at Kariba in 1958 to generate HEP. Focus for water-based recreation, e.g. boat-hire, fishing, water-skiing, heritage, etc.

Mana Pools National Park
A world heritage site. Big game are attracted to waterholes in large numbers in dry season.

Victoria Falls
The Zambezi River flows over 107 metre high basalt cliff. The world's largest sheet of falling water (1.7 km).

Eastern Highlands
Rugged mountains. Cool climate, with trout fishing, horse riding and golf are main attractions.

Hwange National Park
Zimbabwe's largest park. Contains over 100 species of mammals and reptiles, and over 400 species of birds. Visitors are accommodated in three camps, in lodges, chalets, caravans and tents.

Cultural and historical attractions also pull in the tourists. These include Great Zimbabwe, a ruined city that was home to a native civilisation long before the arrival of Europeans. Both Victoria Falls and Great Zimbabwe are World Heritage Sites.

Types of tourism

Zimbabwe has made a deliberate attempt to avoid mass tourism. Instead, tourism is geared to small groups seeking special interest holidays. These so-called **eco-tourists** will include wildlife enthusiasts, bird watchers, botanists and photographers.

Investing in tourism

A successful tourism industry needs a good infrastructure as well as attractive resources. Tourists require decent accommodation, electricity, clean water, roads and airports. Zimbabwe has one of the best developed infrastructures in Africa. There are top-class hotels (mainly owned by international chains) in Harare and Bulawayo. In the national parks, accommodation ranges from luxury lodges to chalets and caravans. There is also an adequate road network and the capital, Harare, has an international airport.

Tourism issues

Managing the resources for tourism

Zimbabwe has plenty of experience in protecting and managing its tourist resources to ensure sustainable development. For example, large areas of the country are set aside as national parks. The government owns these areas and reserves them exclusively for wildlife.

Conflicts between wildlife and local farmers

Subsistence farmers living on the edges of national parks often come into conflict with wildlife. These farmers are poor and, with rapid population growth, land is in short supply. This situation often tempts farmers to grow crops and graze their animals inside the national parks. The result is the destruction of natural habitats and their wildlife. Further conflict also occurs when wild animals (e.g. elephants) destroy crops or threaten the lives of local people.

Aware of these problems, the government reasoned that local people would only protect the wildlife if they received some benefit from it. Thus, in 1984, the government introduced its Communal Areas Management Programme For Indigenous Resources (CAMPFIRE) scheme. Through this scheme, local communities receive money from game hunting fees (e.g. tourists pay up to £4600 to shoot an elephant), and from selling hides and meat. Local people also receive compensation for any crop losses caused by wild animals. As a result, the attitude of local people towards wildlife has changed. Now, because wildlife provides them with an income, they have an interest in protecting it.

Similar conflicts between local farmers and wildlife have arisen in other parts of Africa, including Kenya (Figs. 7.34–7.36). There the government has been notably less successful in resolving the trouble.

EXERCISES

13 Make a list of the ways in which national parks in Zimbabwe differ from those in England and Wales.

REMEMBER
A large proportion of the revenue from tourism in LEDCs 'leaks' overseas, to benefit travel agents, airlines, hotel chains etc. based in MEDCs.

7 Tourism

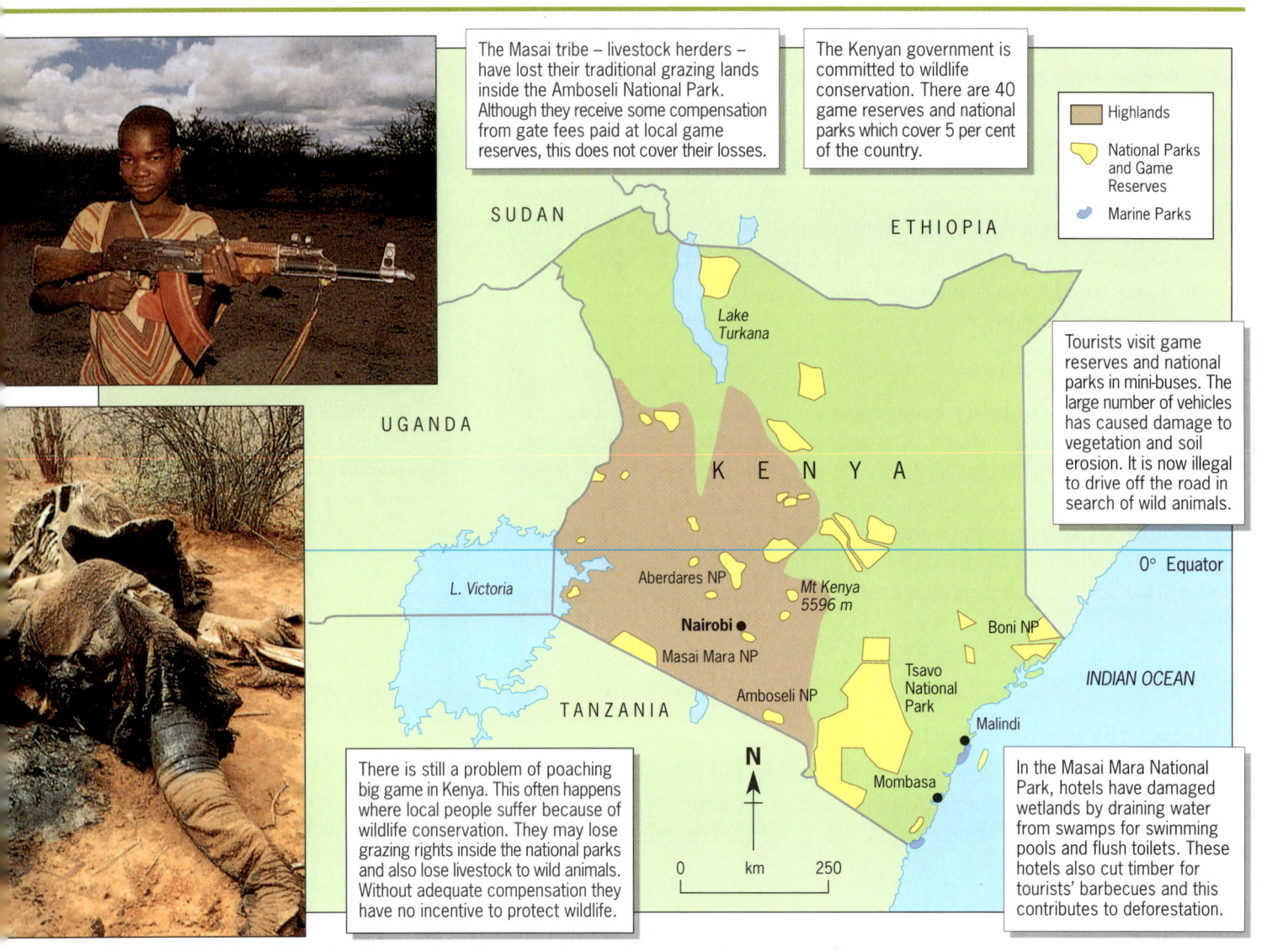

The Masai tribe – livestock herders – have lost their traditional grazing lands inside the Amboseli National Park. Although they receive some compensation from gate fees paid at local game reserves, this does not cover their losses.

The Kenyan government is committed to wildlife conservation. There are 40 game reserves and national parks which cover 5 per cent of the country.

Tourists visit game reserves and national parks in mini-buses. The large number of vehicles has caused damage to vegetation and soil erosion. It is now illegal to drive off the road in search of wild animals.

There is still a problem of poaching big game in Kenya. This often happens where local people suffer because of wildlife conservation. They may lose grazing rights inside the national parks and also lose livestock to wild animals. Without adequate compensation they have no incentive to protect wildlife.

In the Masai Mara National Park, hotels have damaged wetlands by draining water from swamps for swimming pools and flush toilets. These hotels also cut timber for tourists' barbecues and this contributes to deforestation.

Figure 7.34 (above) Eco-tourism and conflict in Kenya

Figure 7.35 (top inset) Shepherd boy with a weapon for use against predatory animals and livestock thieves, northern Kenya

Figure 7.36 (bottom inset) Elephant slaughtered by poachers, Tsavo Park, Kenya

Types of tourism development

In parts of Zimbabwe, there has recently been some concern about the nature of tourist developments. Victoria Falls, visited by two out of every three foreign tourists, has become increasingly commercialised. Bungee jumping, micro-light aircraft flying over the falls, and the sale of cheap trinkets and even drugs are giving the Falls a 'tacky' image. The problem at Victoria Falls highlights the dilemma facing countries wishing to expand their tourism industries. Do they promote exclusive eco-tourism or more downmarket mass tourism? Zimbabwe has chosen the former.

Tourism and political unrest

Zimbabwe's tourism industry faced a major crisis in 2000. Violence in elections and conflict between war veterans and white farmers were widely publicised in the MEDCs. Zimbabwe was seen as an unsafe destination. As a result, foreign tourists, especially those from the UK and the USA, stayed away. The decline in tourism receipts (normally 5 per cent of GDP) hit the economy hard. Many hotels were only half full, and thousands of workers in the tourism industry were laid off. It could be many years before confidence returns. Meanwhile, neighbouring countries, such as South Africa, Kenya and Botswana, which offer similar attractions to Zimbabwe, will benefit.

7 Tourism

TOURISM

Advantages
- Earns valuable foreign exchange.
- Provides direct employment in hotels, transport, guides, wardens, etc.
- Provides indirect employment, e.g. farmers providing food to hotels, making and selling souvenirs, etc.
- Attracts foreign investment – hotels, airports, roads, etc.

Disadvantages
- Provision of infrastructure is costly.
- Much of the profit 'leaks' overseas to MEDCs, e.g. many hotels are owned by foreign chains (Hilton, Sheraton, Holiday Inn, etc.); holidays are organised by foreign tour operators; foreign airlines fly tourists to and from tourist destinations; services and food are often flown in from overseas.
- Creation of a dual economy: local people living in or near a resort may benefit but there may be few benefits elsewhere.
- Environment may be at risk from degradation by mass tourism, e.g. pollution of rivers and coasts; development of ugly resorts, etc.
- Mass tourism may change local cultures and traditions.
- Local people may be excluded from conservation areas without compensation.
- Tourism may be an unreliable source of revenue, influenced by fashion, political unrest etc.

EXERCISES
14* Study Figures 7.34–7.36. What evidence is there that, in Kenya, tourism based on its wildlife is currently unsustainable?

15 With reference to Figure 7.37, describe the benefits and possible disadvantages of tourism in Zimbabwe.

Figure 7.37 The advantages and disadvantages of tourism in LEDCs

REMEMBER
Excessive dependence on tourism has economic risks. Political instability, changes in taste, and the emergence of new tourism destinations can lead to sudden declines in visitor numbers.

CASE STUDY

7.10 Tourism and conservation in the Galapagos

FACTFILE
- The Galapagos Islands lie in the Pacific Ocean, close to the Equator and about 1000 km from the coast of South America (Fig. 7.38).
- The archipelago consists of 13 large islands and six smaller ones.
- All the islands are volcanic in origin.
- The cold southern equatorial current, which sweeps westwards from South America, gives the Galapagos a relatively cool climate (average in January is 20ºC).
- The Galapagos Islands are a World Heritage Site.

Resources for tourism

The tourist attractions of the Galapagos are their remarkable plant and animal life. The islands are home to several land animals found nowhere else in the world. Most famous are the giant tortoises (Fig. 7.39), iguanas (Fig. 7.40), and birds such as the flightless cormorant and red-footed booby (Fig. 7.41). The seas around the islands have more than 300 different species of fish, whales, dolphins, sea lions and seals. Penguins are common.

7 Tourism

Figure 7.38 The Galapagos Islands

Figure 7.39 (above) Watching a giant tortoise

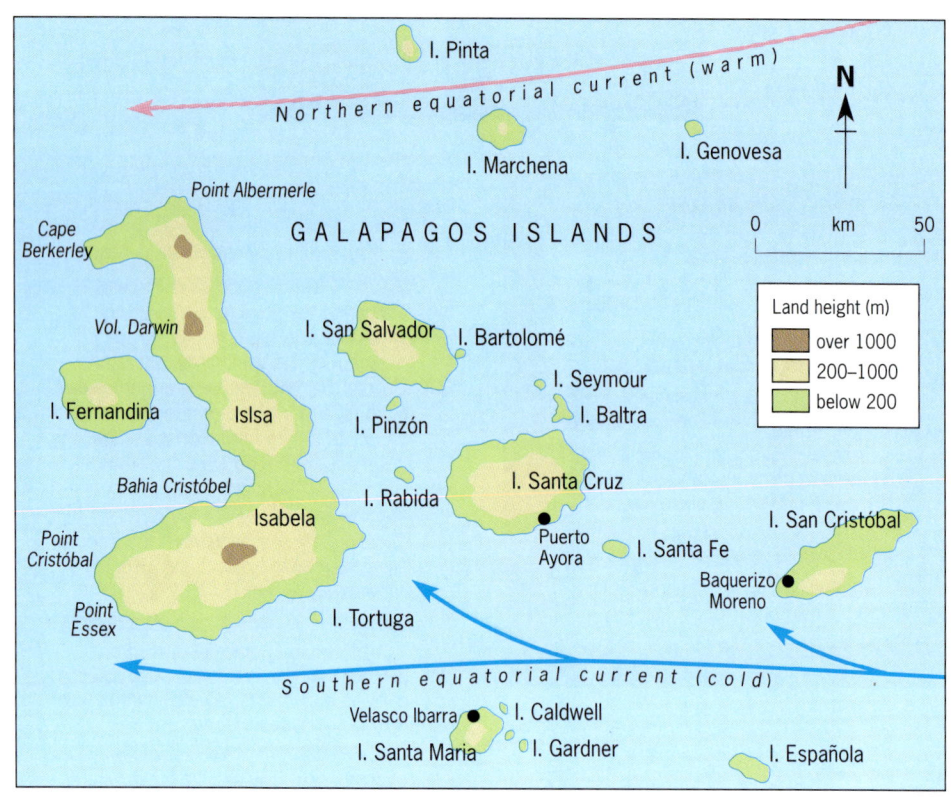

Biodiversity

There are four possible reasons for this extraordinary diversity of wildlife:
- Animals on the islands have evolved in isolation over millions of years.
- The great diversity of habitats.
- The absence of predators, which also accounts for the unusual tameness of many land animals.
- Ocean currents cause an upwelling of cold water and bring nutrients to the surface, providing food for fish, marine mammals and sea birds.

Figure 7.40 (right) Iguanas

7 Tourism

Charles Darwin

Tourists have another reason for visiting the Galapagos: to follow in the footsteps of the famous naturalist Charles Darwin. In 1835, Charles Darwin's epic voyage aboard *HMS Beagle* took him to the Galapagos. He was struck by how closely related species of birds differed slightly from island to island. Many years later, these observations made on the Galapagos inspired his theory of evolution – perhaps the greatest scientific theory of the 19th century.

Tourists on the Galapagos

Recent estimates put the number of tourists visiting the Galapagos at 100 000 a year. Although tourism is still small scale, this represents an increase of almost 50 000 people in ten years. The Galapagos attract small parties of eco-tourists who come to experience its unique environment and wildlife. Around the coast, snorkelling and diving are also popular activities. Visitors fly into the Galapagos from Quito, the capital of Ecuador and then tour the islands by boat. All parties have guides who direct the tourists to some of the official 45 visitor sites where there are concentrations of wildlife.

Figure 7.41 Red-footed booby on its nest of twigs

Conservation measures

Tourism in such a small and fragile environment has to be carefully managed. During their stay on the islands, all tourists must follow a strict code of conduct (Fig. 7.42). Since 1959, 97 per cent of the land area of Galapagos has received full protection as a national park. In 1986, the coastal areas received similar protection when they were designated a marine resources reserve.

Conservation issues

The Galapagos Islands are owned by Equador. Equador is a LEDC with a GDP per capita of $1300 a year, and large foreign debts. Tourism is one of the country's leading industries and a major source of income. Galapagos is the main attraction for tourists to Equador. But sustainable tourism on Galapagos requires careful control of visitor numbers. In the 1980s, numbers were limited to 25 000 a year. However, this limit was never enforced. Today's visitor numbers are estimated to be about 100 000 a year.

Galapagos National Park
Rules for tourists

Do not disturb or remove any native plants, rocks or animals.
Do not transport any plants or animals to the islands or from island to island.
Do not touch or handle the animals.
Do not feed the animals.
Do not startle or chase any animal from its resting or nesting spot.
Do not leave the marked footpaths and trails.
Do not leave any litter.
Do not buy any souvenirs made from native Galapagos products.

Figure 7.42 Code of conduct for tourists on the Galapagos

Environmental problems

Tourism on Galapagos is strictly controlled. Most tourists visit the island by cruise ship and there are just 45 approved visitor sites. Tourists must stay on marked footpaths and be accompanied by a trained naturalist guide.

7 Tourism

> **EXERCISES**
>
> **16a** Summarise the main features of tourism in the Lake District, Zimbabwe and the Galapagos Islands in a table. Use the following headings: resources for tourism; benefits of tourism; problems of tourism; management responses; sustainability.
>
> **b*** 'Tourism has mostly negative effects on environment and culture.' State and explain your view on this observation.

Despite these controls, and the protection afforded by the Galapagos Islands' status as a nature reserve, national park and world heritage site, the environmental impact of tourism is considerable.

- Current rates of population growth on Galapagos are 12 per cent. Most immigrants come to the islands to work in tourism. This increases pressure on scarce resources such as fresh water, beaches and seafood.
- Remote villages have been transformed by discos, hotels, restaurants and souvenir shops – features associated with mass tourism not eco-tourism.
- Local divers have damaged coral formations to make tourists' souvenirs.
- Illegal fishing and overfishing is widespread. Fishing nets do indiscriminate damage to wildlife, killing penguins, dolphins, iguanas, sea lions and flightless cormorants.
- Refuse is dumped overboard from cruise ships.

7.11 Summary: Tourism

> **KEY SKILLS OPPORTUNITIES**
> **C1.1**: Ex. 11; **C1.2**: Ex. 1b, 2a, 5a, 8, 9a, 10a, 10b, 12a, 13; **C1/2.3**: Ex. 6d, 8, 10c, 11, 15; **C2.1**: Ex. 6d, 10c, 11; **C2.2**: Ex. 1b, 5b, 7a, 9a, 10a, 10b; **N1/2.1**: Ex. 2b, 3, 6c, 7b, 8, 9b, 12b, 14; **N1/2.2**: Ex. 6a; **N1/2.3**: Ex. 6b.

Key ideas	Generalisations and detail
Tourism has grown rapidly in the 20th century.	• Mass tourism began in Victorian Britain when seaside resorts for the populations of large industrial centres developed. • Tourism in the late 20th century is a global economic activity. Rising incomes, longer paid holidays and cheap transport explain tourism's growth.
Environments differ in the possibilities they offer for tourism.	• Tourism is based on natural resources and those made by humans. • Natural resources include an agreeable climate, attractive landscapes, clean beaches, wildlife, etc. The mass tourism in Mediterranean countries exploits the region's warm, dry summers. Exotic wildlife sustains tourism in Zimbabwe, Kenya and the Galapagos Islands. • Human-made resources include historic buildings, local customs and cultures, museums, etc. Paris, London and Venice are popular centres of tourism because of their artificial resources.
Tourism creates advantages and disadvantages for tourist destinations.	• Tourism creates employment and brings money and investment to tourist areas. However, in LEDCs, many of the economic benefits of tourism 'leak' overseas to the developed world and trans-national corporations (hotel chains, airlines, tour operators etc.). • Tourism often damages the environment. • Native customs and culture may be degraded by tourism.
Mass tourism is often non-sustainable.	• Large-scale, unplanned tourism may cause permanent damage to environmental resources (e.g. pollution of coastal waters by holiday resorts, damage to coral reefs in the tropics, footpath erosion in national parks, etc.). • In the long term, a sustainable, i.e. green, tourism is needed.
Many countries give areas of high environmental value a special conservation status (e.g. national parks).	• These areas are managed in order to protect their resources (e.g. landscapes, wildlife, etc.) and make their use sustainable. • Examples of management include the Lake District's controversial traffic scheme, Zimbabwe's CAMPFIRE scheme, and the Galapagos code of conduct for visitors.

8 Contrasts in development

8.1 Introduction

We live in a world divided between rich and poor nations. The rich countries account for:

- 20 per cent of the world's population.
- 85 per cent of the world's income.
- 80 per cent of the world's energy consumption.
- 86 per cent of the world's industry (Fig. 8.1).

For many people in LEDCs, life is about survival. One child in every three is malnourished, and each year 12 million children die from causes that could be prevented, often for just a few pence (Figs. 8.2–8.5, see also Figs. 4.18 and 5.3–5.6). Throughout much of the tropics, in Africa, Asia and South America, millions of people live in absolute poverty. This means they lack even the most basic services – decent housing, proper sanitation and clean water – which we in the developed world take for granted.

Figure 8.1 Global inequality and distribution of Gross National Product

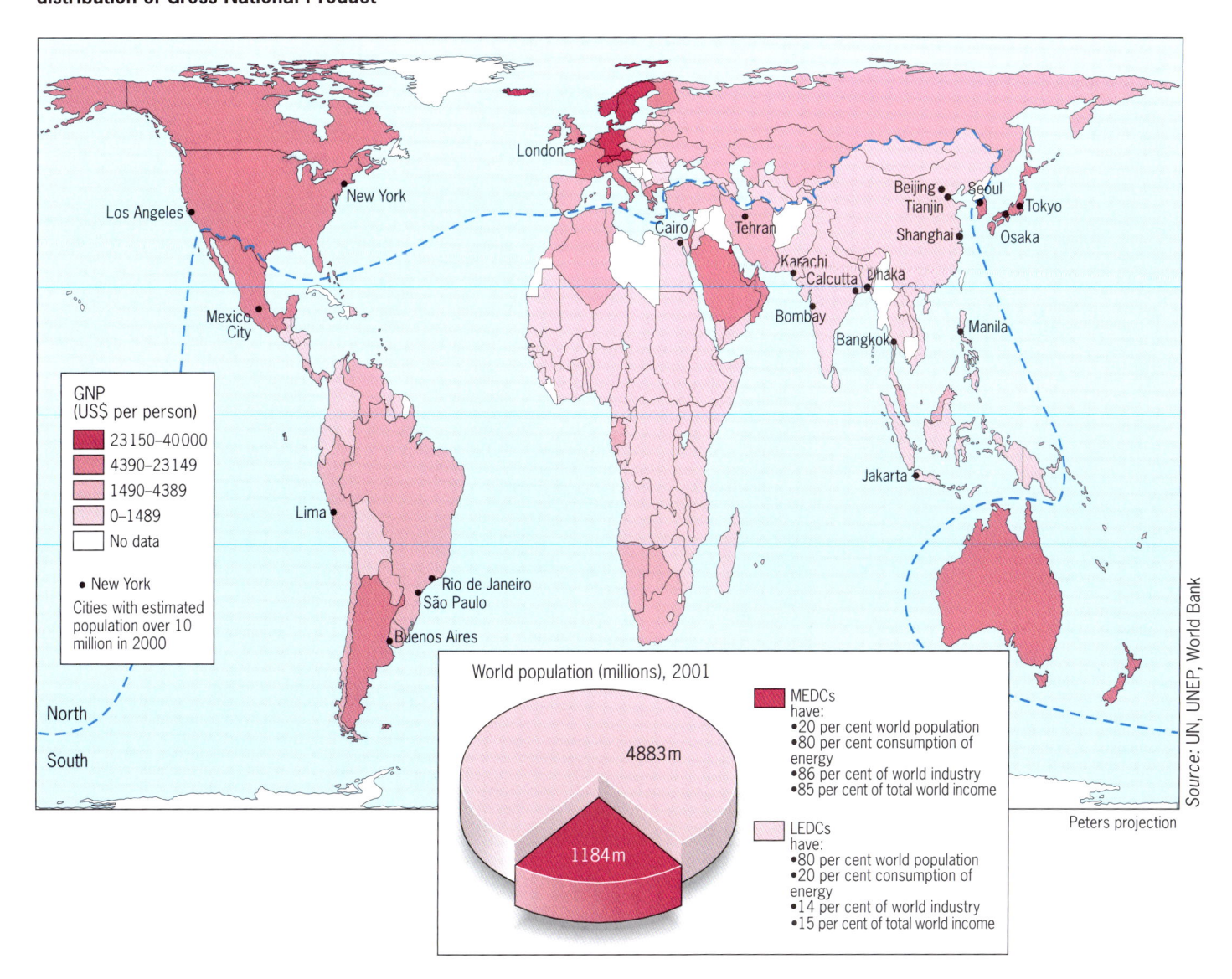

8 Contrasts in development

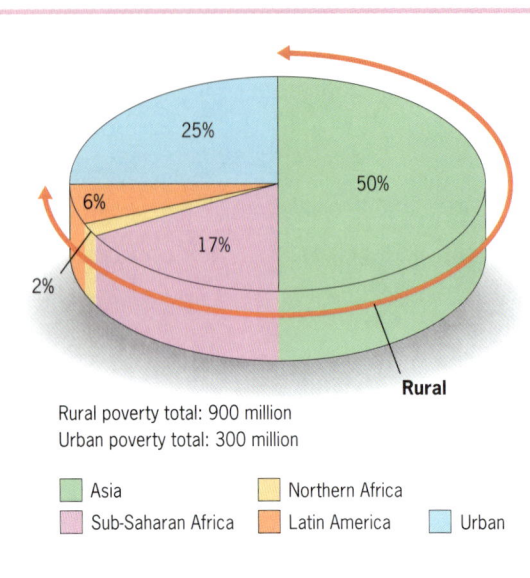

Figure 8.2 (left) Global distribution of poverty, 2001

- 50% Rural
- 25% Urban
- 17% Sub-Saharan Africa
- 6% Latin America
- 2% Northern Africa

Rural poverty total: 900 million
Urban poverty total: 300 million

- Asia
- Sub-Saharan Africa
- Northern Africa
- Latin America
- Urban

Figure 8.3 (below) Children pick their way along open sewers to go to schools in hot tin huts, Port-au-Prince, Haiti

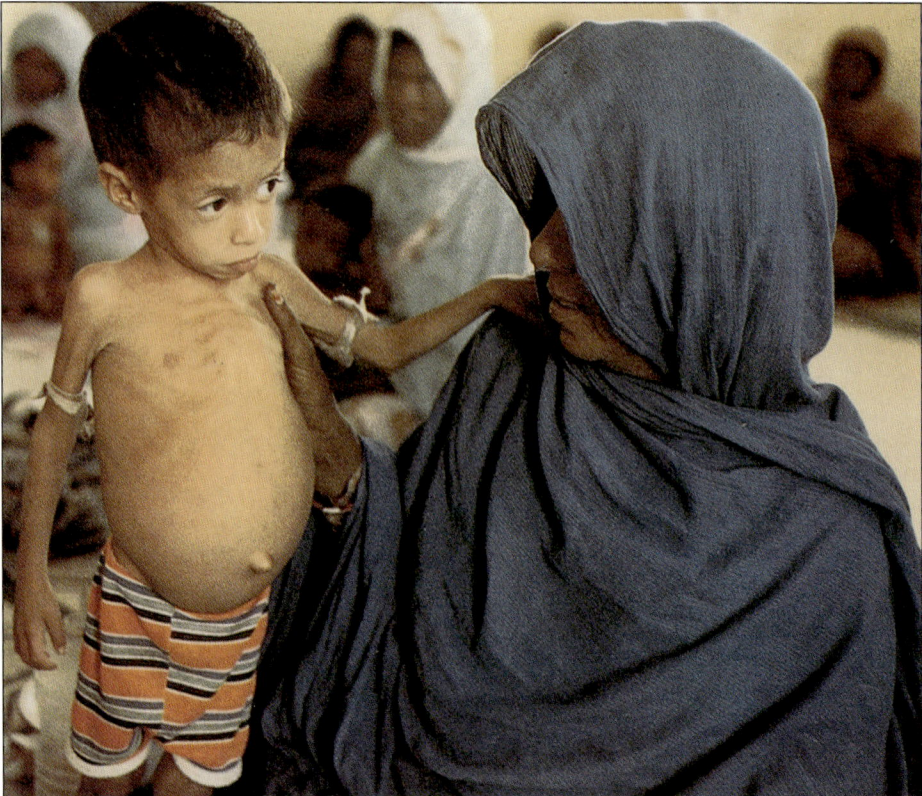

Figure 8.4 Living on the street, Calcutta, India

Figure 8.5 (above) Malnourished child, Nouakchott, Mauritania

8 Contrasts in development

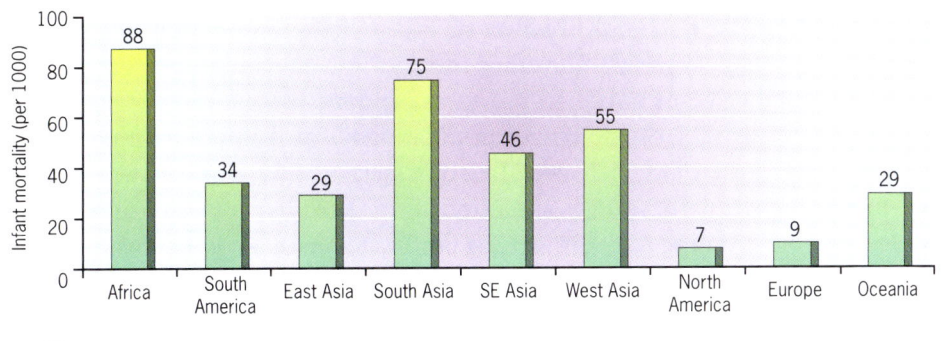

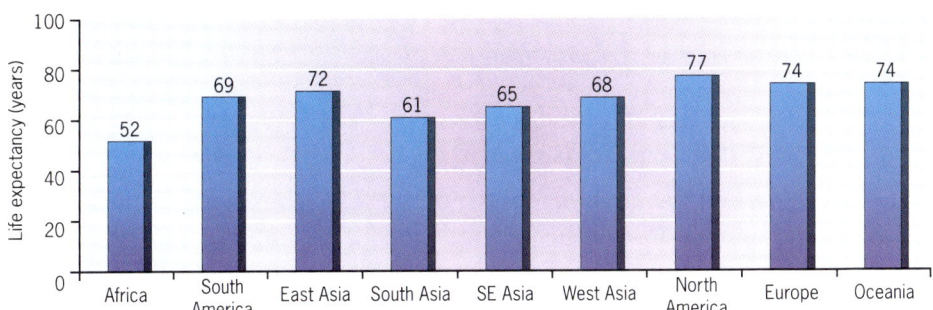

Infant mortality (per 1000) by region:
- Africa: 88
- South America: 34
- East Asia: 29
- South Asia: 75
- SE Asia: 46
- West Asia: 55
- North America: 7
- Europe: 9
- Oceania: 29

Life expectancy (years) by region:
- Africa: 52
- South America: 69
- East Asia: 72
- South Asia: 61
- SE Asia: 65
- West Asia: 68
- North America: 77
- Europe: 74
- Oceania: 74

> **EXERCISES**
>
> **1a** Describe the distribution of rural and urban poverty in Fig. 8.2.
> **b** Describe the global patterns of infant mortality (Fig. 8.6) and life expectancy (Fig. 8.7).
> **c** Suggest how infant mortality and life expectancy might be linked to poverty.

Figure 8.6 (above left) Infant mortality by world regions, 2001

Figure 8.7 (left) Life expectancy by world regions, 2001

8.2 What is development?

Development is one way in which poor countries can tackle the problem of poverty. It is about releasing the natural and human resource potential of a country, region or locality (see Section 6.3).
- Economic development aims to improve farming, industry and transport (Fig. 8.8).
- Social development is about providing services such as schools and clinics (Fig. 8.9), and improving people's skills through education and training.

Figure 8.8 (below left) Workers constructing a road to Abuja, Nigeria's federal capital

Figure 8.9 (below right) Primary school in Nigeria

8.3 Measuring development

Gross national product (GNP) per person

Before we look at geographical patterns of development, we need to find a way of measuring it. There are several possible measures of development.

8 Contrasts in development

EXERCISES

2a What is the difference between economic development and social development?
b Describe the distribution of countries in Figure 8.10 with HDIs of 0.90 and above, and those with HDIs under 0.50.
3 Study Figure 8.11.
a Which countries have a level of social development that is not accurately shown by GNP per person?
b* Using the evidence of Figure 8.11, how useful a measure of human welfare in East and South-east Asia is GNP per capita?

They include **gross national product (GNP)** per person per year, infant mortality, population growth, levels of literacy, housing and health care. GNP (the total value of goods and services produced by a country, including income from overseas, divided by its population) is probably the simplest measure (Fig. 8.1), but it has several weaknesses:

- It ignores prices and takes no account of the purchasing power of money.
- It is only an average measure, so has limited value in a country where a small elite are very wealthy (e.g. Saudi Arabia), but the majority are poor.
- It says nothing about social welfare – education, health care or housing.
- It understates income in poor countries where many farmers grow crops for subsistence rather than for cash (see Book 1, Chapter 8).

Human development index (HDI)

The United Nations has devised its own measure of human well-being. Known as the human development index (HDI), it combines data on life expectancy, education and the purchasing power of incomes, measuring social as well as economic progress. The index ranges from 0 to 1; the higher the score the greater the level of development (Fig. 8.10).

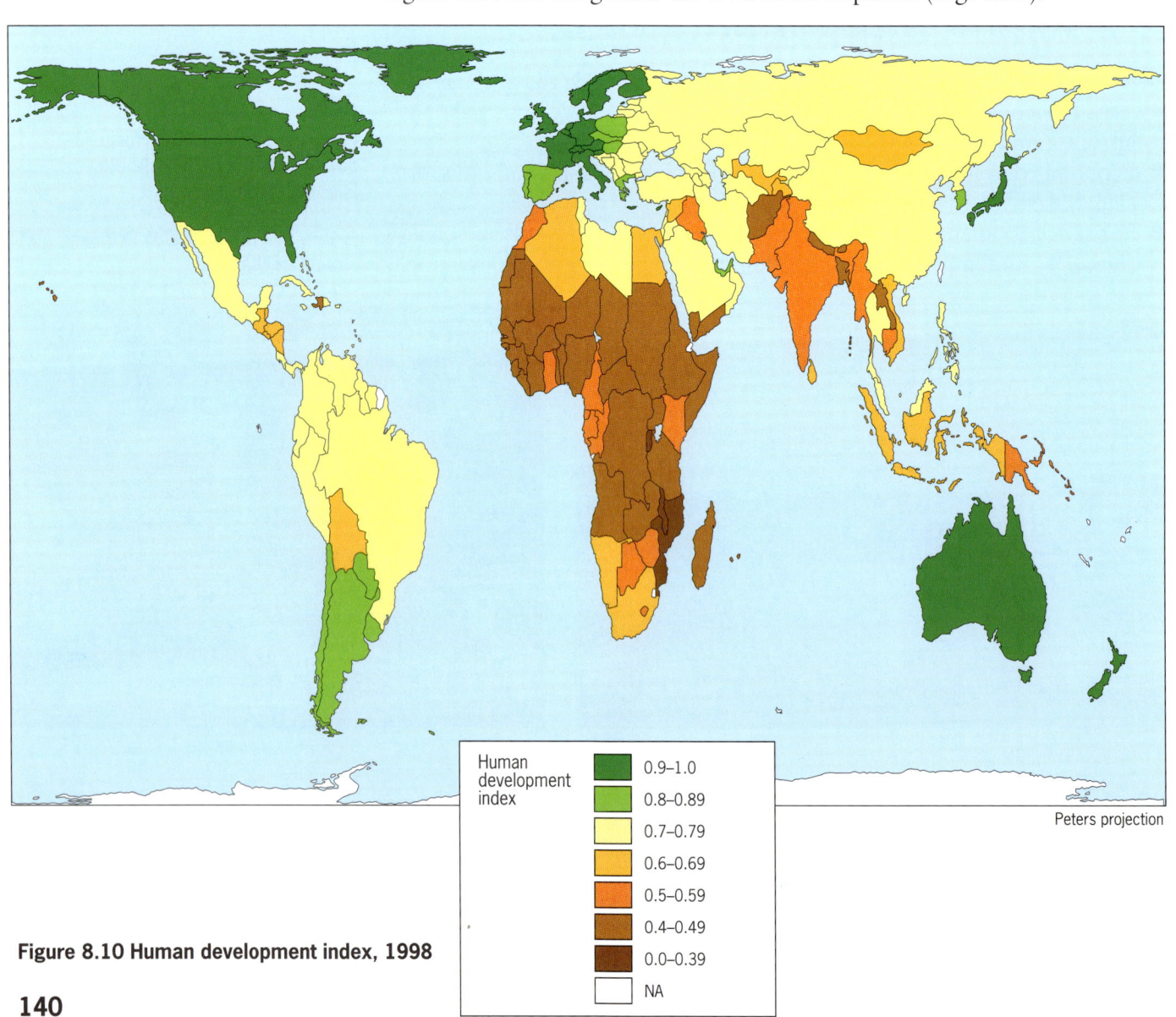

Figure 8.10 Human development index, 1998

Human development index:
- 0.9–1.0
- 0.8–0.89
- 0.7–0.79
- 0.6–0.69
- 0.5–0.59
- 0.4–0.49
- 0.0–0.39
- NA

Peters projection

8 Contrasts in development

Table 8.1 Measures of development in East and South-east Asia

	Population growth (% p. a.)	Urbanisation (%)	HDI
Brunei	1.7	59	0.889
Cambodia	2.3	24	0.422
Indonesia	1.6	40	0.637
Laos	2.5	39	0.420
Malaysia	2.0	51	0.834
Myanmar (Burma)	0.8	28	0.481
Philippines	2.2	51	0.677
Singapore	0.9	100	0.896
Thailand	0.9	19	0.838
Vietnam	1.5	20	0.560
China	0.9	35	0.650
Japan	0.2	77	0.940
North Korea	1.4	63	0.766
South Korea	0.9	86	0.894
Mongolia	1.5	59	0.604

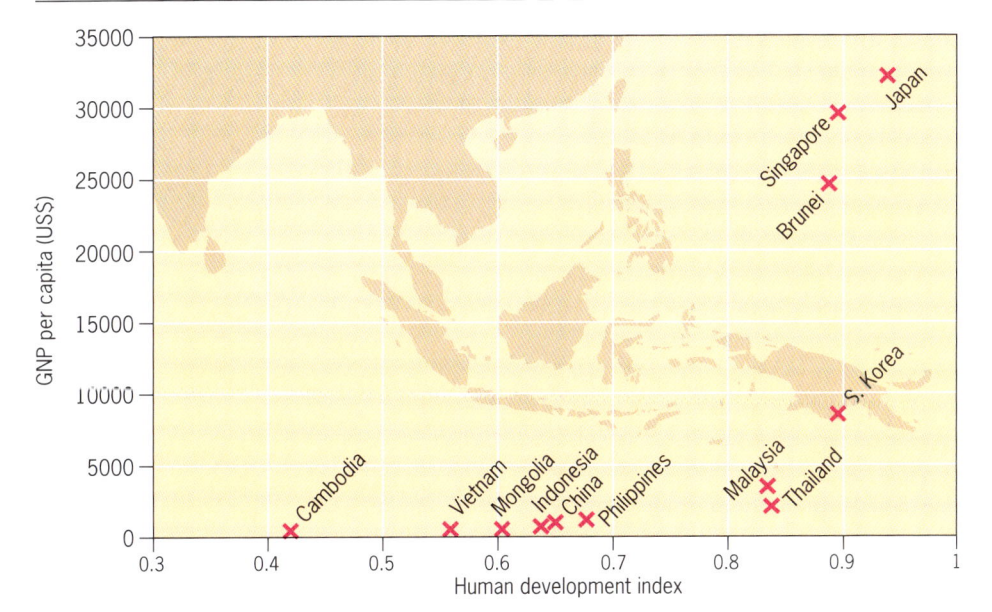

EXERCISES

4a Using the information in Table 8.1, plot two scattergraphs: one to show population growth (y) against HDI (x), and the other to show urbanisation (y) against HDI (x).
b Describe briefly the relationship between development, population growth and urbanisation shown by your graphs.
c* Suggest how development might influence rates of population growth and urbanisation.
d Log on to website www.undp.org/hydro/HDI.html. The website shows the HDIs of 174 countries in rank order. Of the 50 countries with the lowest HDIs, find out how many are in Africa, Asia and South and Central America.
e Present this data as a pie chart (using a spreadsheet) and comment on the results.

Figure 8.11 GNP and HDI in East and South-east Asia

8.4 Global contrasts in wealth

In the last 200 years, economic and social development has allowed countries like the UK, France, the USA and Japan to escape from poverty (see Section 4.5). Unfortunately, the world's rich countries account for only one-fifth of the human population (see Fig. 8.1).

The rest of the population, little affected by development, live in the world's poorer countries. The geographical contrasts between rich and poor can be clearly seen in Figure 8.1.

◆ Apart from Australia and New Zealand, the economically developed world is in the northern hemisphere. It includes North America, Europe, Russia and Japan.
◆ The economically developing world lies to the south in Africa, Asia and South America.

These geographical contrasts are evident when we refer to the rich 'North' and the poor 'South'.

REMEMBER
There is no single measure of levels of development. The most comprehensive measure, which covers both economic and social development, is the UN's human development index.

8 Contrasts in development

8.5 Global contrasts: water supplies

Water is already a serious issue in many LEDCs. Lack of safe drinking water is a major health hazard. In 1995, 2 billion people (40 per cent of the world's population) had no access to clean water. Apart from water quality, there is the problem of an ever rising demand for water throughout the developing world. Rapid population growth, industrial development, expanding irrigation and the growth of tourism could lead to acute water shortages in the developing world early in the 21st century.

CASE STUDY

8.6 Improving water in Moyamba, Sierra Leone

FACTFILE
- In 1999, Sierra Leone had a human development index of just 0.184.
- GNP per capita declined from $160 in 1997 to $130 in 1999.
- Sierra Leone has a national debt of $825 million and is crippled by interest payments to MEDCs.
- Life expectancy in Sierra Leone is 34.7 years; infant mortality is 169 per 1000; and only 30% of adults can read and write.
- In the 1990s, Sierra Leone was devastated by civil war and AIDS.

Figure 8.12 (below left) Sierra Leone and Moyamba

Figure 8.13 (below right) Cross-section of a typical hand-dug well used in Moyamba

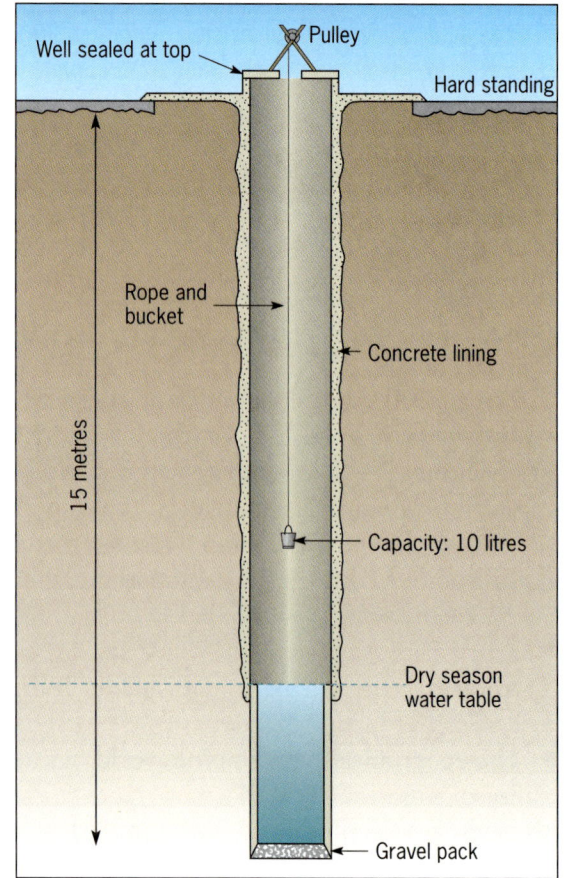

8 Contrasts in development

Moyamba in southern Sierra Leone (Fig. 8.12) shows, on a small scale, how water problems can be tackled in LEDCs. Moyamba is an isolated rural area with a population of 250 000. Most people are subsistence farmers. Most are desperately poor (Fig. 8.14).

Until 1980, few villages had safe drinking water. They relied on streams and pools, which became badly polluted during the dry season. Water-borne diseases were widespread and people's health was poor.

A UK-funded project (undertaken by the charity CARE) to sink wells in the region began in 1980. So far, more than 200 hand-dug wells have been completed (Fig. 8.13). The project aimed to:
- Provide people with safe drinking water.
- Reduce illness and mortality caused by water and hygiene-related diseases.
- Improve health in the villages covered by the project.

In the early years, the project had limited success because, without proper sanitation and health education, even well water became polluted. It became obvious that safe drinking water could be guaranteed only if improvements in sanitation also took place, so the project was modified to provide pit latrines in addition to wells. There has been a notable improvement in health and a reduction in water-borne diseases such as diarrhoea and hookworm infestation.

The development project at Moyamba has been a success for the following reasons:
- It involves local people, using their labour, skills and technology.
- It can be sustained by the local people.
- It is cheap and benefits everyone, so is popular.
- Local people now appreciate the link between disease and polluted drinking water and poor sanitation.

EXERCISES

5 Why do you think it is important that development projects like the one at Moyamba use local labour, local skills and simple technology?

Figure 8.14 Rural poverty in Sierra Leone

8.7 Global contrasts: feeding a hungry world

A healthy adult needs to consume at least 2500 kilocalories (Kcals) a day. Many countries in the economically developing world fall well below this level. For example, the average sub-Saharan African consumes only 2050 Kcals a day. This is just three-quarters of the average food intake of someone in the UK. Even more worrying is the fact that average food consumption per person in Africa actually fell by 3 per cent between 1995 and 2000 (Fig. 8.15).

Food shortages and ill health

Lack of sufficient food, which we call **undernutrition**, eventually leads to death by starvation. More common is **malnutrition**. Malnutrition results from an unbalanced diet. A healthy diet includes carbohydrates (e.g. cereals), protein (e.g. meat), fats and vitamins. Unfortunately, poor people often cannot afford to eat meat and fat. Instead, they survive on cheap, starchy foods, such as rice, cassava and sweet potatoes. This is part of the reason why so many children in the economically developing world suffer from malnutrition (see Fig. 8.5). Malnutrition causes stunted growth,

REMEMBER
Small scale development schemes – which are low cost, use local skills and simple technology, and are sustainable – are often the most beneficial to poor people in LEDCs.

143

8 Contrasts in development

EXERCISES

6 Study Figure 8.15.
a By how much did food production increase/decrease in • the world • Asia • Latin America and the Caribbean • sub-Saharan Africa between 1980 and 2000?
b Suggest possible explanations for the rapid growth of food output per person in Asia (see Chapter 6).

	1980	1985	1990	1995	2000
World	93.1	98.8	100.7	101.5	106.3
Asia (developing)	79.4	91.8	100.6	114.9	124
Sub-Saharan Africa	100.8	98.1	98.1	100.6	98.6
Latin America & Caribbean	96.4	100	99.9	106.5	112.9

Figure 8.15 (right) Per capita food production in the economically developing world, 1980–2000

Figure 8.16 (below) Child suffering from kwashiokor, Somalia

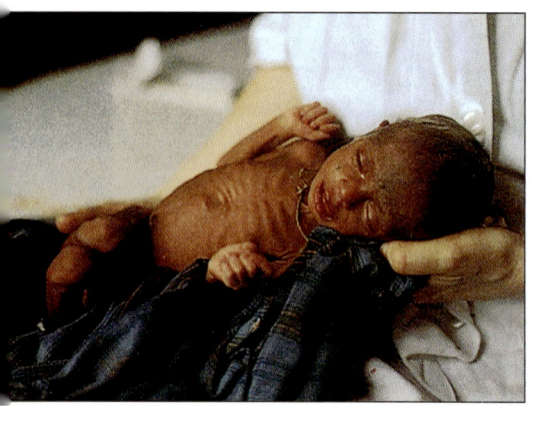

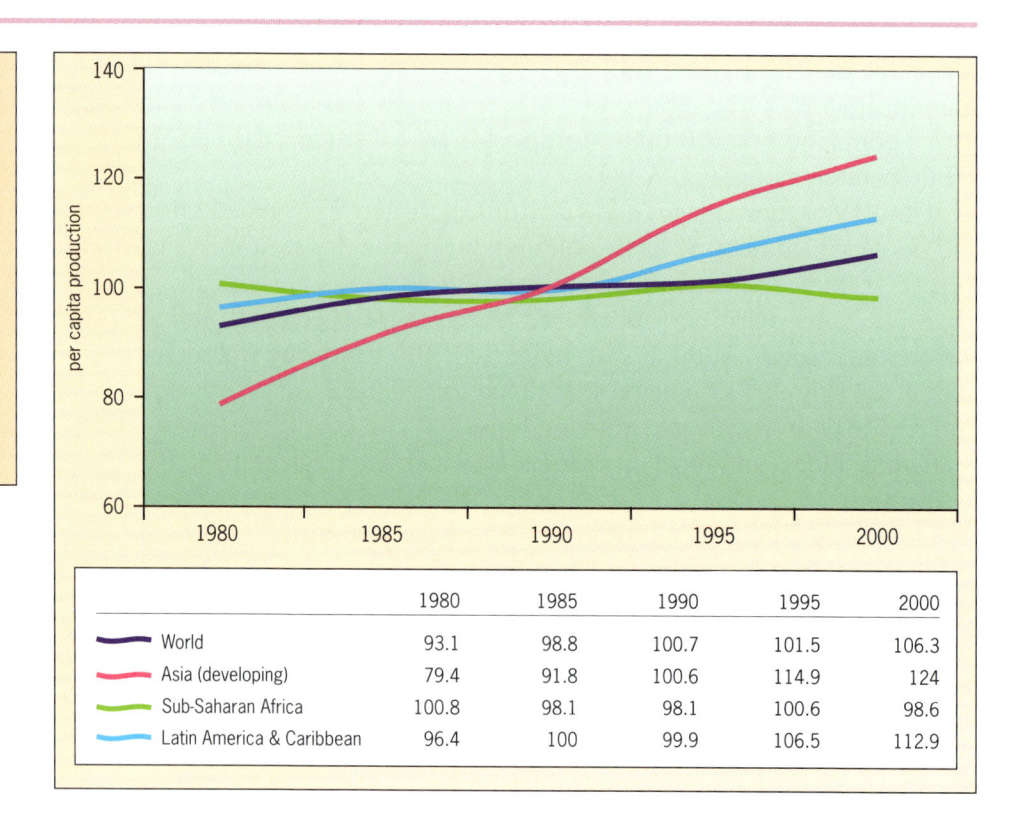

impairs mental development and, in extreme cases, results in diseases such as kwashiokor (Fig. 8.16) and marasmus.

Famine and food security

Famine occurs when food shortages are acute. Although common in Europe before the 18th century, famines are confined today to the economically developing world. They take place because millions of people in LEDCs cannot rely on their food supplies. We say that they have little food security. The poor are the most vulnerable. Even when food is available, they may starve simply because they cannot afford it.

In Book 1, Chapter 9, we saw how the green revolution has resolved food problems in many LEDCs. However, the green revolution is not always appropriate. In the next section, we will see what has been done in one such country – Nepal – to raise food production and improve food security.

CASE STUDY

8.8 Increasing food production: Koshi Hills, East Nepal

FACTFILE
- Nepal is a mountainous country in South Asia, about the size of England.
- Nepal is most famous for its 8000 m peaks, which include Mount Everest.
- In 1999, its human development index (0.351) meant that Nepal was ranked the 152nd (out of 174) poorest country in the world.
- GDP per capita in 1999 was just $220.
- 77 per cent of adult females are illiterate.

8 Contrasts in development

The causes of poverty

The Koshi Hills is one of the poorest regions in Nepal (Fig. 8.17). Here, most people depend on farming. Below 2000 m, farmers grow cereal crops on small terraces cut into steep valley sides (Fig. 8.18). They use ox-drawn ploughs and hand tools. Above 2000 m, only livestock farming is possible.

Physical conditions are not ideal for farming. Not only have powerful rivers like the Arun and Tamur and their tributaries carved steep valleys, but between the valleys the land rises to more than 3000 m. Meanwhile, rapid population growth has reduced the average farm to less than one hectare in size. Such tiny farms are too small to be self-sufficient, so many farmers also have to work as part-time labourers to survive. Given these conditions, it is not surprising that three out of five families live in absolute poverty.

Changing production

Since the early 1980s, attempts have been made to increase food production and improve levels of nutrition in Koshi.

- ◆ New, higher-yielding varieties of maize, wheat, rice and soyabean have been introduced, increasing food output and food security for many families (see Book 1, Chapter 9).
- ◆ Outputs of milk and meat have also risen, thanks to the introduction of new breeds of cattle, sheep, goats and pigs, and to better animal health care. Surplus milk production has made it possible to set up a milk marketing scheme involving 800 farmers. The scheme supplies milk for

Figure 8.18 Cultivation terraces in Nepal

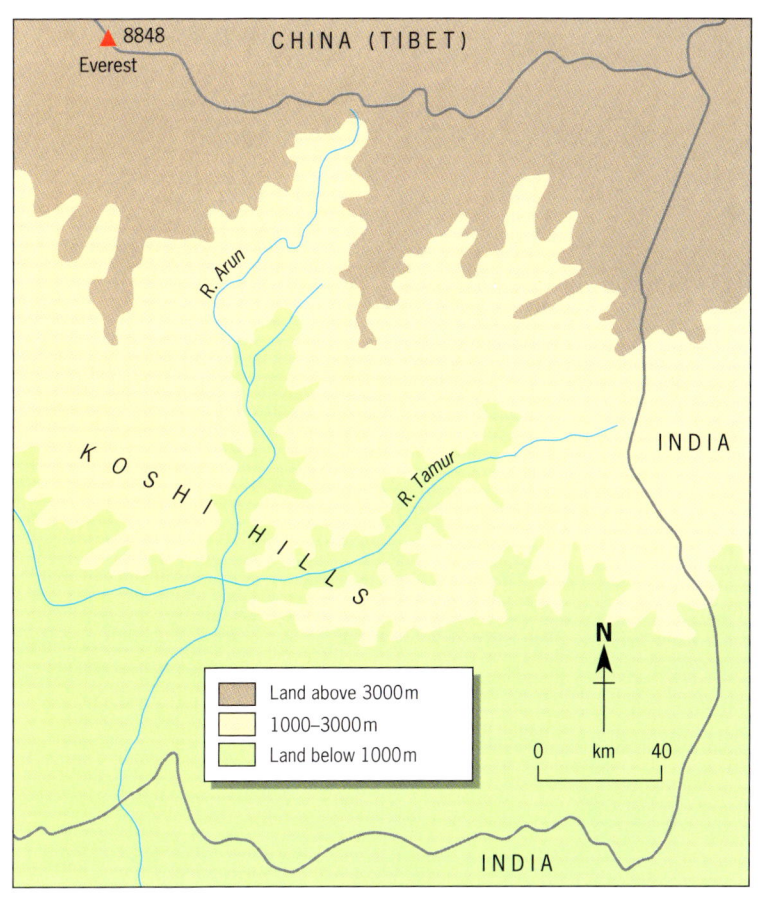

Figure 8.17 (above) Koshi Hills, East Nepal

EXERCISES

7 Study Figure 8.18.

a Suggest two environmental problems that farmers might face in this area.

b* Why do farmers cultivate such steep slopes?

8 Contrasts in development

urban areas and generates extra money for farmers. The farmers receive cash payments every 15 days, which gives them a regular income.
- Reafforestation schemes, covering 7000 hectares, have been established. Planting trees helps to stabilise steep slopes, protect cultivation terraces, and reduce soil erosion. In addition, it gives local people a sustainable source of timber and fuelwood.

These initiatives have raised farm incomes; improved people's health, diet and well-being; and helped to regenerate the environment.

8.9 Explaining global contrasts: economic problems

Economic obstacles to development:

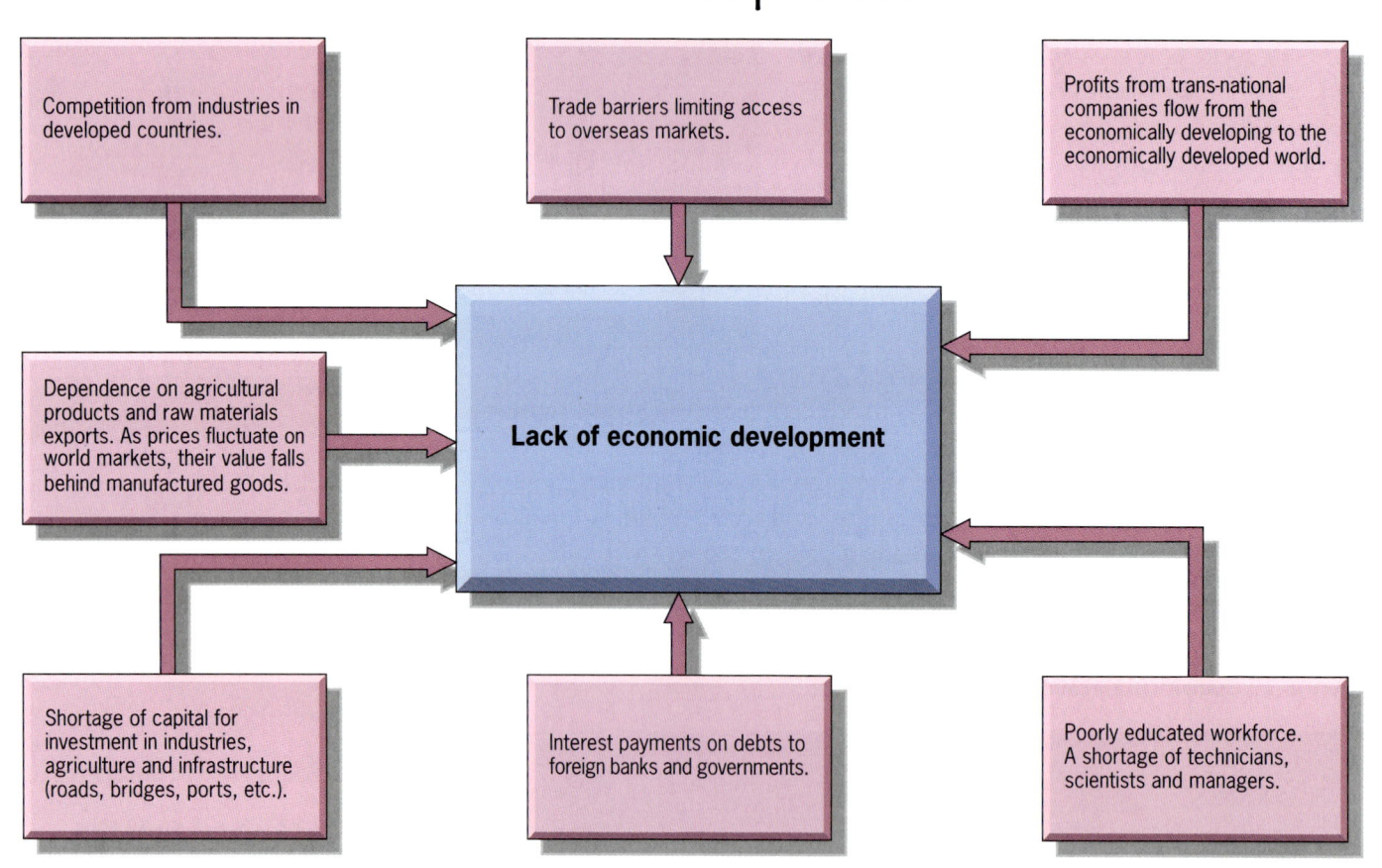

- Competition from industries in developed countries.
- Trade barriers limiting access to overseas markets.
- Profits from trans-national companies flow from the economically developing to the economically developed world.
- Dependence on agricultural products and raw materials exports. As prices fluctuate on world markets, their value falls behind manufactured goods.
- Shortage of capital for investment in industries, agriculture and infrastructure (roads, bridges, ports, etc.).
- Interest payments on debts to foreign banks and governments.
- Poorly educated workforce. A shortage of technicians, scientists and managers.

→ Lack of economic development

Figure 8.19 Economic obstacles to development

Economic problems act as obstacles to development in poor countries (Fig. 8.19). In this section we shall investigate some of these problems and see how they contribute to poverty.

Shortages of capital

Lack of money for investment (capital) is a major barrier to development in many poor countries. Capital is essential if these countries are to exploit their natural resources, improve their agriculture, develop new industries or pay for roads, airports, schools and hospitals.

Where will this capital come from? There are two possibilities:
- Investment from large foreign companies or **trans-national corporations**

8 Contrasts in development

(TNCs) like Ford, BP, Unilever, Samsung, Toyota, Honda and Nike (Fig. 8.20).
- Loans from governments and banks in the economically developed world.

Profits leaking overseas

Although foreign investment creates jobs in LEDCs, many of the resulting financial benefits 'leak' overseas. For example, if a foreign TNC invests in a country, a large slice of its profits return to its investors, who are located overseas. Moreover, shortages of skilled labour in poor countries often mean that foreign workers get the best paid jobs. Some experts even argue that foreign TNCs hold back development.

Figure 8.20 Nike workers in Jakarta, Indonesia

Borrowing and debt

In the 1980s and 90s, many poor countries borrowed huge sums of money from the developed world. A sudden rise in interest rates meant that they were unable to pay the interest on their loans. As a result, they fell into ever increasing debt. Today, 32 of the 54 poorest countries in the world have a severe debt problem. Most of these countries are in Africa. Guinea-Bissau in West Africa is a typical example. The annual interest payments on its foreign debt are two and a half times the value of its exports. Such crippling debts stifle development. Money that might have gone into hospitals, schools or housing has to be paid, instead, to the developed world.

In 2000, the UK and other MEDCs agreed to forego interest payments on loans to 41 of the world's poorest countries. In return, the debtor countries had to agree to use the money to tackle poverty.

> *EXERCISES*
>
> **8** Study Figures 8.21 and 8.22. Nearly all of the leading cocoa producers are countries in the developing world.
>
> **a** What were the highest and lowest prices of cocoa in the period 1984–2000?
>
> **b** Using the evidence in Figure 8.22, what factors appear to influence price?
>
> **c*** Suggest possible problems that swings in world cocoa prices might cause for the producing countries.
>
> **d*** What could the cocoa producers do to reduce these swings in world prices?

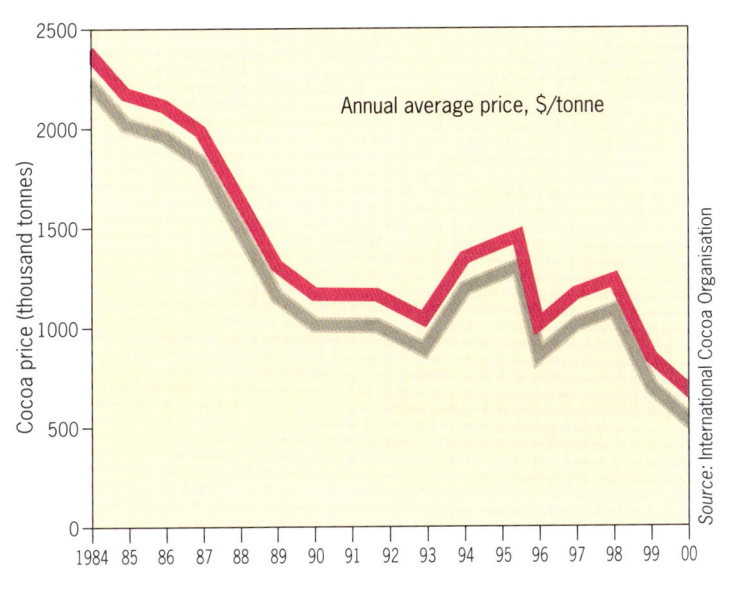

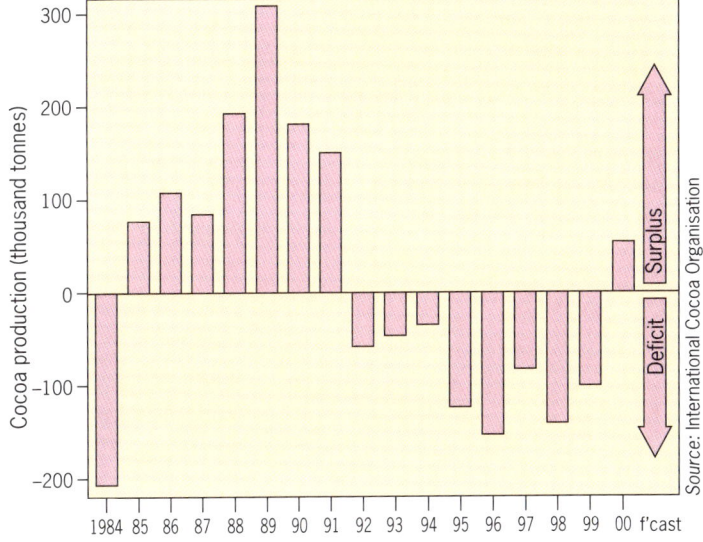

Figure 8.21 (below left) Changes in world cocoa prices, 1984–2000

Figure 8.22 (below right) Changes in world cocoa supply balance, 1984–2000

8 Contrasts in development

> ### EXERCISES
> **9** Read the newspaper article on Tanzania's foreign debts (Fig. 8.23).
> **a** What is the size of Tanzania's foreign debt?
> **b** Which countries and institutions have made loans to Tanzania?
> **c** What impact does foreign debt have on the people of Tanzania?
> **d*** Should the developed world write off debts owed to it by the world's poorest countries? Present and justify your view to your class.

Trade

Most poor countries export **primary products**, e.g. food (Fig. 8.22) and raw materials, and import manufactured goods. This creates two problems. First, the value of these primary products tends to fluctuate wildly on world markets (Fig. 8.21). And second, the prices of primary products have failed to keep up with those of manufactured goods (see Chapter 9). Thus, poor countries get less for their exports and have to pay more for their imports (Fig. 8.23). Finding markets for exports is also difficult. Rich countries and economic groups like the European Union often protect their own industries from foreign competition by setting barriers to trade, such as import taxes (tariffs) and import limits (quotas).

Skills shortages

It is difficult for poor countries, which lack technology and skills, to compete with rich countries in making manufactured goods. In many LEDCs, fewer than one in three adults can read and write. Yet without skilled workers, scientists, technicians and managers, development has little chance of success (see Section 6.3).

Government finances are under pressure because of Tanzania's foreign debt. It owes $8 billion (£4.8 billion) to its international creditors – £163 for every man, woman and child in the country. The government must spend nine times as much on debt repayments as on basic health care, and four times as much on debt as on primary education.

Since the government was forced to cut back on spending, school buildings have fallen into disrepair and there are shortages of equipment.

A recent Oxfam report estimated that, on average, there was one desk per seven pupils and each textbook was shared by four children. Classes of more than 50 pupils are not uncommon.

At Lusala school the children are comparatively lucky. The parents built the school themselves in the days when the government still provided roofs for free. There are seven classrooms for 450 pupils and almost every child has a desk.

But the introduction of fees has made the school unaffordable for some.

The government hopes that some of the pressures on its budget will be relieved soon through the World Bank's Heavily Indebted Poor Countries (HIPC) initiative.

The main lenders – Western governments and international financial institutions such as the IMF and World Bank – have agreed to write off up to 80 per cent of the loans for developing countries with unsustainable debts.

Tanzania should be the next in line for relief under the HIPC initiative – provided its debts still meet the definition of "unsustainable" stipulated by the IMF and World Bank.

Their criteria are strictly financial. If a country spends more than 20-25 per cent of its export earnings on servicing its debts, it qualifies for relief.

Debt relief alone cannot reduce poverty but it is a necessary first step, says Simon Levine, Christian Aid's programme officer for Tanzania.

"Tanzania will never be able to provide any kind of reasonable living conditions for its citizens without debt relief," he says.

Figure 8.23 The real cost of shouldering the debt burden, *The Guardian*, 19.02.99

8.10 Population problems

In MEDCs, population growth is slow: just 0.1 per cent in 2001. Contrast this with countries in the economically developing world, where population growth is, on average, more than 15 times faster (Fig. 8.24). In many poor countries, population growth is faster than economic growth. Thus, while there are always more mouths to feed, more jobs to provide, and more demands for education and health care, there is less money per person to pay for them. The end result is that people get even poorer (Fig. 8.25). Burkina Faso in West Africa is a country caught in this poverty trap. Its population growth is 2.8 per cent a year; half the population is under 15 years; and

8 Contrasts in development

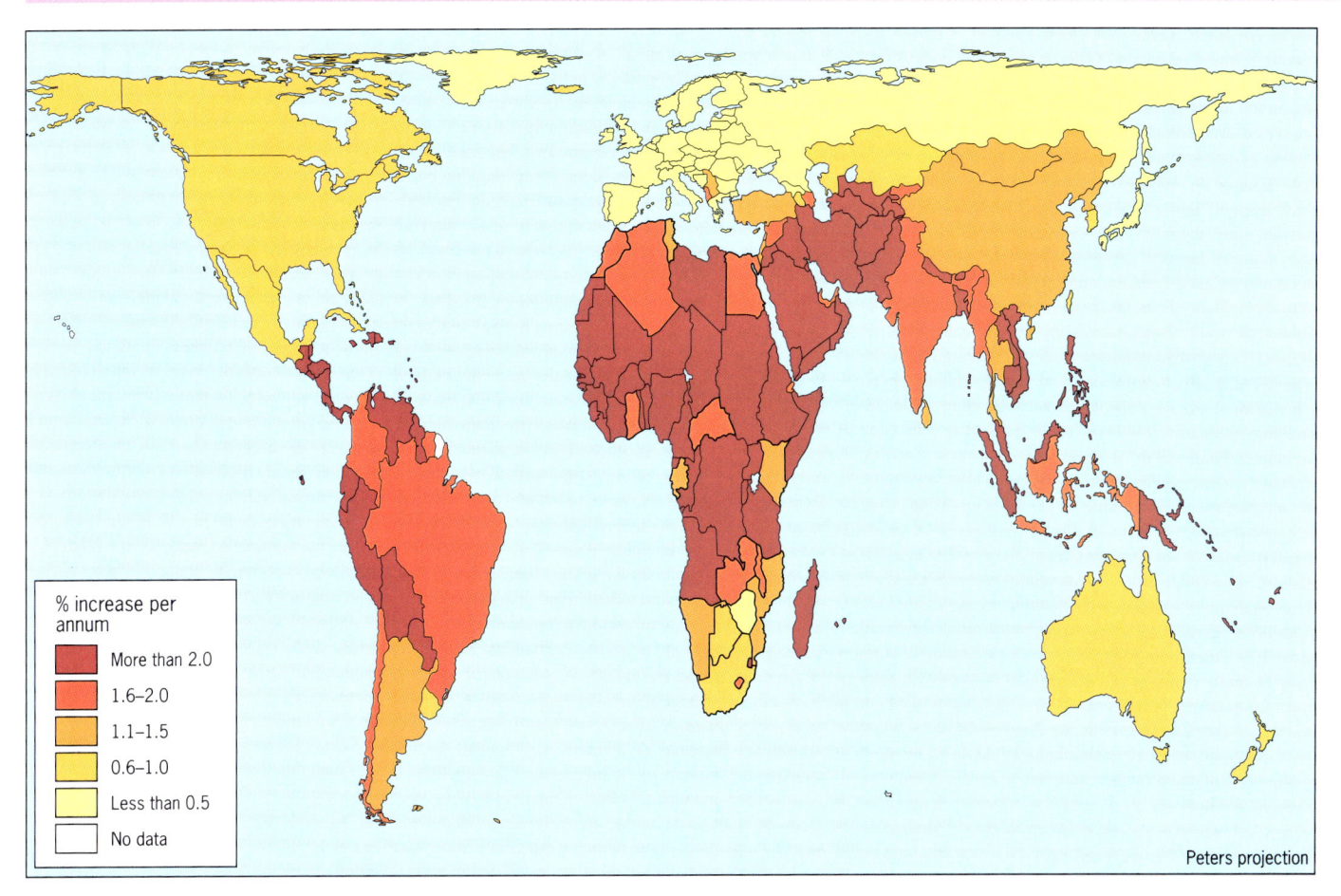

Figure 8.24 (above) Population growth rates, 1999

% increase per annum
- More than 2.0
- 1.6–2.0
- 1.1–1.5
- 0.6–1.0
- Less than 0.5
- No data

Peters projection

Figure 8.25 (right) The links between rapid population growth and poverty

women have an average of 6.4 children. Only 28 per cent of adults in Burkina Faso are literate and there is only one doctor to every 57 000 people. Unless countries can control their population growth, development will do little to improve the lives of most people.

8.11 Social problems

The rapid and successful development of several **newly industrialising countries (NICs)** in East Asia (e.g. Taiwan and South Korea) is due partly to investment in education. Today, these NICs have workforces that are as highly skilled and educated as those in MEDCs.

Literacy

The contrast between Asia's NICs and other Asian countries, such as Pakistan, is striking. Pakistan has one of the lowest literacy rates in the world. Only 3 out of 10 boys and 2 out of 10 girls have had even a primary school education. Lack of education reduces the chances of employment and

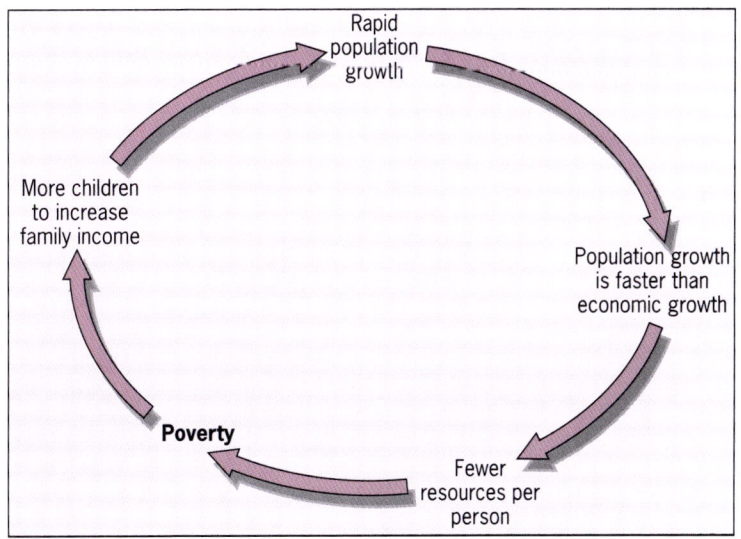

EXERCISES

10a What evidence in Figure 8.24 suggests that AIDS is beginning to have an effect on population growth in Africa?

b Study Figure 8.25 and draw a similar diagram to show the possible effects of family planning on poverty.

8 Contrasts in development

> **EXERCISES**
>
> **11a** Draw a bar chart to represent male and female illiteracy in South Asia.
> **b** Outline the problems posed by illiteracy to economic and social development in LEDCs.

Table 8.2 Male and female illiteracy in South Asia, 1999

	% Males	% Females
Afghanistan	49.6	79.9
Bangladesh	48.3	70.7
India	32.2	55.5
Nepal	42.0	77.2
Pakistan	41.1	70.0
Sri Lanka	5.7	11.4

Figure 8.26 Women at heavy labour in a brickworks, Bengal, India

of escaping poverty. In the past, religion has discouraged the education of women and girls. Parents were reluctant to spend money educating daughters who would only leave home when they married. Such attitudes towards women are common in the economically developing world. But with half the workforce made up of women (Fig. 8.26), can these countries afford to ignore them any longer?

Not only are women less educated than men in many parts of the developing world, but they have little control of their own lives. Where women have greater independence, social development is often more advanced. In Kerala state in southern India, women have an unusually high status. It is no coincidence that here, adult literacy and life expectancy are higher, and infant mortality is lower than in any other part of India.

8.12 Political problems

Political problems also hold back progress in much of the economically developing world. For example, many governments do not use their income to benefit the people. Smuggling and corruption are all too common. Meanwhile, civil wars destroy essential infrastructures like roads, pipelines and hospitals. Money that might have been used to build schools, houses and factories, goes instead on weapons. War also disrupts food production, forces people off the land and creates millions of refugees. Once again, the situation is worst in Africa. In the 1990s, there were civil wars in Somalia, Sudan, Rwanda, Sierra Leone (Fig. 8.27), Mozambique and Liberia. Elsewhere in Africa, violent changes of government have occurred. Political instability makes these countries extremely risky and therefore unattractive places for foreign investment.

8 Contrasts in development

Rich pickings but empty coffers in Sierra Leone
Claudia McElroy

FROM his veranda, Chief Abu Mbawa Kongorma views with bitterness the neglected buildings and potholed roads of Sierra Leone's diamond capital, which he says has seen little gain from more than 60 years of mining.

Digging around the town of Koidu, in the diamond-rich Kono district, has turned the land into a moonscape, its dug-up wealth far away on the European diamond market. Despite the civil war, which wrecked the country's mining-based economy and killed 15 000 people, foreign investors and native miners are still attracted there. 'This area is the promised land in terms of mining ... Sierra Leone should be one of the richest countries in the world. Instead it is one of the poorest,' the chief said. So why are state coffers empty when official export statistics show that in 1995, 213 000 carats reached the European market? One reason may be that since rebels launched a guerrilla campaign from Liberia in 1991, much of the east and south cannot be entered by security forces, and so illegal mining and smuggling has spread. It costs an estimated £147 million a year in lost foreign exchange.

Some argue that the war is an excuse. They claim that state authorities have had a direct stake in the trade for 60 years, making themselves rich, not the treasury.

Others blame foreigners for controlling the profits. The South African mining company De Beers dominates the world market, fixing supplies and boosting prices. On a national level, buying is largely monopolised by about 30 licensed Lebanese dealers and a few unlicensed Guineans. Then there are foreign mining companies with seemingly risky investments.

Branch Energy Ltd of Britain has invested £9 million in six mining and exploration projects since 1995. 'Sierra Leoneans will benefit, principally through the employment opportunities,' the company's head, Alan Patterson, said. 'What we're doing will provide a massive boost to the economy.'

Such confidence may seem surprising because of Sierra Leone's political unrest. Four years of military rule, which saw two coups, ended in March when Ahmed Tejan Kabbah was sworn in as president, after the first democratic election in almost 30 years.

Figure 8.27 Article about Sierra Leone: one of the poorest countries, *The Guardian*, 11.9.96

EXERCISES

12 In spite of its rich diamond deposits, Sierra Leone was ranked 174th out of 174 in the UN's 1999 HDI league table. Read Figure 8.27.
a What benefit does Sierra Leone get from its diamond resources?
b State and explain three reasons why Sierra Leone gets so little money from its diamonds.
c* Suggest a plan of action for Sierra Leone, to enable the country to benefit more from its diamond resources.

REMEMBER
Many factors contribute to poverty in LEDCs. They include illiteracy, disease, political instability, debt, unfair terms of trade and rapid population growth.

CASE STUDY

8.13 Environmental problems in the Dominican Republic

The Dominican Republic (Fig. 8.28) occupies half of the island of Hispaniola in the Caribbean (the other half belongs to Haiti, see Chapter 6). By the standards of the world's least developed countries, the Dominican Republic is not badly off, but compared to the economically developed world, it is poor (Table 8.3).

Table 8.3 Levels of development in the Dominican Republic and the UK

	Dominican Republic	UK
Population growth (% p.a.)	1.8	0.1
Infant mortality (per 1000)	40	6
GNP per capita ($US)	1910	22640
Human development index	0.720	0.932

Spiralling debt

Poverty in the Dominican Republic is due partly to the misuse of the country's natural resources. This goes back to the 16th century, when Spanish conquerors exploited the island's rainforest for timber. More recently, sugar cane plantations have replaced much of the forest (Fig. 8.29).

Heavily dependent on sugar cane, disaster hit the Dominican Republic in the early 1980s, when world sugar prices slumped. To make matters worse, the country's largest market, the USA, introduced quotas on imports of sugar. This meant that the country could not sell as much sugar as it needed to meet its debt repayments.

8 Contrasts in development

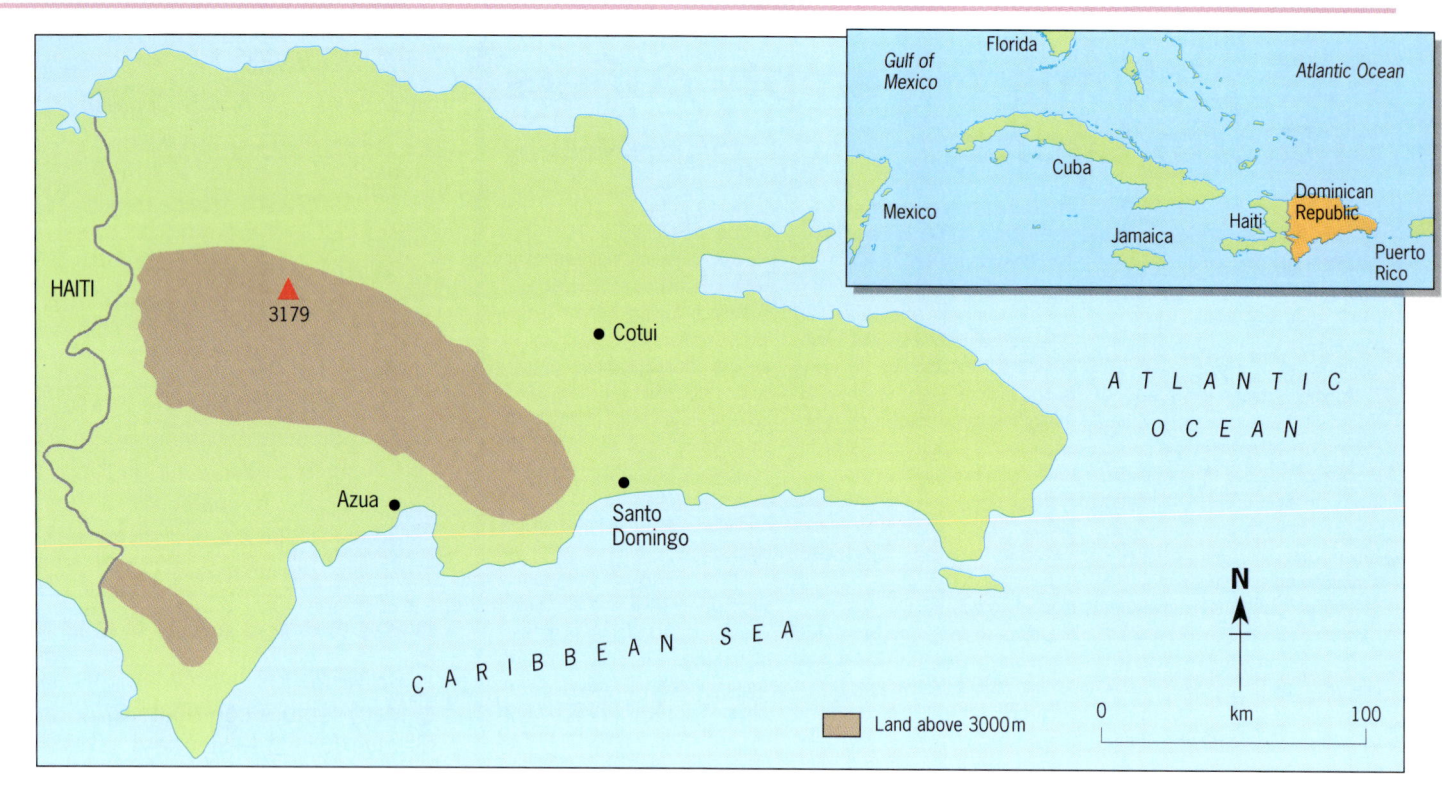

Figure 8.28 The Dominican Republic

Development brings problems

The government responded by moving away from its old dependence on sugar. It encouraged foreign TNCs to invest in agribusiness (see Book 1, Chapter 9), which is high-tech farming based on chemical fertilisers and pesticides. It also tried to promote tourism.

Problems of tourism

Development has caused problems, both for the environment and people. Tourists have eroded beaches and damaged coral reefs. Large areas of mangroves have been cleared (they harbour malarial mosquitoes) to make the country more attractive to tourists. Apart from degrading the environment, these developments have caused the extinction of hundreds of native species. While tourism has created small areas of prosperity, it has brought few benefits to the rest of the country. In fact, its main impact has been to widen the gap in living standards between rich and poor.

Problems of agribusiness

The introduction of agribusiness, controlled by foreign companies and based on export crops such as melons, coffee and citrus fruits, has widened inequalities. For instance, just 2 per cent of land owners control 55 per cent of the land. Agribusiness has also caused environmental problems.

- In Azua and Cotui (see Fig. 8.28) many peasant farmers have lost their land to agribusiness. No longer able to farm the land, these people now work as labourers for the huge agribusiness enterprises.

Figure 8.29 Harvesting cane on a sugar plantation, the Dominican Republic

8 Contrasts in development

Figure 8.30 The possible effect of tourist development on the environment

- Intensive rice and cattle farming have polluted rivers, lakes and soils. Contaminated drinking water has damaged people's health, while the accumulation of agro-chemicals in soils has made some farmland useless.
- To accommodate tourists, golf courses have been built on prime agricultural land.
- Deforestation has led to soil erosion and has interrupted the water cycle, changing local climates and causing rivers to dry up. Deforestation has also produced shortages of fuelwood.

> *EXERCISES*
>
> **13a** How does the approach to rural development in the Dominican Republic differ from that in Moyamba in Sierra Leone (Section 8.6)?
> **b*** Draw a flow diagram to show how the Dominican Republic's trade problems in the 1980s led to environmental degradation.
> **14*** In what way could poverty and lack of development in the economically developing world be blamed on rich countries?

8.14 Diseases and pests

Compared to people living in rich countries, those living in the economically developing world have relatively poor health (Table 8.4). Much of their ill health is due to poverty.

Diseases

Malnutrition, overcrowded living conditions and poor sanitation encourage the spread of diseases (Figs. 8.3–8.5). Poverty also means that individuals cannot afford essential medicines, and governments cannot afford to train enough doctors or build enough hospitals and clinics.

Disease and poor health are not only a sign of a lack of development, they also contribute to it. Malaria, for example, affects more than 300 million people. People who suffer from malaria are constantly tired. Weakened and lacking energy, they have little strength to work. When malaria strikes young adults who have children to support, it can lead to severe economic effects. River blindness (Fig. 8.31) is a disease common in West Africa that has similar effects on the economy (Fig. 8.32). It often blinds young adults who should be at their most productive time of life.

Figure 8.31 Teenage girl in Mali, Africa leading her father who has been blinded by onchocerciasis (river blindness)

153

8 Contrasts in development

> **EXERCISES**
>
> **15** Study Table 8.4.
> **a** Which diseases kill most people in LEDCs?
> **b** What are the main causes of the diseases listed in Table 8.4?
> **c*** How does poverty and lack of development contribute to ill health in LEDCs?

Other diseases are major killers. Ten million children a year die from measles; diarrhoea kills 5 million; and TB between 2 and 3 million. Most of these deaths could be prevented with vaccination programmes, clean drinking water, and the use of drugs.

Pests

We call animals that eat and destroy crops pests. It is estimated that each year, pests such as locusts, weevils and rats destroy up to half of all the world's food crops (Fig. 8.33). Most of these losses are in poor countries that cannot afford to protect crops with pesticides or store food securely.

In sub-Saharan Africa, wooded and grass savannas are home to one of the world's greatest pests – the tsetse fly (Fig. 8.34). The tsetse fly is a blood-sucking insect that transmits sleeping sickness to people, and a similar disease called nagana to cattle. The problem is so bad that huge areas of tropical Africa cannot support cattle. If the tsetse fly could be eradicated, cattle farming would be profitable. This would help to raise living standards in what is probably the poorest region in the world. So far, the use of pesticides and drugs has had little success.

Table 8.4 Major diseases in the economically developing world

Disease	Causes	Treatment and control	Number of people affected
Malaria	The most significant tropical disease. 300 million people infected. Malarial parasite is spread by mosquitoes (see Fig. 6.12).	Spray breeding sites (stagnant water) with insecticides. Difficult to control because insects become drug-resistant.	107 million new cases a year. 1–2 million deaths.
Bilharzia	Parasitic infection. Caused by infected larvae of water snails. Larvae penetrate the skin as people swim or work in irrigated fields. Poor sanitation spreads the larvae.	Very effective drugs exist, but often do not reach the people who need them.	200 million cases a year. About 200 000 deaths.
Sleeping sickness	Parasites spread to humans by tsetse flies, which suck blood from infected cattle and wild animals. Causes nagana in domestic cattle.	50 million people in Africa are at risk. It is difficult to eliminate the tsetse fly. Biological solutions that restrict the fly's breeding offer some hope.	25 000 cases a year.
Filariasis	Parasitic disease caused by worms and their larvae. Spread by mosquitoes and other blood-sucking insects. One form of the disease causes blindness.	Drugs are available to cure river blindness. Insecticides can be used to destroy the insects.	More than 100 million cases a year.
Tuberculosis (TB)	Bacteria spread in the air when infected people cough or sneeze. Highly infectious in overcrowded living conditions.	Risk and death toll is increasing as bacteria become drug resistant. TB is also a growing problem in MEDCs.	The world's largest killer of adults. 2–3 million deaths a year.

8 Contrasts in development

The bite spreads parasitic worms. They breed under your skin and produce hordes of microscopic worms called microfilariae which infest your body for years. You itch, you get skin problems, you become debilitated.

Then the microfilariae invade your eyes. And you slowly become blind.

It's called River Blindness. You'll find it in a host of fast-flowing rivers in West and Central Africa. And there you will find most of the victims – the million or so who have already lost or are losing their sight and seventeen million who have the symptoms and are at risk of blindness; their sight could be saved – quickly and easily.

Consider what happens if you lose your sight in a poor African country. No disability benefit, no regular medical facilities, little prospect of remedial work.

There is just your family living at the harshest subsistence level. You become a burden on them. You will need to be led from place to place. You can do little to contribute to their welfare.

For this is the particular agony of River Blindness. It takes years to gestate. When blindness strikes, you are likely to be in your thirties or forties. You are likely to be the family breadwinner. Without you, the family will lose their hard-won self-sufficiency and become dependent on the charity of neighbours.

And it is all avoidable.

THE BLACK SIMULIUM FLY

Mectizan: a drug to halt River Blindness. It is a simple-to-take white tablet and it has been thoroughly tested in Africa. It not only prevents blindness but it tackles the painful symptoms as well.

SIGHT SAVERS is committed to distributing Mectizan in Nigeria, Ghana, Mali, Guinea, Uganda and Sierra Leone. We intend to expand the programme to include Cameroon.

If this sounds easy, it isn't. A distribution programme on this scale will strain Sight Savers' resources to the utmost. Then again, you have to allow for the reality of medical work in poor countries. In some of those countries there are village health workers or committees who we can train to distribute Mectizan. In others we are relying on nurses who go from village to village. We shall be very reliant on local people, village chiefs in particular.

And there are transport problems. The dry flatlands of Mali make a 50cc motor-cycle a reasonable means of travel; in the hilly, forest regions of Guinea, we'll need a tougher 125cc machine. A lot of Mectizan will be distributed by simple bicycle.

And the programme will go on for years. In each community at risk, Mectizan will have to be distributed annually for at least ten years to break the cycle of the disease. Our aim is to ensure that none of today's children will lose their sight through River Blindness.

SIGHT SAVERS INTERNATIONAL
Sight Savers International
13 Cheap Street, Frome,
Somerset BA11 1BN
Telephone: 01373 452272

Figure 8.32 (left) Advertisement by the charity Sight Savers

EXERCISES

16 Study Figure 8.32.
a What insect is responsible for transmitting river blindness?
b How many people have lost their sight to river blindness, and how many are at risk?
c Describe the problems faced by people who have lost their sight in LEDCs.
d Name the drug that is effective in fighting river blindness.
e* Although river blindness is easily treated by drugs, there are many obstacles to helping people who suffer from the disease. Give a brief account of these problems.

Figure 8.33 Locust swarm, Ethiopia

REMEMBER
Diseases have a particularly severe economic impact when they strike the young adults on whom children and the elderly depend. For this reason, the effect of AIDS in southern Africa in the next 20 years could be disastrous.

8 Contrasts in development

EXERCISES

17 Study Figure 8.34.
a Describe the distribution of cattle in Africa.
b How does the distribution of the tsetse fly affect the distribution of cattle in Africa?
c* Suggest one other factor that appears to influence the distribution of cattle in Africa.
d* Suggest possible reasons why it is especially difficult to control pests in LEDCs.

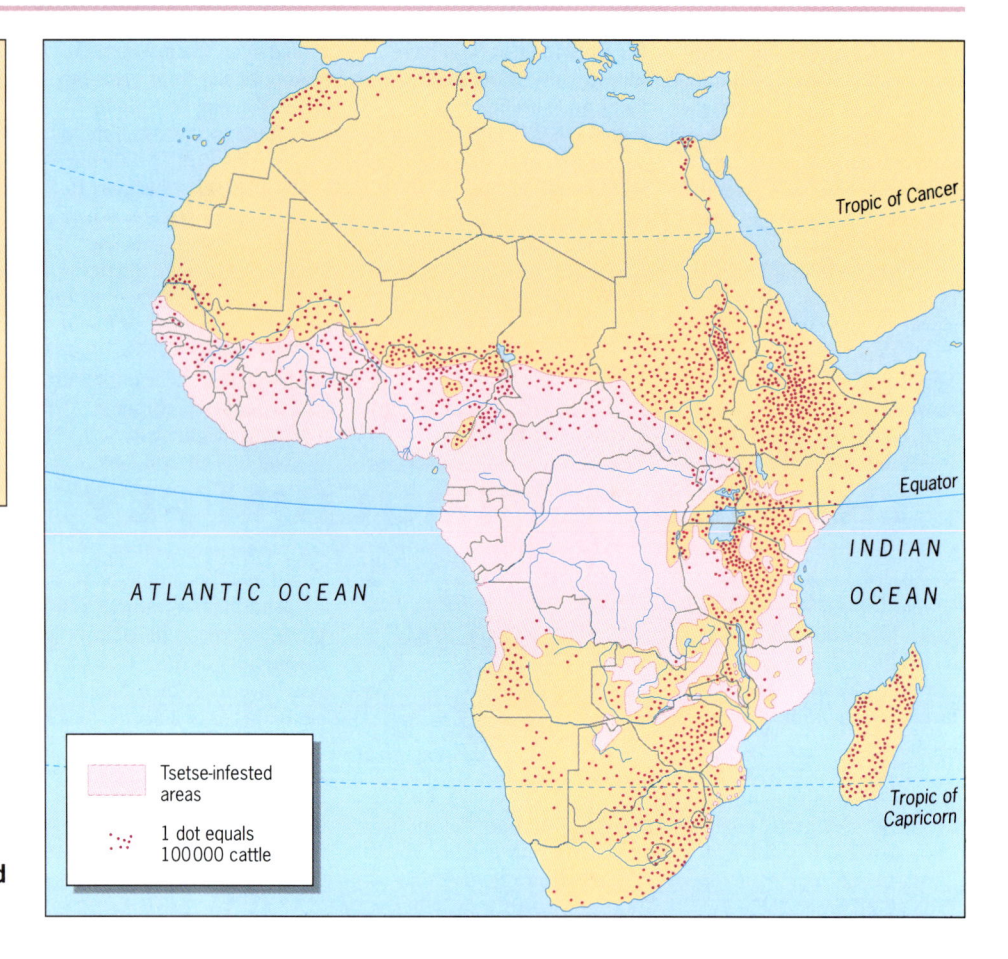

Figure 8.34 Distribution of cattle and the tsetse fly in Africa

8.15 Regional contrasts in development

Just as there are contrasts in wealth between countries, we find similar differences between rich and poor regions *within* countries. Such regional contrasts are not, as you might think, found only in the economically developing world. Italy is a rich country, but its regional contrasts in wealth are as great as any country in the world.

CASE STUDY

8.16 Italy: a divided country

Italy is a country split between the rich, developed North and the poor, backward South (Fig. 8.35). People in the North are as prosperous as their neighbours across the Alps in Austria and Switzerland (Fig. 8.36). In contrast, in the South, standards of living are much lower – so low that southern Italy could almost be a different country (Fig. 8.37). In almost every respect, the South is worse off than the North (Table 8.5).

Since 1950, the Italian government (and more recently the EU) has tried to reduce these differences. Huge sums of money have been poured into the South to improve agriculture, build roads and develop new industries, such as steel and petrochemicals. But despite all these efforts, the contrasts in wealth between North and South are as strong as ever.

8 Contrasts in development

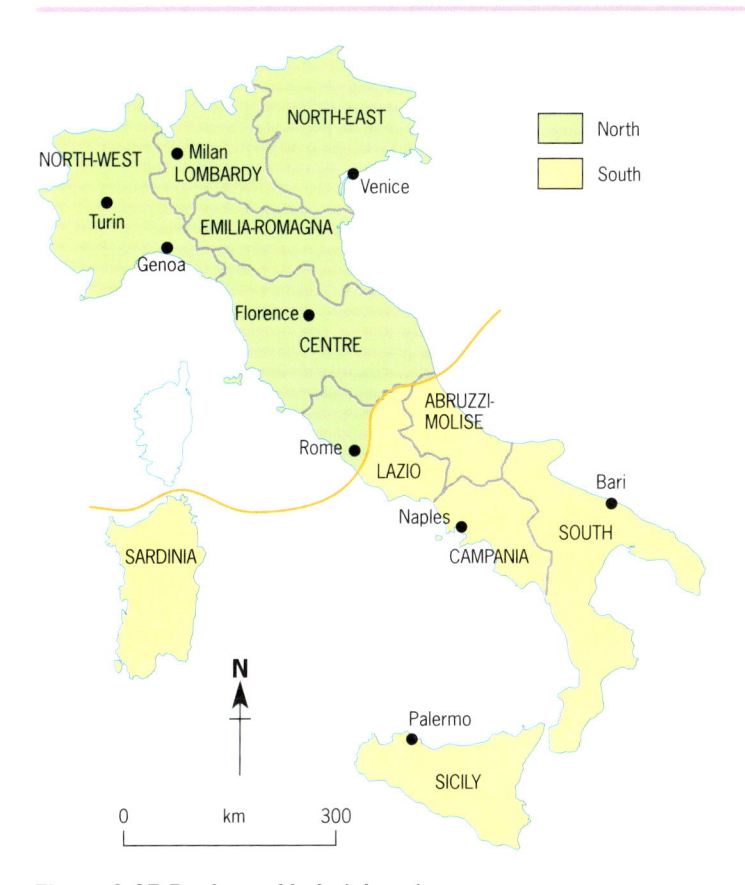

Figure 8.35 Regions of Italy (above)

Explaining Italy's regional contrasts

The Italian South is on the edge, or **periphery**, of the EU. This location is one reason why its development lags so far behind the North. Other peripheral regions in the EU, such as the Scottish Highlands, western Ireland and the Greek islands, have similar problems. Generally, these peripheral regions are poorer than regions at the centre or **core** of the EU.

Figure 8.36 (top) Productive vineyards, Tuscany, northern Italy

Figure 8.37 (above) Poor, hilly, dry, rural landscape in southern Italy

Table 8.5 Regional contrasts in Italy: 2000

	Unemployment (%)	GDP per capita (EU= 100)	Employment in agriculture (%)	Average wheat yields 100 kg/ha
North-west	9.3	118	3.8	46
Lombardy	5.7	131	2.1	58
North-east	5.0	124	5.3	56
Emilia-Romagna	5.7	131	6.7	57
Centre	7.9	107	3.8	41
Lazio	12.3	113	2.9	33
Campania	24.9	65	7.5	32
Abruzzo-Molise	11.2	87	7.7	30
South	22.7	65	12.0	26
Sicily	25.6	67	9.1	22
Sardinia	21.5	65	8.1	20
Italy	12.3	102	5.4	49

EXERCISES

18a Using the outline of Italy (Fig. 8.35) and the data in Table 8.5, draw choropleth maps (proportional shading maps) to show: • the regional pattern of unemployment • GDP per person • % unemployment • employment in agriculture • wheat yields in Italy.
b* Study the maps you have drawn and summarise the regional contrasts of development in Italy.

157

8 Contrasts in development

> **EXERCISES**
>
> **19*** Draw a similar diagram to Figure 8.38 to show a possible vicious circle of decline in the Italian South.

Virtuous circle of growth

Figure 8.38 provides an explanation for the prosperity of the core. Past investment in industry has created jobs and a large, skilled workforce. People have migrated to work in the core, and as its population and wealth grew, an attractive market for industry developed. Service industries have set up, and motorways, railways, ports and airports have improved. This success has led to more investment, thus allowing the spiral of growth to continue.

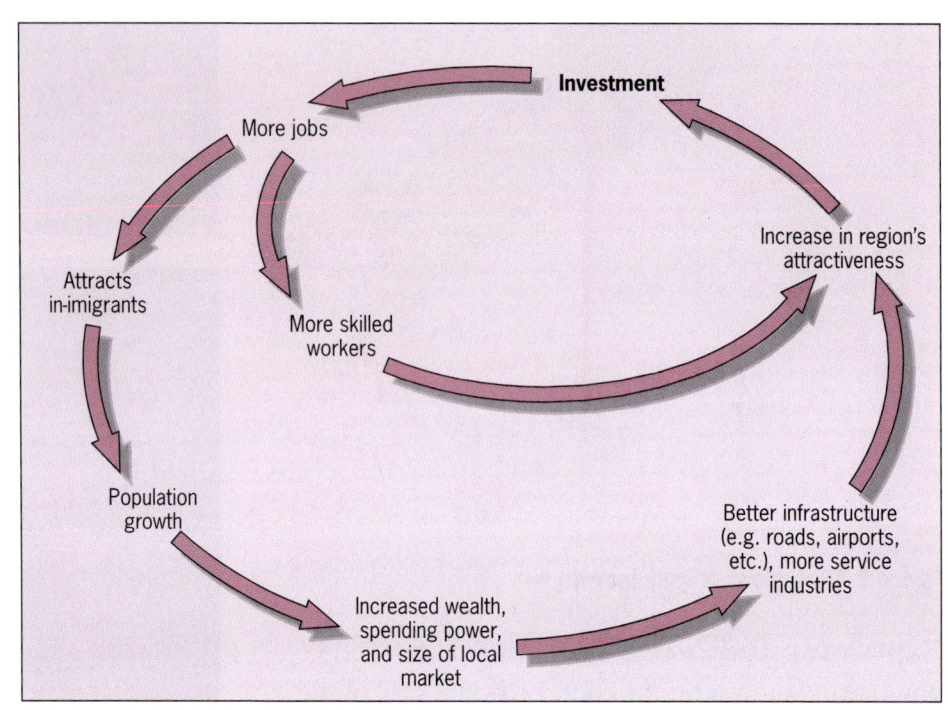

Figure 8.38 (right) Virtuous circle of growth in a core region

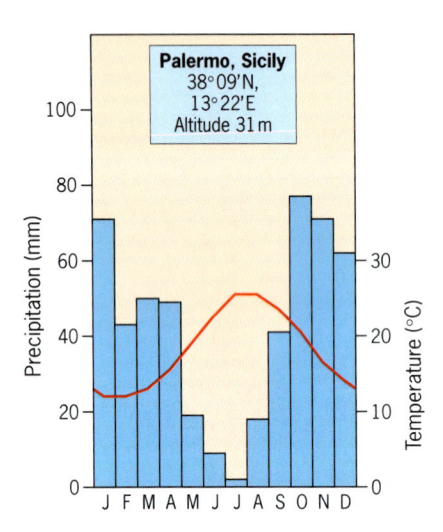

Figure 8.39 (above) Palermo: mean monthly temperature and precipitation

Geographical disadvantages of the South

Apart from its location, the South has other disadvantages that have hindered its development.
- In summer, its climate is hot and dry (Fig. 8.39). This makes cultivation without irrigation difficult.
- Overcultivation and overgrazing have led to land degradation and eroded and exhausted soils.
- The region's mountainous relief has often limited resources for farming.
- Massive out-migration from the South has occurred in the past 100 years. Driven by the search for jobs, people have flocked to northern cities like Milan and Turin, and further afield to Europe and the USA.

The future

Attempts to transplant manufacturing industry in the South in the past 50 years have largely failed, and are not likely to succeed in the future. Tourism offers the best hope for economic development. Hitherto, the region's resources for tourism – warm, sunny climate, coastline and cultural/historical heritage – have hardly been touched. All this could change in the next 20 years; massive new investments in tourism are already underway in several areas, including Naples and Palermo.

8.17 Summary: Contrasts in development

KEY SKILLS OPPORTUNITIES
C1.1: Ex. 5, 14; **C1.2**: Ex. 1a, 3a, 9a, 9c, 9d, 12a, 15a, 15b, 16a, 16b, 16d; **C1/2.3**: Ex. 1b, 5, 12c, 16e; **C2.1**: Ex. 9d; **C2.2**: Ex. 2b, 10a, 12b, 13a, 14, 16c, 17a; **N1/2.1**: Ex. 1c, 8c, 15c, 17c, 17d; **N1/2.2**: Ex. 4a, 6a, 8a, 11a, 18a; **N1/2.3**: Ex. 4b, 4c, 8b, 18b; **IT1/2.1**: Ex. 4d; **IT1.2**: Ex. 4e.

Key ideas	Generalisations and detail
The world is divided into rich and poor countries.	• Rich countries are economically developed. The economically developed world (apart from Australia and New Zealand) is in the northern hemisphere. • Poor (economically developing) countries are in the tropics, in the southern hemisphere. • The developed world is the 'North'. The developing world is the 'South'.
Development includes both economic and social progress.	• Economic development is about realising the physical and human resource potential of a country. It includes investment in economic activities (e.g. industry and agriculture). • Social development concerns the provision of services (e.g. health care, sanitation, education, etc.). It is a direct attempt to improve the quality of life and human well-being.
There is no simple measure of development.	• GNP per capita considers only economic development. • The human development index is the best single measure. It takes account of economic and social factors (incomes, life expectancy and education).
There are global contrasts in the provision of food and water supplies.	• There are severe shortages of food in many parts of the economically developing world. Famine, malnutrition and undernutrition affect millions of people, especially the poorest. • Problems of water availability and water quality are considerable in many LEDCs. Lack of clean water is a major cause of ill health in these countries. • Small-scale projects can help to overcome localised food and water supply problems (e.g. Moyamba, Koshi Hills).
Economic factors are major obstacles to development in many poor countries.	• LEDCs suffer from shortages of capital and skilled workers, and the burden of debt owed to the developed world. • Many of the economic benefits of investment by TNCs in poor countries go to MEDCs. World trade operates against many poor countries, which rely on exporting primary products and importing manufactured goods from the developed world.
Rapid population growth hinders development.	• In many LEDCs, population growth is faster than economic growth. As a result, people on average get poorer. • The remedy is birth control and family planning.
Adult illiteracy and the low status of women are barriers to development in many LEDCs.	• Without an educated and skilled workforce, economic development on a large scale is impossible. In many poor countries, barely one-third of adults can read and write. For religious and social reasons, the education of women and girls is often ignored. • The economic success of Asia's NICs would not have been possible without enormous spending on education.
Political problems often hold back development.	• Political instability is a problem in many poor countries (especially in Africa). Violent changes of government, civil wars, corruption and money spent on weapons hinder development. Wars sometimes cause the breakdown of farming, creating famine and millions of refugees.
Environmental problems in the economically developing world contribute to poverty.	• The development of natural resources has often been unsustainable. • Deforestation, agribusiness, the rapid growth of tourism, etc. damage forests, soils, and landscapes (e.g. Dominican Republic). This increases the poverty of local people.
Diseases and pests are serious hazards in many LEDCs.	• Diseases are widespread in LEDCs in the tropics. They often reflect poor living conditions, inadequate diets and contaminated water supplies. Some of the most serious diseases are water-borne (e.g. diarrhoea, river blindness). Insect pests such as locusts destroy crops; others like the tsetse fly make cattle farming impossible over large areas.
Contrasts in wealth exist at a regional scale.	• Regional contrasts in wealth exist in both MEDCs and LEDCs. Sometimes these contrasts result from differences in resources between regions. Sometimes they result from relative location. • Core regions (e.g. northern Italy) develop a virtuous circle of growth bringing prosperity. • Peripheral regions often suffer from their remoteness. This may limit investment and cause growth to lag behind the core.

9 Trade and aid

Table 9.1 The growth of world trade ($bn), 1950–99

1950	113
1960	232
1970	637
1980	4095
1990	6865
1993	7480
1999	11330

9.1 Introduction

Since 1970, there has been a huge expansion of world trade (Table 9.1). Ideally, such expansion should benefit everyone (Fig. 9.1), and in the long run it should help to narrow the gap between rich and poor countries. Unfortunately, not all countries have shared in this rapid growth of trade. In particular, it is the poorest countries that have gained least. Why is this? And what can be done to help these countries? These are questions we will try to answer in this chapter. But first we need to know a little more about trade.

EXERCISES

1a Draw a line graph to show the growth of world trade from 1950 to 1999 (Table 9.1).
b From your graph, try to predict the level of world trade in 2000 and 2010.

Figure 9.1 Trade and prosperity

Top left: Lifting of trade barriers at Heathrow Airport's duty free, London

Top right: Manufacturing trainers in East Asia for export to MEDCs

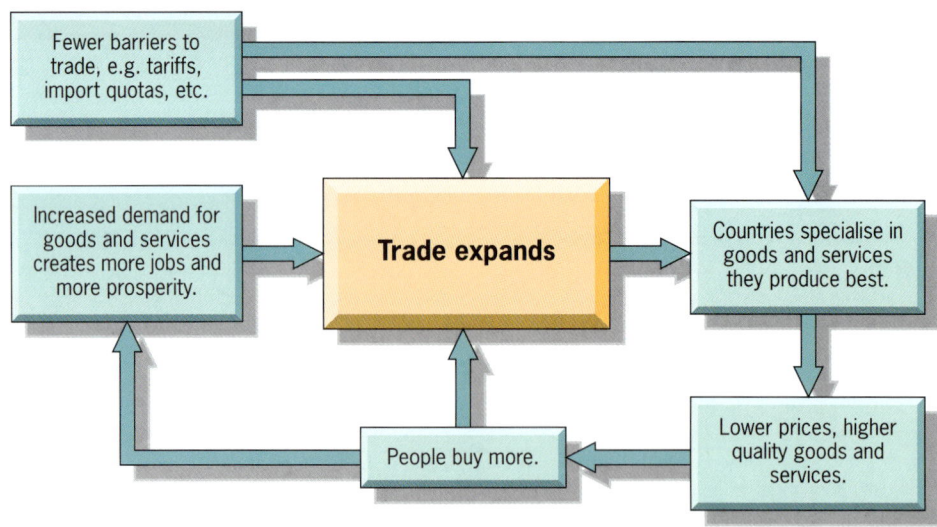

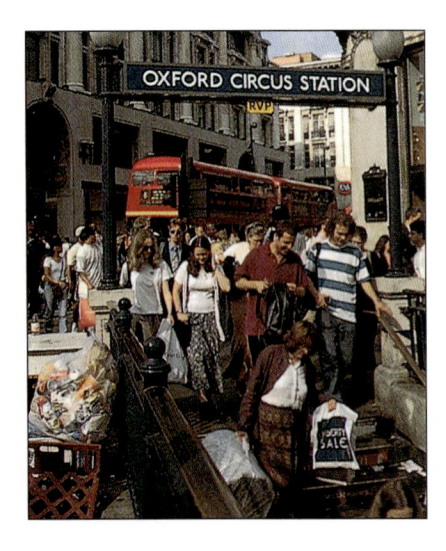

Bottom left: Increased demand for goods means more shoppers, London

Bottom right: Lower prices result in more imports, Nigeria

9 Trade and aid

Figure 9.2 Value of trade per person, 1999

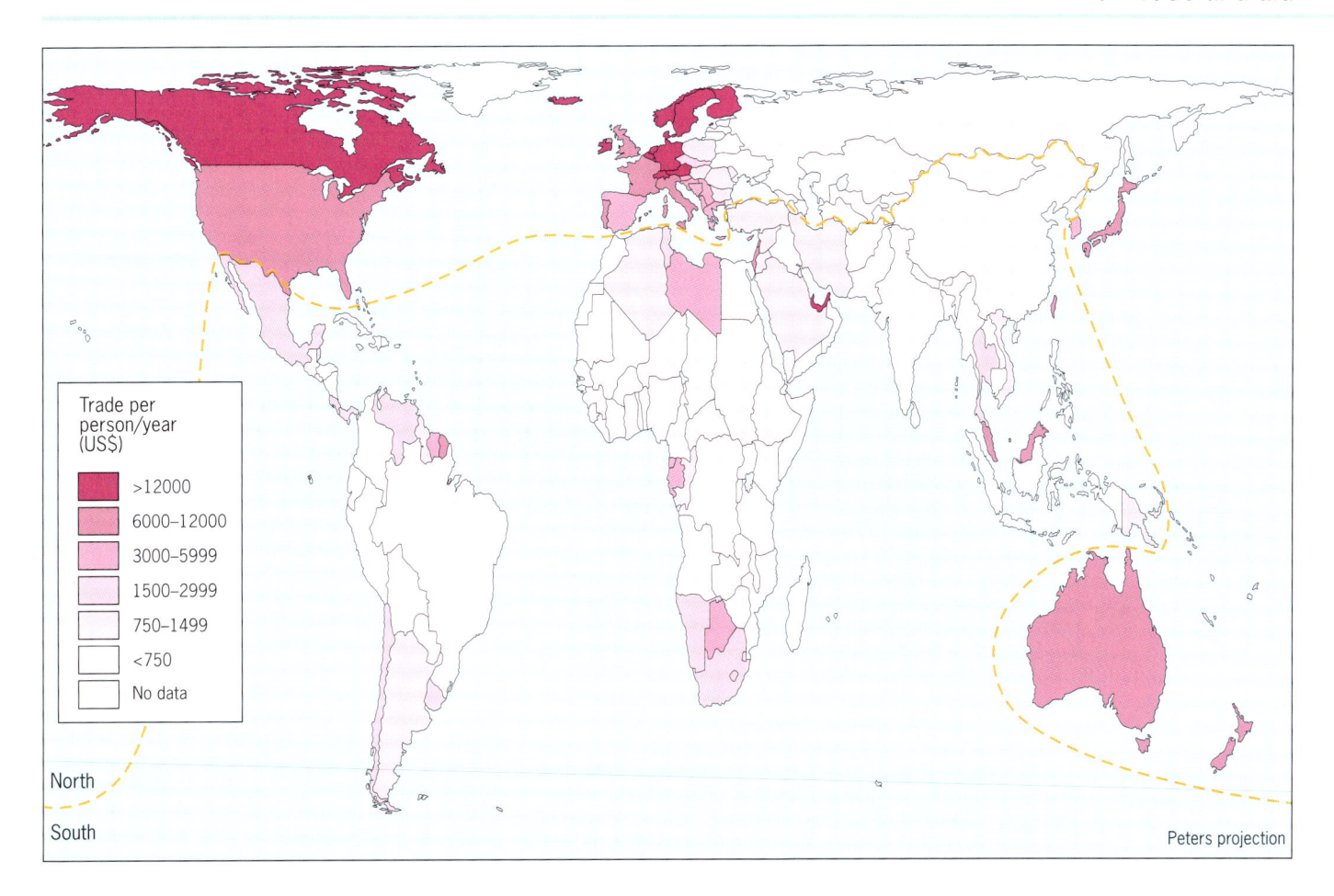

9.2 What is international trade?

International trade is the exchange of goods and services between countries. The goods and services that countries sell abroad are **exports**: those bought from other countries are **imports**. The difference in value between a country's exports and imports is known as the **balance of trade** (Table 9.2).

Table 9.2 International trade of three countries

	UK	Sweden	Indonesia
Imports: total value ($bn)	318	68.5	24
Imports: value per person ($)	5345	7784	107
Exports: total value ($bn)	268.2	84.8	49
Exports: value per person ($)	4507	9545	218
Trade: total value per person ($)	9852	17329	325
Population (millions)	59.5	8.8	224.8

9.3 Influences on international trade

There are many influences on international trade (Fig. 9.3). Some, such as foreign investment and free trade agreements, help to increase the volume of trade. Others, such as tariffs and quotas, have the opposite effect. In this section we will look at some of these influences in more detail.

EXERCISES

2* Study Figure 9.2. Describe how, on a global scale, the amount of trade per person is related to economic prosperity (see Figure 8.1).

3 Study Table 9.2.
a Calculate the trade balances for the UK, Sweden and Indonesia.
b Using the data in Table 9.2, state and explain which country belongs to the economically developing world.
c* Which country appears to benefit most from international trade?
d* Why is the 'value per person' data more useful than the 'total value' data when assessing the importance of a country's trade?

161

9 Trade and aid

Figure 9.3 Influences on international trade

Diagram: Factors pointing to "International trade": Transport infrastructure, National government policies, Free trade agreements, Policies of trans-national corporations/globalisation, Tariffs, Commodity price agreements, Economic and trade groups

> **REMEMBER**
> The promotion of free trade, and the removal of tariffs, export subsidies etc. expands the volume of global trade, and benefits both MEDCs and LEDCs.

Trade barriers

Tariffs are taxes on imported goods. There are two main reasons why countries use tariffs.

- Tariffs make imports more expensive, and so protect a country's own industries from competition.
- Tariffs reduce the demand for foreign goods, and so cut imports and strengthen the balance of trade.

Since 1947, the General Agreement on Trade and Tariffs (GATT) has reduced tariffs and promoted **free trade** between countries. This has given world trade an enormous boost. In 1995, the World Trade Organisation (WTO) replaced the GATT. It has 140 member countries. All of them are committed to free trade and further reductions in tariffs.

Even the total abolition of tariffs will not automatically lead to free trade. Countries use other ways of blocking imports. These include imposing fixed limits, i.e. quotas, on specific goods; giving subsidies to home industries; and insisting that imports, e.g. cars, meet strict technical standards.

Trading groups: the European Union

Many countries have formed trading groups to promote trade between themselves (Fig. 9.4). The European Union (EU), with its 15 member states, is the largest and best-known group (Fig. 9.5). Within the EU, there is a single market. This means there are no tariff barriers to the movement of goods and services between member states. People and capital can also move freely within the EU.

As well as being a single market, the EU is also a customs union and, in 1999, 11 of its 15 states adopted a single currency (the Euro). All EU states have agreed a common tariff on goods and services imported from outside the EU. Why is this needed? The simple answer is to protect EU industries from foreign competition. Without protection, many industries could go bankrupt and tens of thousands of jobs would disappear.

9 Trade and aid

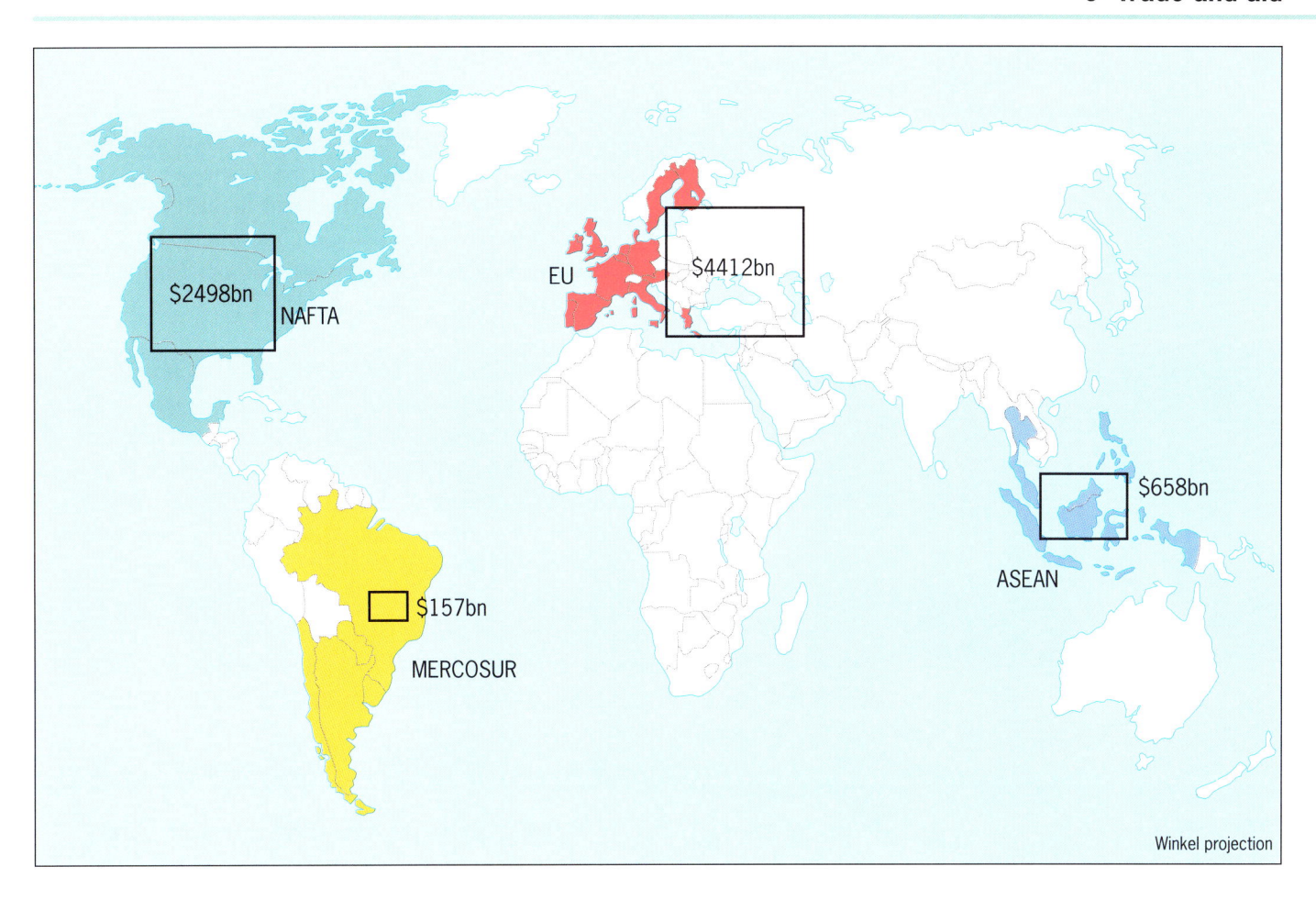

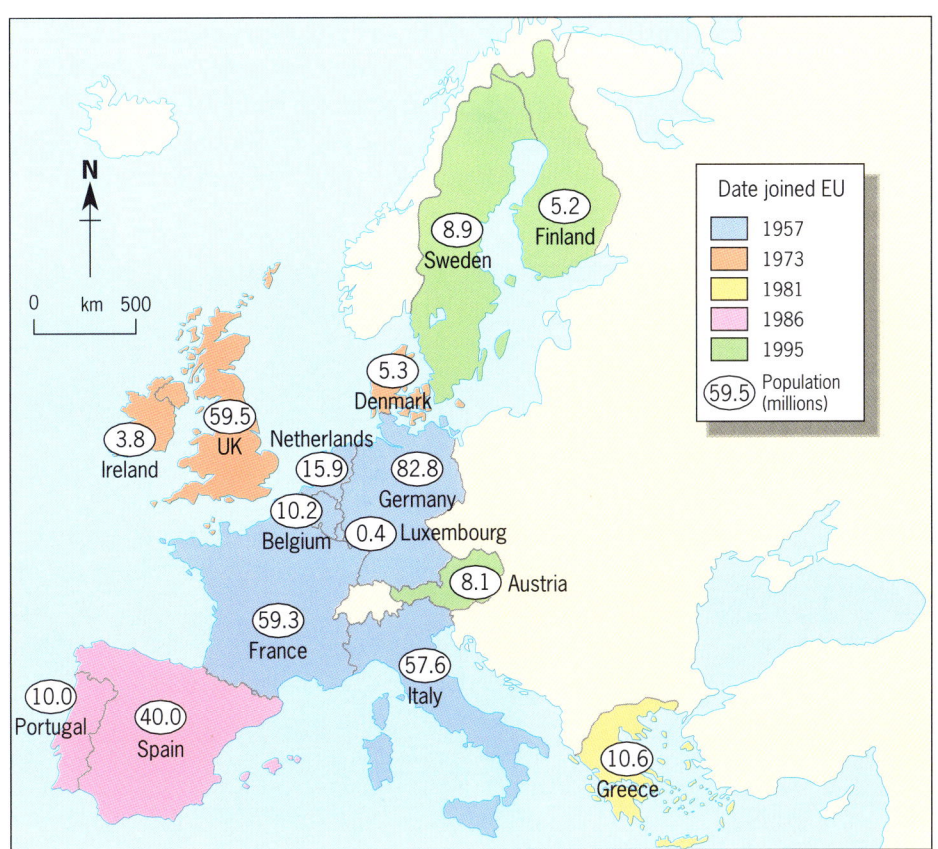

Figure 9.4 (above) Main regional trading groups and value of imports and exports, 1999

Figure 9.5 (left) The European Union

EXERCISES

4 Study Figure 9.5.
a In terms of population, how big is the common market in the EU?
b Draw a bar chart to show how the size of the common market grew between 1957 and 1995. (NB assume that the populations of EU countries have not changed during this period.)
c* Suggest one possible advantage of large markets to industry.

163

9 Trade and aid

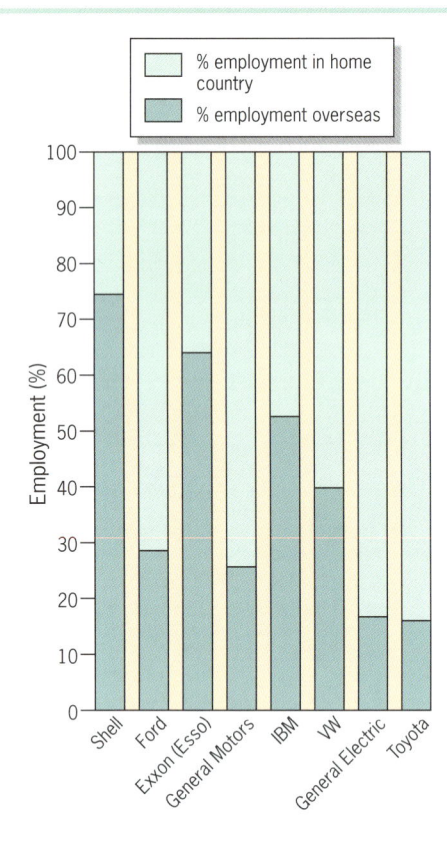

Figure 9.6 Top 8 TNCs: overseas employment

Trading groups: Mercosur and NAFTA

There are several other trade groups that are similar to the EU, including the North American Free Trade Association (NAFTA) and Mercosur (the common market of the South) (see Fig. 9.4). The USA, Canada and Mexico formed NAFTA in 1994. Established in 1991, Mercosur includes Brazil, Argentina, Uruguay, Paraguay and Chile (an associate member). Both NAFTA and Mercosur will eventually become free trade zones for member states. Mercosur hopes to go a step further and, like the EU, agree a common tariff on all imports.

Already, NAFTA and Mercosur have expanded trade between member states. Because of NAFTA, the USA and Canada have access to the Mexican market of 90 million people. Also, because Mexico's labour costs are only a fraction of those in the USA, the border region of Mexico with the USA has become a magnet for US-owned industries, as well as other trans-national corporations (TNCs) (Fig. 9.6). Located on the Mexican side of the border, these firms can easily supply the US market. In return, Mexico gets badly needed investment and thousands of jobs. Of course, there are some disadvantages. For example, Mexican maize farmers now have to compete with more efficient growers in the USA. And there are worries that millions of jobs in industry in the USA and Canada might disappear if firms decide to shift to cheaper locations in Mexico.

Globalisation

More and more TNCs are locating their factories worldwide, outside their country of origin. This **globalisation** of industry has a number of advantages for TNCs.
- They can profit from lower labour costs or cheaper materials.
- They can expand production and reduce costs (economies of scale).
- They have direct access to markets that might otherwise be protected by tariffs and other trade barriers.

Because of globalisation, about 40 per cent of all foreign trade takes place within these TNCs today.

EXERCISES

5a List the possible advantages and disadvantages of trading groups like the EU for member states.
b* State and explain the possible disadvantages of such trading groups for non-member states.
6a* The economic activities of Shell and Exxon (Esso) differ from the company shown in Figure 9.7. Suggest how.
b* Suggest one possible reason why these two TNCs depend so heavily on overseas investment.

Figure 9.7 Branch factory for manufacturing audio equipment on the Mexico-USA border

9 Trade and aid

Foreign direct investment

Most foreign investment by TNCs has gone to countries in the economically developed world. Europe, for example, took 40 per cent of all foreign investment in the early 1990s. None the less, in 1998, the rich countries still invested more than $170 billion in the economically developing world. This sort of investment is vital to poor countries. It spreads technology and skills, encourages economic development and provides jobs (Table 9.3).

Thanks to foreign investment, the importance of manufacturing will continue to grow in LEDCs. Nowhere is this more evident than in South America. In 1999, overseas companies invested $29 billion in Brazil. Much of this went to the automobile industry. Between 1997 and 2001, GM, Ford, Honda and Daimler Chrysler all opened new factories in Brazil.

Yet, while Brazil has attracted huge foreign investment, other countries, especially in Africa and South Asia, lag far behind. Lack of investment in these countries is one reason why they are the poorest in the world (Fig. 9.8).

Table 9.3 World shares of global manufacturing (percentage)

	MEDCs	LEDCs
1970	88	12
1980	83	17
1990	84	16
2000	80	20
2005	71	29

9.4 The pattern of world trade

The rich countries of the northern hemisphere dominate global trade (Fig. 9.9). Most of this trade takes place between rich countries. Yet, although trade between rich and poor countries is much smaller, it is growing rapidly.

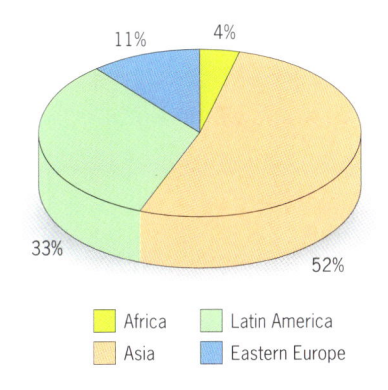

Figure 9.8 (above) Foreign direct investment in LEDCs, 1999

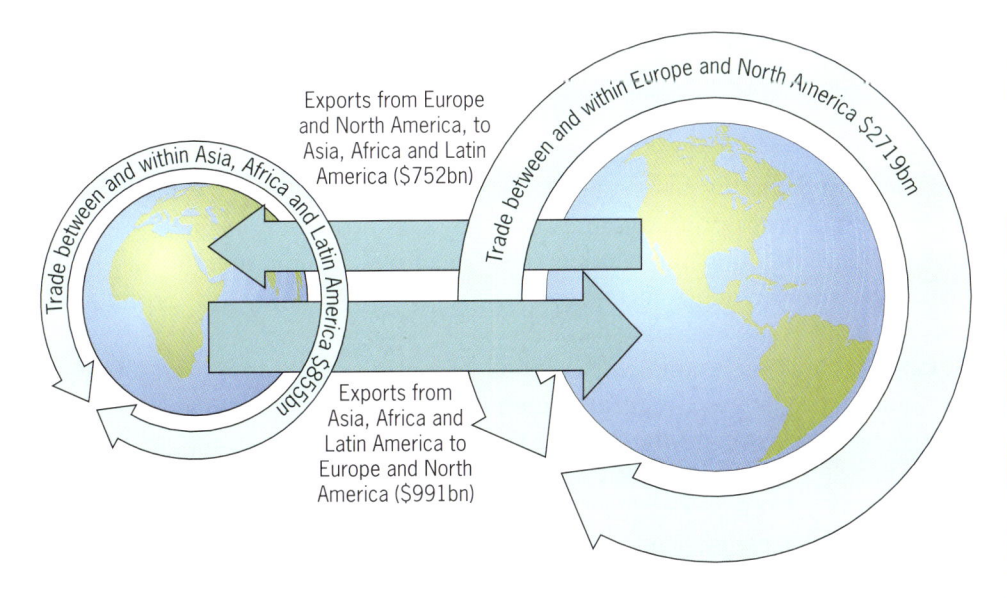

Figure 9.9 The pattern of international trade, 1999

Terms of trade: a fair exchange?

Recent industrial growth in South America, India and the Pacific Rim countries means that today, manufactured goods are important exports for a number of LEDCs. Even so, many of the world's poorest countries still depend heavily on exports of primary products, such as food and raw materials (Fig. 9.10). This has three main drawbacks.

EXERCISES

7 Refer back to Sections 8.7 and 8.10 and explain why some countries in the economically developing world are more attractive for foreign investment than others.

8* Suggest why the pattern of international trade in Fig. 9.9 does not give an entirely accurate picture of trade between and with the economically developed and economically developing world.

9 Trade and aid

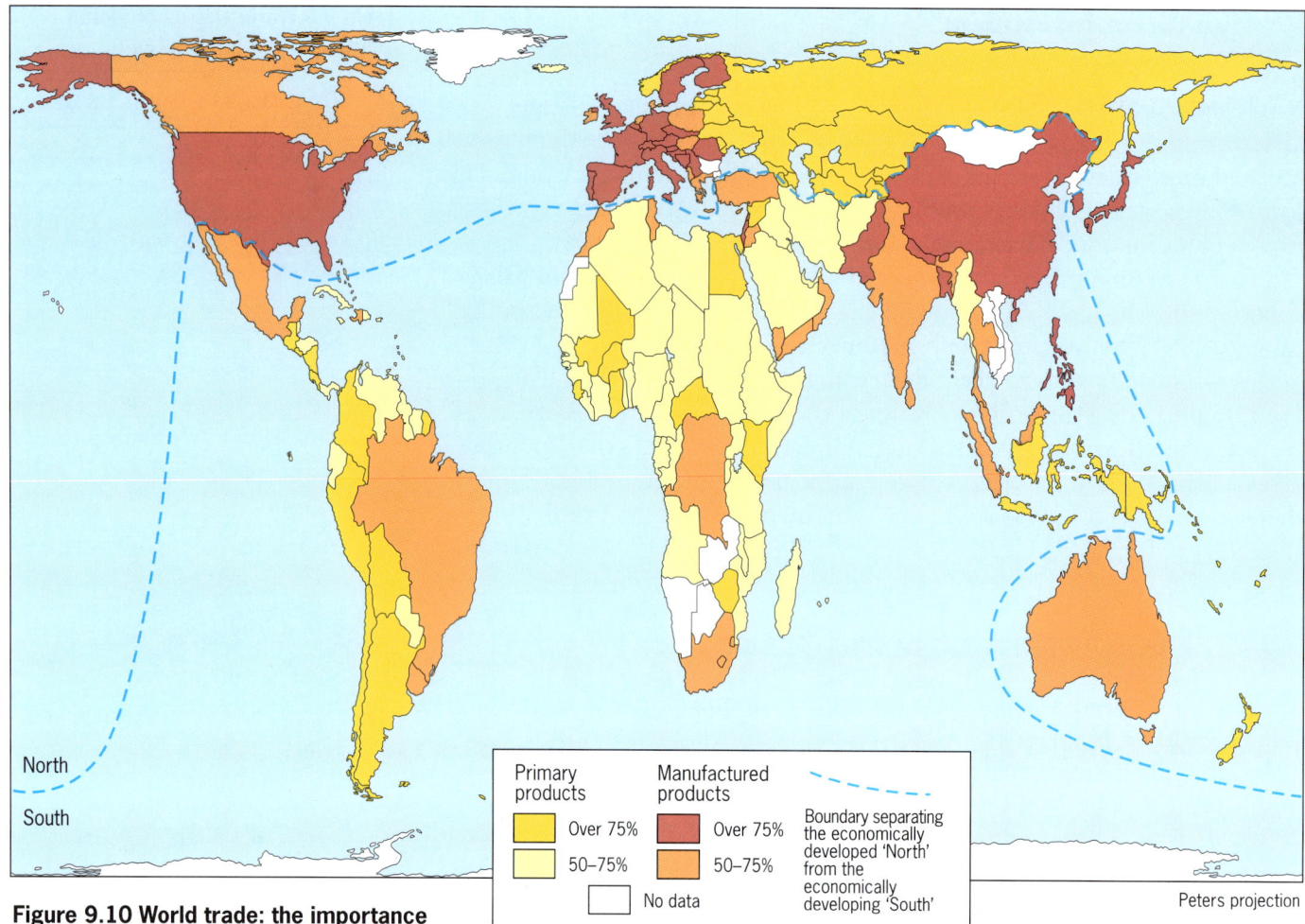

Figure 9.10 World trade: the importance of primary and manufactured goods

Legend:
- Primary products: Over 75% (yellow), 50–75% (pale yellow)
- Manufactured products: Over 75% (dark red), 50–75% (orange)
- No data (white)
- Boundary separating the economically developed 'North' from the economically developing 'South'

Peters projection

REMEMBER
Many LEDCs depend heavily on a narrow range of primary products (unprocessed food and raw materials) for export. Compared to manufacturing, the value of these goods has fallen, and their prices fluctuate wildly on the world market.

EXERCISES
9a Study Table 9.4 then match up each primary product to the appropriate value-added manufactured good.
b* Why do you think that the processing of raw materials occurs mainly in the economically developed world?

Value added

One problem is that primary products have little '**value added**'. Value is added when primary products are processed. This processing mainly takes place in MEDCs. Thus, the benefits of value added, and the millions of jobs it creates, go to the rich consuming countries rather than the poor producers (Table 9.4).

Coffee is a classic example. It is grown as a cash crop in the tropics. Only here can its specific growing requirements be met. These include an absence of frost, no month where the average temperature falls below 11°C, and rainfall within the range of 200–900 mm a year.

Table 9.4 Primary products and value-added goods

Primary products	Value-added goods
bauxite	petroleum
rutile	shoes
cocoa	aluminium
crude oil	chocolate
leather	titanium
rubber	wood veneer
timber	cigarettes
tobacco	tyres

9 Trade and aid

Figure 9.11 Picking coffee beans, Kenya

Figure 9.12 Value added on a jar of coffee

Coffee

Coffee is the second most valuable commodity that enters international trade (after petroleum). Poor countries, like Uganda, Ethiopia and El Salvador, depend on it for half of all their export earnings (Figs. 9.11, 9.14 and 9.17). And yet, for every jar of instant coffee sold in the UK, the producers receive less than one-third of the price we pay our local supermarket (Fig. 9.12). Food processing, food packaging, advertising and retailing (more than two-thirds of the value of the product) are all done in MEDCs.

World commodity prices

There is another problem for countries that rely heavily on primary products. World prices for these commodities often fluctuate wildly on international markets (Table 9.5, Fig. 9.13). If prices fall, countries get less for their exports. This makes it difficult for them to pay for both imports and the interest on foreign loans.

Cost of imports

Relying on the export of primary products has one other disadvantage. Compared to manufactured goods, the value of primary products has fallen sharply in the last 20 years. For instance, in 1986 a kilo of coffee bought almost twice as many manufactured goods as in 2000. As most imports to poor countries are manufactured goods, we can see how the terms of trade have worked against LEDCs.

Table 9.5 Coffee price changes, 1980–2000

Year	Price (US cents/kg)
1980	331
1981	254
1982	275
1983	282
1984	311
1985	293
1986	376
1987	237
1988	255
1989	202
1990	157
1991	147
1992	117
1993	136
1994	296
1995	305
1996	257
1997	407
1998	286
1999	227
2000	198

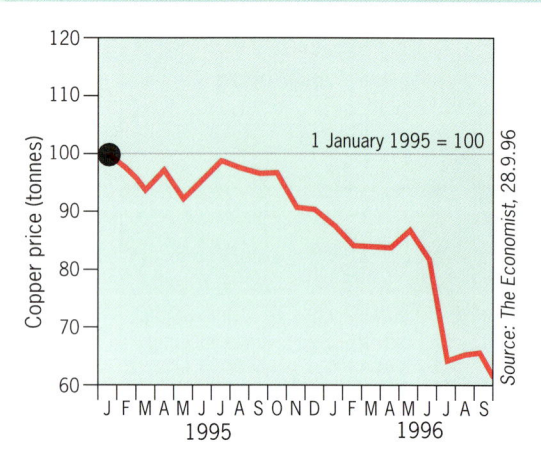

9 Trade and aid

Figure 9.13 (above) Copper prices, January 1995–September 1996

Figure 9.14 (right) Exports from selected African countries

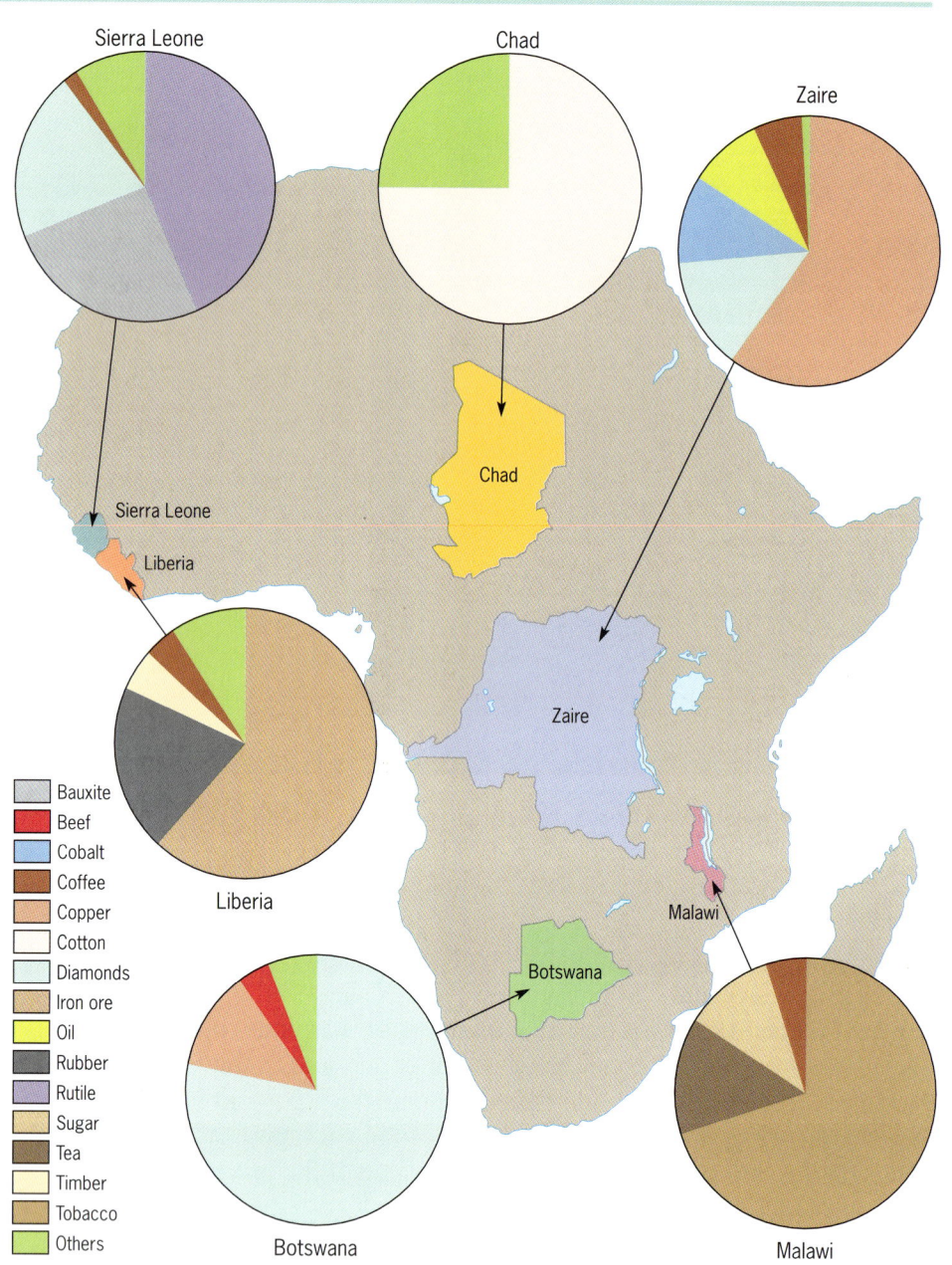

EXERCISES

10a Describe the price changes for copper between January 1995 and September 1996 (Fig. 9.13).

b* Suggest two possible effects of these price changes on a country like Zaire (Fig. 9.14) that relies heavily on copper exports.

c Plot the changing price of coffee from 1980–2000 (Table 9.5) as a line graph.

d Suppose a country's only export was coffee. How many times greater would its export earnings be in the best year, compared to the worst year, in the period 1980–2000?

EXERCISES

11a Describe the pattern of exports of the six African countries in Figure 9.14.

b* Explain how this pattern of exports could possibly create disadvantages for these countries.

The exports of the African countries in Figure 9.14 are not typical of all countries in the economically developing world (see Fig. 9.10). In fact, manufactured goods account for more than half of the value of exports from LEDCs. Although low-tech manufactured products, such as textiles, clothing and metals, feature prominently, high-tech exports are becoming more important. Brazil, for example, has modern export industries that include chemicals, cars, electronic equipment and even aircraft (Table 9.6).

9.5 Foreign aid

For a variety of reasons, including poverty and contrasts in levels of development (see Chapter 8), many countries in the economically developing world depend on aid. The main purpose of this aid is to help reduce inequalities.

9 Trade and aid

Foreign aid is the transfer of money, food, equipment and technical assistance from rich countries to poor countries. The United Nations (UN) recommends that rich countries should give 0.7 per cent of their GNP in foreign aid to poor countries. In fact, few countries achieve this target. Take the UK: although in 1998 it gave foreign aid worth £3 billion, this was only 0.28 per cent of its GNP.

Table 9.6 Brazil's leading exports

Industry	65%
Motor vehicles	14.9%
Metal products	10.6%
Chemicals	7.2%
Other industry	32.3%
Non-industry	**35%**
Soya beans	9.3%
Coffee	7.0%
Metal ores	7.0%
Other non-industry	11.7%

Development aid given by the richest countries (Dollars per person/year; figures in brackets show aid as a % of GNP):

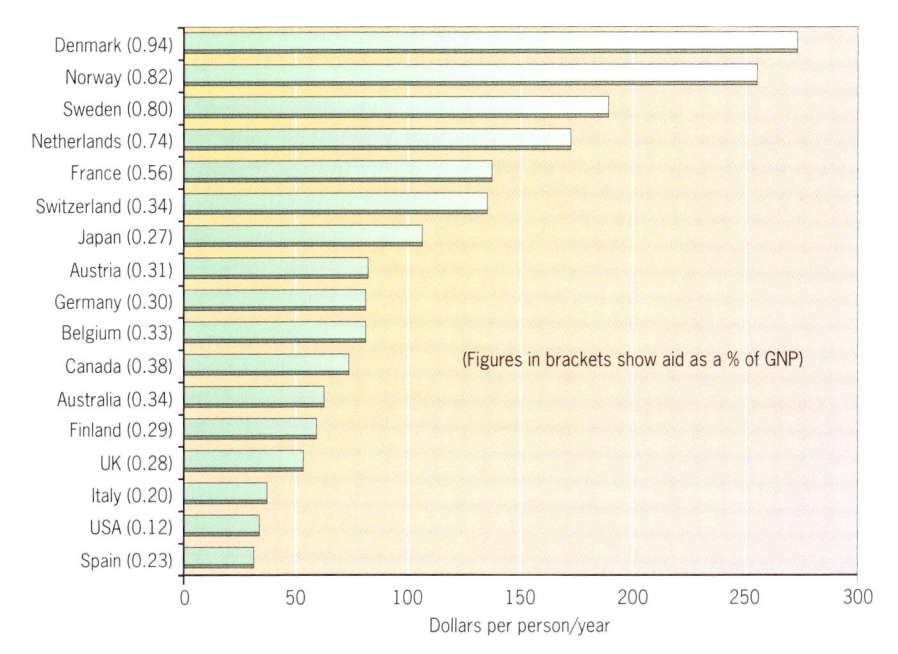

- Denmark (0.94)
- Norway (0.82)
- Sweden (0.80)
- Netherlands (0.74)
- France (0.56)
- Switzerland (0.34)
- Japan (0.27)
- Austria (0.31)
- Germany (0.30)
- Belgium (0.33)
- Canada (0.38)
- Australia (0.34)
- Finland (0.29)
- UK (0.28)
- Italy (0.20)
- USA (0.12)
- Spain (0.23)

Figure 9.15 Development aid given by the richest countries

9.6 Types of foreign aid

Foreign aid often hits the headlines after natural disasters, such as famines, earthquakes and floods. In such emergencies, governments, charities and international organisations like the World Bank and the Red Cross provide **short-term aid**. Typically, this aid consists of money, medicines, food, blankets, tents and other equipment to help people through the immediate crisis (Fig. 9.16).

Other types of aid are **long term**. Such aid is for economic development, and involves encouraging industry, agriculture or energy production. Long-term aid may also be for social development, e.g. health care, family planning, education, etc. The purpose of long-term aid is to improve the quality of life of people living in the economically developing world.

EXERCISES

12a Draw a pie chart to show the information in Table 9.6.
b Which exports in Table 9.6 have significant value added, and which have little or no value added?
c* Compare Brazil's exports with those of the six African countries in Figure 9.14.
13 Study Figure 9.15.
a Which countries achieved the UN target for development aid?
b Which countries are most and least generous in donating aid?
c* Draw a scattergraph of aid given in dollars per person per year, and aid given as a percentage of GNP.
d Comment on what your graph shows.

REMEMBER
Very few countries achieve the UN target for foreign (development) aid – 0.7% of GNP. Foreign aid is often more beneficial for donor countries than for recipient countries.

Figure 9.16 Red Cross workers give emergency supplies and medical aid to earthquake survivors, Mexico 1985

9 Trade and aid

Figure 9.17 Advertisement for a non-governmental organisation

Types of long-term aid

- Aid given by one country directly to another is known as **bilateral aid**. The British government's support for improvements in water supply in Sierra Leone and agriculture in Nepal (see Chapter 8) are both examples of bilateral aid.
- **Multilateral aid** is assistance given to poor countries through international bodies such as the UN, the World Bank, and the International Monetary Fund. Each organisation has its own aid programme. Funding comes from economically developed countries.

Charities such as Oxfam, Christian Aid and Save the Children are also important sources of aid (Fig. 9.17). These are **non-governmental organisations** that rely on private donations and gifts from businesses. Apart from giving emergency aid, their aid programmes are often small scale. There is an emphasis on low-cost schemes, based on simple technology and using local knowledge and skills (see Book 1, Section 9.2).

9.7 Foreign aid: who benefits?

Why do rich countries give foreign aid? The most obvious answer is for humanitarian reasons: to help people who live in poverty and misery (Fig. 9.18). However, this is not the only reason.

Tied aid

It may surprise you to know that about half of all foreign aid is given for selfish reasons. For example, if the UK gives money to a poor country, it may insist that the money is spent only on British goods. This is **tied aid**. Such aid often benefits the donor country.

- It boosts exports and provides jobs for its workers.
- It is a way of supporting the donor country's own firms or farmers.
- It may subsidise the donor's own industries in order to protect jobs.
  Tied aid can also cause problems for LEDCs.
- It increases the **dependency** of countries on MEDCs.
- It may discourage poor countries from developing their own industries, and may undermine existing industries and agriculture.
- Donors often pay for inappropriate schemes that require their exports and expertise (e.g. expensive dam projects), whereas smaller, low-technology projects would be more effective.

Corruption and inefficiency

If foreign aid does not always benefit poor countries, neither does it always reach the poorest people in the developing world. Misuse of foreign aid in LEDCs is widespread. All too often, aid finds its way into the pockets of politicians, government officials and the better-off. Thus, millions of poor people get little benefit from foreign aid.

Foreign aid is often wasted by incompetent governments. Giving aid to these countries may do little to help the poor. Rather than give cash donations, the poor in these countries are better served by small-scale projects that promote literacy, appropriate technology and so on.

9 Trade and aid

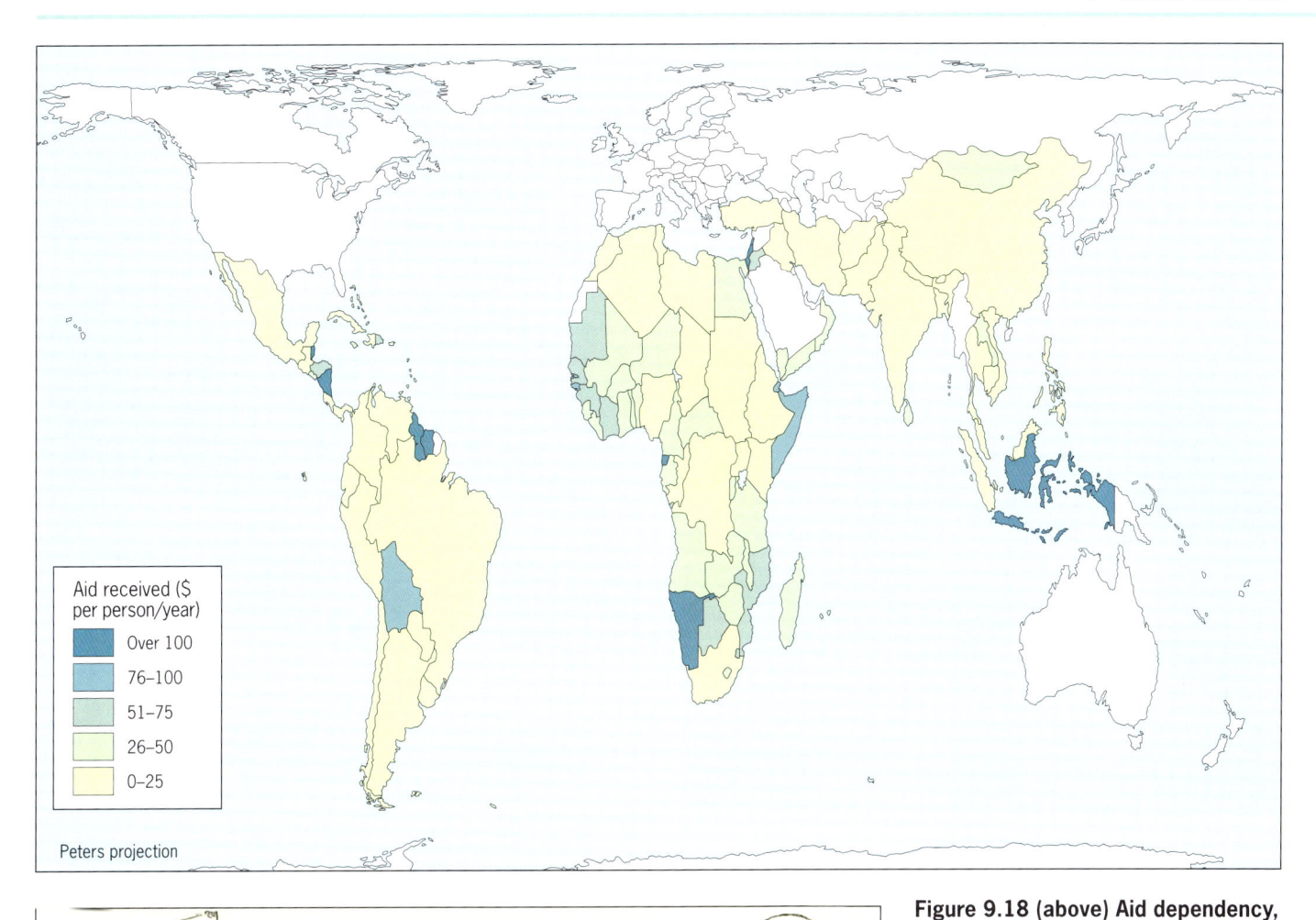

Figure 9.18 (above) Aid dependency, 1998

Figure 9.19 (left) What price development?

9.8 Long-term aid projects

In this section, we will look at two contrasting aid projects funded by the Department For International Development (DFID) – the UK government agency that is responsible for the country's official aid programme.

EXERCISES

14 Study Figure 9.18 and compare it with Figure 8.10.
a Does foreign aid go to the poorest countries?
b* Suggest possible reasons why some poor countries may be refused foreign aid.
15a Suggest two reasons why the farmer in Figure 9.19 might benefit little from the donation of foreign aid.
b Explain how the donor country might benefit from the aid given to the farmer in Figure 9.19.

9 Trade and aid

CASE STUDY

9.9 Large-scale aid: Ilisu dam project

The Ilisu dam in Anatolia, South-east Turkey, is an example of tied aid. The project will generate millions of pounds worth of business for engineering and constructional companies in MEDCs.

FACTFILE

- The Ilisu dam is sited on the River Tigris, 65 km from the border of Iraq and Syria, in South-east Turkey (Fig. 9.21).
- The dam is due to be completed in 2006 and will cost £1.25 billion.
- It will create a reservoir 200 km long, with a surface area of 300 sq km.
- The Ilisu dam will generate 1300 MW of hydroelectric power.
- The Ilisu is just one scheme in a huge project to develop the hydroelectric and agricultural resources of South-east Anatolia (Fig. 9.21). When complete, the project will provide 7000–8000 MW of electricity and irrigate 17 600 sq km of farmland.
- The Ilisu project relies on the financial backing of several MEDCs, including the UK, Germany, the USA and Japan.
- The UK government will give export guarantees worth £200 million to underwrite the project.

Figure 9.20 (above) Medieval city of Hasankeyf, Turkey, destined to be flooded when the Ilisu dam is built

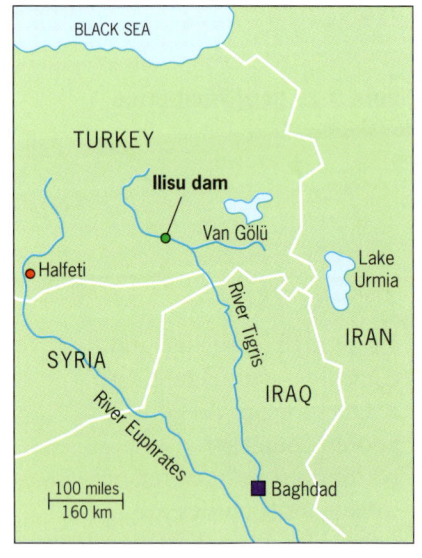

Figure 9.21 (above) Location of Ilisu and Halfeti in South-east Turkey

Figure 9.22 (right) Villagers left high and dry by the tide of progress, *The Guardian*, 13.07.00

Villagers left high and dry by tide of progress

"They could have waited until we had picked the pistachios one more time', saya Ahmet Erdogan, "it would have been a good crop this year". He is standing by a new man-made lake, where the rising waters have just submerged the five domes of Halfeti's historic Turkish bath.

Down the road his neighbours are hard at work knocking down their house for salvage, to sell wood and concrete for scrap. The pistachio groves disappeared weeks ago and soon half of this village, which has been a functioning settlement since 1000BC, will be under water.

As controversy mounts over the British government's plans to back the construction of the Ilisu dam, other dams are already rearranging the topography of south-eastern Turkey and forcing thousands of people out of their homes.

Halfeti is the village where the Kurdish rebel leader Abdullah Ocalan was born. It clings to the side of a valley which the Euphrates river cuts through the region. People who have been displaced are being resettled a few miles away in a new village on the barren plain.

Most people did not want to move at all, but that was not an option. A random sample of villagers produces complaints of a lack of consultation, of insufficient compensation and of a general feeling that they are losing much more than they stand to gain.

"It's not going to be like this anymore," says Celal Guneri, casting his eye over the old village. "There are nice gardens and trees here. Up there it's dry, there's no work – it's all artificial".

The new Halfeti is indeed a regimented place – row after row of identical concrete structures in a neat grid on the stony brown earth. Perhaps it just needs time to develop, but at the moment the jumbled feeling of a real riverside community has been replaced by a village without a soul.

The state rejects suggestions that old Kurdish communities are being deliberately broken up, to allow central government to gain better control over an unruly region.

Many Kurdish pressure groups are convinced the allegation is true, but local villagers have more pressing economic needs.

Supporters of the big dam projects argue that the disappearance of old village communities is a necessary price of progress, and that some things have to be sacrificed for the greater good. With or without Ilisu, south-eastern Turkey is already in a cycle of irreversible change.

9 Trade and aid

Who benefits?

Large-scale aid projects in LEDCs based on the construction of dams, are often controversial. Ilisu is no exception. Critics of the project point out that Ilisu has major environmental, economic, social and political disadvantages.
- The dam will reduce downstream flows of the River Tigris by 10 per cent.
- The River Tigris has its headwaters in Turkey, but the downstream states of Iraq and Syria, which rely on its water, will lose valuable resources. This could lead to international conflict.
- There is a danger that the reservoir behind the dam could increase the spread of water-borne diseases such as malaria and leishmaniosis.
- Large areas of productive farmland in the Tigris Valley will be flooded by the reservoir.
- Up to 78 000 people living in the Tigris Valley will be displaced when the reservoir floods 80 villages and towns. Many of these people are landless and will receive no compensation. It is argued that the forcible eviction of these people (who are minority Kurds, currently fighting for a homeland in South-east Turkey) is a violation of human rights.
- Important archaeological sites will be flooded, including the town of Hasankeyf, which is the centre of Kurdish culture.

> *EXERCISES*
>
> **16** In your view, who will benefit most and who will benefit least from the Ilisu project? Justify your answer.
>
> **17** Search the internet for the latest up-date on the Ilisu dam. Then prepare a statement on the project for debate in class that represents the views of one of the following interested parties: the Turkish government, the governments of Iraq and Syria, the Kurdish minority.
>
> **18** Read the newspaper article (Fig. 9.22) on the dam at Halfeti on the Euphrates River in South-east Turkey, near Ilisu.
> **a** Describe the impact of the dam on the lives of the people in Halfeti.
> **b** Explain why local people resent being resettled.
> **c** What arguments are put forward to justify big dam projects?

CASE STUDY

9.10 Small-scale aid: Ghana

Unlike Ilisu, most of the aid given by the DFID is for small projects that use modest amounts of money. Typical examples are the small-scale development projects currently being supported by the DFID in Ghana. These projects cost about £18 million a year (Fig. 9.23).

Local involvement

Ghana is a poor country in West Africa (Fig. 9.23). In the 1970s and 1980s, it suffered a steep decline in its trade. Although its economy improved during the 1990s, not everyone has benefited. The DFID's aid programme for Ghana specifically targets the poorest groups in the country and includes health care, education, agriculture and industry. The emphasis is on the involvement of local people. The projects aim to make small but significant improvements in people's quality of life. This so-called 'bottom up' approach, is favoured by charities (see Book 1, Chapter 9). Local people work to improve roads, repair public buildings, dig wells and so on. The technology used is simple, and the skills needed are those that the people already have.

> **REMEMBER**
> Foreign aid is often most effective (i.e. most benefits the poor) when it promotes small-scale projects that involve local people and are sustainable. Large-scale projects (e.g. dams) often impose unacceptable costs on ordinary people in LEDCs.

9 Trade and aid

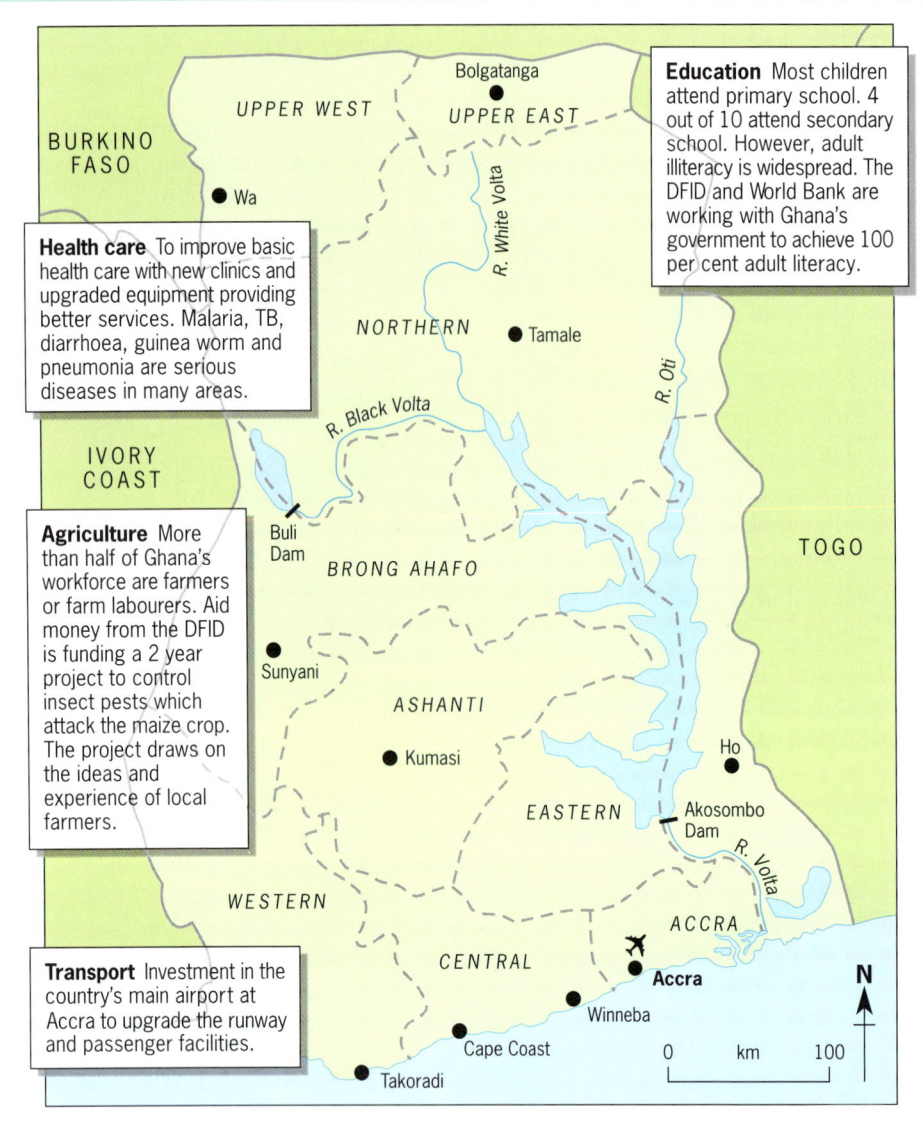

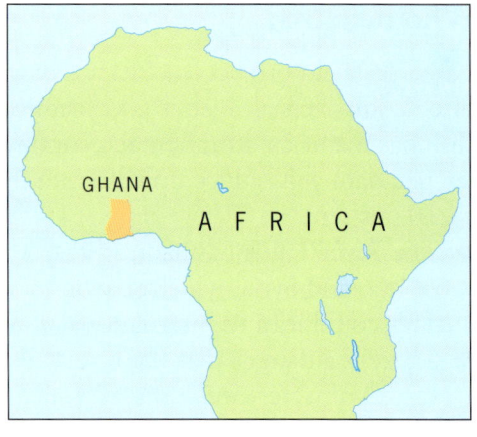

Health care To improve basic health care with new clinics and upgraded equipment providing better services. Malaria, TB, diarrhoea, guinea worm and pneumonia are serious diseases in many areas.

Education Most children attend primary school. 4 out of 10 attend secondary school. However, adult illiteracy is widespread. The DFID and World Bank are working with Ghana's government to achieve 100 per cent adult literacy.

Agriculture More than half of Ghana's workforce are farmers or farm labourers. Aid money from the DFID is funding a 2 year project to control insect pests which attack the maize crop. The project draws on the ideas and experience of local farmers.

Transport Investment in the country's main airport at Accra to upgrade the runway and passenger facilities.

Training

Education is an important part of the Ghanaian aid programme. One scheme sponsored by the DFID aims to provide basic literacy for 15 000 people (Fig. 9.24). Foreign aid provides educational experts and local training for teaching staff, as well as vehicles, materials and equipment.

The DFID's literacy programme is part of a much larger education project undertaken by the Ghanaian government and sponsored by the World Bank. It aims to make basic literacy universal in Ghana by the early 21st century.

Figure 9.23 (above) The DFID's aid projects in Ghana

Figure 9.24 (right) Literacy lesson in a boys' school, Koko Village, Ghana

EXERCISES

19 Read Sections 9.5 to 9.8.
a Make a list of the advantages and disadvantages of foreign aid for rich donor countries and for poor recipient countries.
b* Should rich countries give aid to poor countries? What is your opinion? Support your view with carefully reasoned arguments.

9 Trade and aid

9.11 Interdependence

Increasingly, the developed countries of the North and the developing countries of the South depend on each other. We have seen how trade, investment, foreign aid, loans, interest payments and international migration (Chapter 5) link rich and poor (Fig. 9.25). But who benefits from this **interdependence**? And is the relationship fair or one-sided?

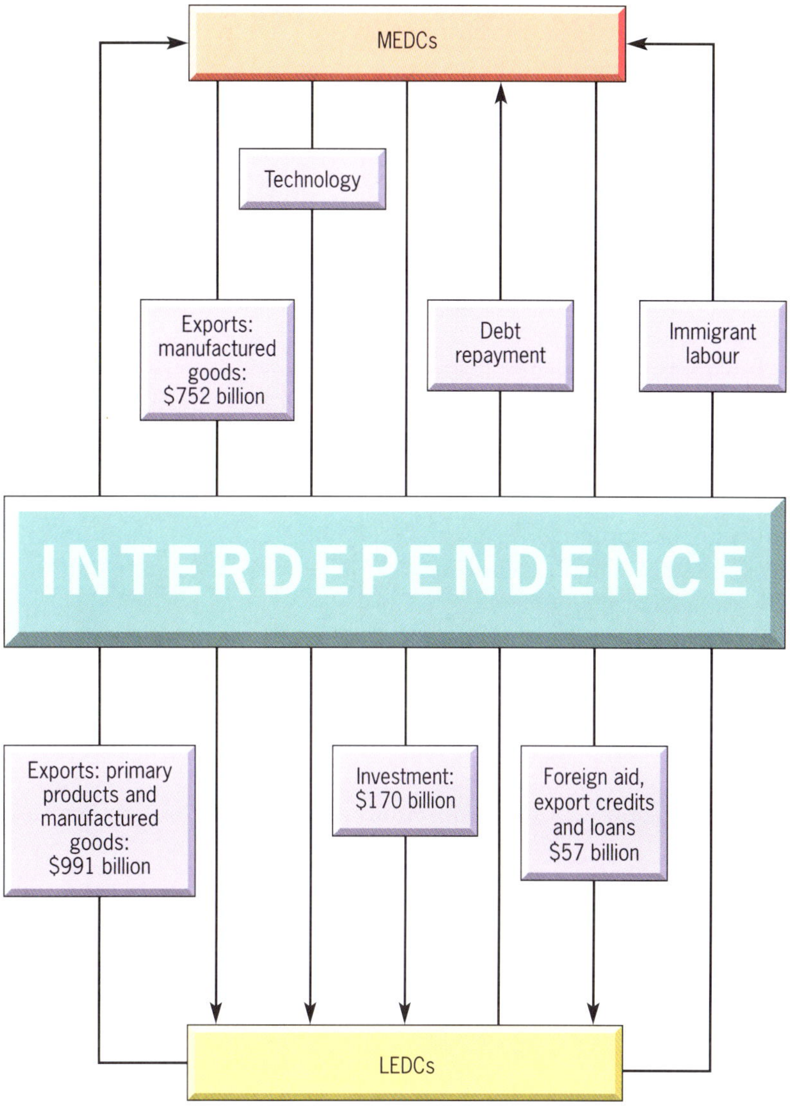

Interdependence and MEDCs

Advantages

MEDC industry depends on raw materials, many of which come from the developing world. These imported materials have become cheaper compared to the manufactured goods exported by the MEDCs. Moreover, as we have seen, most of the profit made through the processing and manufacturing of these materials (i.e. value added) goes to the rich countries. Further profits flow to the developed world from investments in economically developing countries, and from interest payments on loans. Even foreign aid benefits rich countries: tied aid boosts exports and secures thousands of jobs in MEDCs.

Disadvantages

None the less, interdependence does have some disadvantages for rich countries. As TNCs invest in the developing world, jobs may be lost in rich countries. And competition from new industries in the developing world could destroy many industries in rich countries.

Interdependence and LEDCs

Advantages

By exporting goods to the developed world, poor countries receive hard currency to pay for imports and loans. Investment by foreign TNCs provides much-needed jobs, improves skills, allows poor countries to exploit their advantage of cheap labour and brings in modern technology. Poor countries also receive foreign aid from rich countries and from the UN, World Bank, International Monetary Fund and other global institutions.

Disadvantages

However, interdependence is not always a good thing. The terms of trade often work against poor countries and, in the past, the exploitation of raw materials by foreign companies has sometimes been at the expense of the environment.

Figure 9.25 Interdependence between the developed and developing worlds, 1998

EXERCISES

20* Interdependence suggests that changes in the economically developed world will have knock-on effects in the economically developing world, and vice versa. Suppose there is a downturn in the economies of rich countries. Study Figure 9.25 and suggest what effects this might have on countries in the economically developing world.

9 Trade and aid

EXERCISES

21 Select one of the following roles: a trade minister for the government of an economically developing country; the chief executive of a TNC that has invested heavily in the economically developing world; the director of a charity such as Oxfam; an official for the World Bank that makes cheap loans to poor countries. In your chosen role, state whether you believe that the balance of interdependence between rich and poor countries is fair or one-sided. Support your view with detailed arguments.

Shell Nigeria and interdependence

FACTFILE

- Shell produces 2 million barrels of oil a day in Nigeria, most of it for export.
- Shell makes $5 on a typical barrel of oil worth $17.
- $12 out of every $17 for a typical barrel of oil goes to the Nigerian government.
- Shell contributes 40 per cent of Nigeria's export earnings.
- Shell employs 10 000 people in Nigeria, 95 per cent of whom are Nigerians.
- The Nigerian government has a 55 per cent stake in Shell Nigeria.
- Shell spends $20 million a year in Nigeria on community development, health, education, infrastructure projects and training schemes.

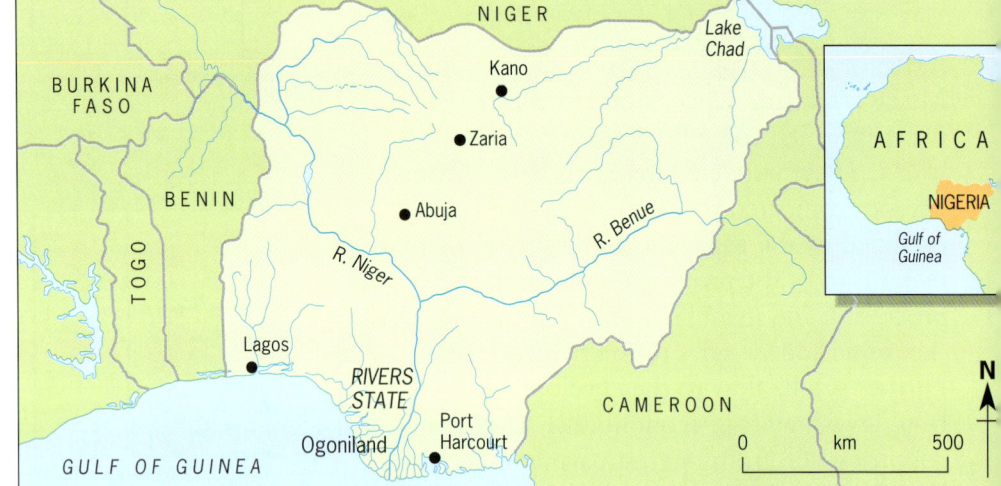

Figure 9.26 (above) Nigeria and Ogoniland

Environmental damage

In Nigeria, massive environmental damage occurred in the oil-rich province of Ogoniland in the Niger Delta in the 1990s (Fig. 9.26). Much of this damage has been blamed on the oil industry and, in particular, on Shell, the Anglo-Dutch oil company that produces half of Nigeria's oil.
Oil is vital to Nigeria's economy. It provides 80 per cent of the government's revenues. For this reason, it has been suggested that the government has tolerated oil spillages and gas flaring in Ogoniland.

The oil industry has also been blamed for the decline in fish stocks, soil acidification and water pollution. On the other hand, a World Bank report in 1995 said that some environmental problems in the Niger, such as loss of biodiversity, the depletion of fish stocks and deforestation, were due to rapid population growth and the construction of dams, rather than oil pollution.

Oil and the Ogoni people

The Ogoni people receive few benefits from the oil produced on their land. Since the late 1980s, they have waged a campaign against Shell and the Nigerian government. About half of all oil spillages in the past five years were the result of sabotage. However, the oil revenues from Ogoniland go direct to the Nigerian government. The Shell company has no say in how or where these revenues are spent.

Figure 9.27 Oil leak destroys a mangrove swamp, Niger Delta, October 1995

9.12 Summary: Trade and aid

KEY SKILLS OPPORTUNITIES
C1.2: Ex. 13b; **C1/2.3:** Ex. 5, 18, 21; **C2.1:** Ex. 16, 17, 19b, 21; **C2.2:** Ex. 2, 10a, 19b; **N1/2.1:** Ex. 7, 8, 10b, 14a, 19, 20; **N1/2.2:** Ex. 1a, 3a, 4a, 4b, 10c, 12a, 13c; **N1/2.3:** Ex. 1b, 3b, 3c, 10d, 13d; **IT1/2.1:** Ex. 17.

Key ideas	Generalisations and detail
International trade is the exchange of goods and services between countries.	• Goods and services sold abroad are exports. Goods bought from other countries are imports. • The balance of trade is the difference in value between imports and exports.
The volume of world trade has expanded massively since 1970.	• This growth of trade has not been shared equally between countries. • The MEDCs dominate world trade. More than 80 per cent of world trade is between these rich countries.
Some factors encourage trade, others discourage it.	• Trade expands when barriers such as tariffs, duties and quotas are removed. The free movement of goods and services between countries is known as 'free trade'.
In the last 50 years, there has been a general reduction in barriers to trade.	• Trade barriers, such as tariffs, protect a country's industries from competition. • Since 1947, the GATT has worked to reduce barriers to trade. Its successor, the WTO, is also committed to tariff reduction and the promotion of free trade.
Many countries have formed large trading groups.	• Large trading groups, such as the EU, NAFTA and Mercosur, aim to remove all internal barriers to trade between member states and create single markets. The EU also has a common tariff on goods and services imported from non-EU states.
Two-fifths of all international trade takes place within large trans-national companies (TNCs).	• Overseas investments by TNCs, e.g. Ford, IBM, are a major influence on global trade. • Globalisation of industry means that firms locate production worldwide to take advantage of lower costs (e.g. cheaper labour in LEDCs).
The terms of trade work against poor countries.	• Most poor countries rely heavily on exports of primary products (food and raw materials), and import manufactured goods. This pattern of trade has several disadvantages: primary products have little 'value added'; prices for primary products fluctuate wildly; the value of primary products, compared to manufactured goods, has fallen sharply in the last 25 years.
Foreign, or development, aid is the transfer of money, food, equipment and technical assistance from rich to poor countries.	• Rich countries normally give between 0.2 and 1.2 per cent of their GNP in foreign aid. • Few rich countries give the UN's recommendation of 0.7 per cent.
There are several different types of foreign, or development, aid.	• Bilateral aid is given directly from one country to another. • Multilateral aid is given through an international organisation, e.g. UN or World Bank. • Non-governmental organisations are charities that give aid from private donations. • Foreign aid may be short term (providing immediate disaster relief) or long term. • Long-term aid aims to promote economic development (encourage industry, agriculture, energy production, etc.) and social development (education, health care, family planning, etc.). • An important part of foreign aid is export credit guarantees.
Foreign, or development, aid does not always benefit the most needy.	• Political and economic factors (e.g. non-democratic governments, failure to pay debts) may mean that some of the poorest countries do not receive significant foreign aid. • Corruption, government inefficiency and misuse of aid in LEDCs sometimes means that aid does not reach its intended target, i.e. the poorest people. • Tied aid (e.g. Ilisu dam) may benefit donor countries more than recipients. • Tied aid may have significant environmental and social disadvantages for recipient countries.
There is increasing interdependence between the rich developed North and the poor developing South.	• MEDCs and LEDCs depend on each other for trade. There are important flows of investment capital from rich to poor countries. • Loans to the economically developing world result in flows of money through interest payments in the opposite direction. • International migration from poor to rich countries and foreign aid strengthen global interdependence. • Interdependence has both advantages and disadvantages for both MEDCs and LEDCs.

Revision section: chapter summaries from Book 1

Book 1, Chapter 1 summary — Tectonic activity

The Earth's interior is made up of several concentric layers.	• The outermost layer is the crust. This thin rocky layer includes oceanic and continental crust. • The lithosphere. • The mantle, which accounts for most of the Earth's mass. • The core. • Density and temperature increase with depth in the Earth's interior.
The Earth's crust and lithosphere are broken into large fragments, or plates.	• There are seven major plates and about 12 minor ones. • The plates are continually moving. This movement occurs along the plate margins and is responsible for most earthquakes and volcanic activity, and a wide range of landforms (e.g. fold mountains, ocean trenches, rift valleys etc.).
Important tectonic processes occur at plate margins.	• There are three types of plate margin: constructive, destructive and conservative. • New oceanic crust is formed by sea-floor spreading at constructive margins. • Old oceanic crust is destroyed at destructive plate margins. • Crust is neither formed nor destroyed at conservative margins. • The movement of plates (plate tectonics) explains continental drift and a variety of tectonic processes and landforms.
Earthquakes are the sudden release of energy from within the Earth in the form of seismic waves.	• Earthquakes result from friction between, and stretching and compression of, rocks within the crust and lithosphere. • Most earthquakes occur close to plate margins. • The energy released in earthquakes is measured on the Richter scale. The damage caused by earthquakes is measured on the Mercalli scale.
Earthquakes are major natural hazards.	• Earthquakes cause major loss of life, injury and damage to property each year. • Death and injury are caused not only by collapsed buildings, but also by fire and disease. • The human impact of earthquakes is often greater in LEDCs than in MEDCs. • In many countries that are vulnerable to earthquakes, earthquake-proof buildings help to minimise loss of life and damage to property.
Volcanoes are points of weakness in the crust where molten rock from the mantle reaches the Earth's surface.	• Like earthquakes, most volcanoes are found at plate margins. • The shape of volcanoes and the violence of eruptions depend on the type of magma and the amount of steam present.
There are several types of volcano.	• Strato-volcanoes, with alternating layers of ash and lava. • Shield volcanoes, with wide bases and low-angled slopes. • Lava domes. • Fissures, which lead to the formation of lava plateaus.
Volcanoes are major natural hazards.	• Eruptions may produce: lava flows, pyroclastic flows and avalanches, ash falls, lahars, nuées ardentes etc. • Eruptions can cause huge loss of life, injury and damage to property. • LEDCs are often hit hardest by volcanic eruptions.
Volcanoes can be used as important resources.	• Volcanoes provide: fertile soils, which allow intensive cultivation e.g. in Java; geothermal energy e.g. in Iceland; resources for recreation and tourism.

Revision section: chapter summaries from Book 1

Book 1, Chapter 2 summary — Rocks and landscapes

Rocks are made from a mixture of minerals.	• Granite is made up of quartz, feldspar and mica; limestone of calcium carbonate.
There are three main types of rock: igneous, sedimentary and metamorphic.	• Igneous rocks such as granite and basalt are formed from molten rock, or magma. • Sedimentary rocks result either from the breakdown of pre-existing rocks (e.g. sandstone) or from the build up of plant and animal remains on the sea floor (e.g. limestone). • Metamorphic rocks (e.g. slate and marble) have been altered by great heat and/or pressure.
Rocks have a structure.	• Rocks contain lines of weakness such as joints and bedding planes. Sedimentary rocks were deposited in layers, or strata.
Rocks are changed by the process of weathering.	• Physical weathering by frost or the sun breaks down rocks into smaller fragments. • Chemical weathering destroys rocks by altering their mineral composition.
Some rocks produce distinctive landscapes.	• Granite often produces upland landscapes (e.g. Cairngorm plateau) with features such as tors, blockfields and screes. Land use is mainly restricted to rough grazing and recreation. • Minor igneous intrusions, such as sills and dykes, form important local features such as vertical scars and waterfalls. • Limestone produces upland landscapes known as karst. They include pavements, scars, dry valleys, shake holes, caves and caverns. Soils are thin and not suited to cultivation, though recreation is often important. • Chalk is associated with gentle uplands or escarpments. These consist of a steep scarp slope and a gentle dip slope. There is little surface drainage today. • Clay gives rise to gentle lowland landscapes of great value to agriculture.
Rocks have great economic value.	• Hard rocks such as dolerite and granite may be used as roadstone. Clay baked into bricks is a valuable building material, and so is sandstone. Limestone is a vital raw material used in the chemical, steel and agricultural industries. Weathered granite forms china clay, which is used for pottery manufacture (Fig. 2.42).
Rock structure and scenery is influenced by folding and faulting.	• Pressure caused by plate movements produces simple upfolds, or anticlines (e.g. the Weald), and downfolds, or synclines (e.g. the London Basin), as well as major fold mountain ranges, such as the Himalayas. When rocks fracture instead of folding, they form rift valleys (e.g. East African Rift Valley) and fault scarps.
Fold mountains present both advantages and disadvantages for human activities.	• The possible advantages of fold mountain areas include: resources for tourism (snow, scenery, lakes), hydro-electric power, pasture for livestock, water supply, forestry, mineral ores etc. The disadvantages include: steep slopes, thin soils, cold climate and a short growing season, which limit farming and settlement; poor accessibility; and fragile ecosystems easily damaged by tourism and farming.

Book 1, Chapter 3 summary — Weather and climate

Weather is the day-to-day state of the atmosphere.	• Weather in the British Isles and north-west Europe is very variable. In other parts of the world (e.g. around the Equator) the weather is very constant.
Climate is the long-term (seasonal) pattern of weather.	• The main feature of climate is seasonal change in temperature and precipitation. Outside the tropics, climates have a warm and cold season. Within the tropics, seasonal differences in precipitation (wet and dry seasons) are more significant.

Revision section: chapter summaries from Book 1

Weather has an important impact on human activities.	• Transport movements are disrupted by snow, ice and fog. Summer droughts affect agriculture and water supplies. Heavy precipitation causes rivers to flood in winter, as happened in The Netherlands in January 1995.
On the global scale, there are broad climate regions corresponding with belts of latitude.	• Latitude is the main influence on temperature. It determines the sun's angle in the sky and the amounts of solar radiation received by a place. From the Equator to the poles, climate changes from equatorial to tropical continental, hot desert, Mediterranean, cool continental/maritime, cold continental and polar.
The British Isles have a mild damp climate.	• The main influences on climate in the British Isles are: latitude, distance from the ocean, the North Atlantic Drift, the prevailing westerlies, altitude and aspect.
The climate of the British Isles has significant regional differences.	• The west is milder and wetter than the east; the south is warmer than the north. Highland Britain is both wetter and colder than Lowland Britain.
Precipitation – the moisture that falls from clouds.	• Precipitation includes rain, drizzle, snow, sleet and hail.
Precipitation occurs in three situations: when air is forced to cross mountains; in depressions along fronts; when air is heated and rises by convection.	• Precipitation in mountainous areas is called relief, or orographic precipitation. It gives high precipitation in British uplands. Frontal precipitation occurs in depressions when air is forced to rise at warm and cold fronts. Most precipitation in the British Isles is frontal. Convectional precipitation follows intense heating of the ground by the sun. It causes showers and thunderstorms and is particularly important in the tropics.
Weather charts provide a daily summary.	• Weather charts summarise temperature, precipitation, cloud cover, wind direction/speed, pressure etc. These charts are essential for making forecasts. They are updated every six hours. The main features on these charts are isobars and fronts.
Depressions and anticyclones dominate Atlantic weather charts.	• Depressions are mid-latitude storms bringing mild, wet, cloudy conditions to north-west Europe. • Anticyclones are areas of high pressure bringing usually dry, settled weather. Temperatures are often extreme (cold in winter, warm in summer) with very variable amounts of sunshine.
Weather forecasters rely increasingly on satellite images.	• Both visible and infra-red images are used. They provide information on cloud patterns and temperature.
Tropical cyclones are violent storms that form over warm oceans.	• Tropical cyclones bring very strong winds and heavy rain. Every year they cause immense loss of life and damage to property in the tropics and sub-tropics.
Tropical cyclones hit poorer countries hardest.	• Hurricane Andrew (Florida 1992) was responsible for 22 deaths. The less-powerful cyclone that struck Bangladesh in 1991 killed 125 000 people. Poverty, lack of early warning, lack of shelters and remoteness mean greater destruction in LEDCs.
Climatic hazards can also have a severe impact on MEDCs.	• The drought in the USA in the first half of 2000 was responsible for the worst forest fires for more than 50 years.

Book 1, Chapter 4 summary — Ecosystems

Ecosystems comprise plants, animals, decomposers and the physical environment.	• The living and non-living parts of ecosystems are linked together by a complex web of relationships. • Ecosystems are powered by sunlight. They have flows of nutrients. • Sunlight is trapped by plant leaves and converted to sugar and starch by photosynthesis. • Energy 'flows' through ecosystems along food chains and food webs.
Ecosystems vary in scale from local to global.	• Moorland is an example of a local ecosystem. The tropical rainforest, savanna grasslands and northern coniferous forest are global ecosystems.

Revision section: chapter summaries from Book 1

The tropical rainforest is the most productive and most diverse ecosystem.	• The tropical rainforest is found in lowland areas within 10 degrees of the Equator and contains 90 per cent of all living species. • The rainforest climate is warm and humid. These conditions are ideal for plant growth.
The rainforest trees are adapted to the equatorial climate.	• Trees are evergreen (they don't lose their leaves once a year). Some trees have leathery leaves with driptips, and buttress roots.
Rainforest soils have little fertility.	• Forest soils are acidic and contain few nutrients. • The forest is sustained by the rapid cycling of nutrients. • Permanent agriculture is not sustainable in the rainforest.
Exploitation of forest resources is rapidly destroying the rainforest.	• Deforestation is caused by agriculture, settlement, road building, mineral extraction, logging and HEP projects. • Most logging in the rainforest is not sustainable.
There are arguments for and against the destruction of the rainforest.	• Arguments centre on the conflict between environmentalists who want to conserve the rainforest, and economists who want to develop its resources.
Savanna vegetation is adapted to climate.	• Rainfall rather than temperature determines the seasons in the savanna. Plants are adapted to the long dry season. Perennial plants, e.g. trees, are often deciduous, fireproof and have leaves modified to reduce moisture loss.
The recent exploitation of agricultural resources in the savanna has often been unsustainable.	• Overstocking of pastures and overcultivation of arable land (both caused by rapid population growth), and the effects of prolonged drought, have led to widespread land degradation (desertification) in the savannas.
Degraded land in the savannas can be restored to achieve future sustainability.	• Restoration of degraded savannas can be achieved by: afforestation, enclosure of farmland, diversification of crops, drilling of boreholes to increase water security, education of farmers etc.
The coniferous forest is found between the temperate deciduous forest and the Arctic.	• Climatic conditions set severe limits on plant growth in the coniferous forest. • Productivity and biodiversity in this type of forest are low.
Coniferous trees are adapted to the severe continental climate.	• Trees are evergreen and have needle-shaped leaves. • Trees have conical shapes.
The coniferous forest is the world's main source of softwood and pulp.	• Canada, Sweden, Finland, Norway and Russia are the leading exporters of softwood timber and pulp.
Coniferous forest once covered the Highlands of Scotland.	• Attempts are being made to re-establish the ancient Caledonian forest in Scotland using seed from the trees in the remaining fragment of the original forest.

Book 1, Chapter 5 summary — Settlement patterns

Settlements can be divided into rural and urban types.	• We recognise farms, hamlets and villages as rural: towns and cities as urban. However, there is no clear division between rural and urban types. Population size, employment in non-rural activities, population density and function are all used to distinguish urban from rural settlements.
Rural settlement patterns may be nucleated or dispersed.	• Nucleated patterns are dominated by villages. They are often associated with localised resources e.g. water, and a communal system of agriculture e.g. open field agriculture in medieval Europe. • Dispersed patterns consist of scattered isolated farms and hamlets. They are associated with pastoral farming, poor resources for farming, and a tradition of individuality. • In Britain, nucleated patterns are more common in the Lowland Zone; dispersed patterns are typical of the Highland Zone.

Revision section: chapter summaries from Book 1

Rural settlement may have uniform, random or clustered distributions.	• Clustered distributions are most common. They result primarily from the influence of physical factors such as relief, climate, soils, water supply etc. These factors can either attract or repel settlement. • Uniform distributions often indicate an even spread of resources e.g. on a lowland plain.
The characteristics of site, situation, shape and function are important features of individual settlements.	• Site refers to the land on which a settlement is built. In the past, sites were chosen to provide resources (e.g. water, soil etc.) that would satisfy the basic needs of farming communities. • Situation is the location of a settlement in relation to the surrounding region. • Settlement shape is influenced by both physical factors (e.g. relief, drainage) and human factors (e.g. roads, planning). • Settlements have a variety of functions, which increase with settlement size. The most important functions are residential, industrial and commercial.
Large settlements are central places or service centres.	• Settlements form hierarchies based on their importance as central places. Large settlements have many functions and serve large trade areas. They support high order functions (comparison goods/services, theatres, hospitals etc.), which require high threshold populations and have a large range.
The functions of many rural settlements are undergoing change.	• Car ownership (giving greater mobility) and new retail formats (e.g. edge-of-town superstores) are responsible for the decline of retailing in many market towns. Smaller places, such as villages, are losing shops (also schools, GPs etc.) to larger centres. • There is a general decline of services in rural areas in the British Isles. This decline is due to a reduction in local demand caused by a) depopulation, b) rural commuters, c) retired people and d) second homes.

Book 1, Chapter 6 summary — Urbanisation and urban structure

Cities first appeared about 5500 years ago.	• The first cities were in the Middle East and Egypt. They were supported by food surpluses from irrigation agriculture in fertile river valleys.
Urbanisation is an increase in the proportion of people living in towns and cities.	• Urbanisation has increased in the last 200 years. By 2000, nearly half the world's population lived in urban areas. • Rapid urbanisation occurred in Europe and North America in the 19th century. • Today, urbanisation is concentrated in the economically developing world.
Today, urbanisation in the economically developing world is leading to the growth of mega cities.	• Mega cities have populations of 5 million and above. Many of these cities are primate cities and dominate industry, commerce and investment in their respective countries.
Most urban dwellers live in LEDCs.	• Two out of three urban dwellers live in LEDCs. Most of them are poor.
Urbanisation results from natural population increase and rural-urban migration.	• Rural-urban migration is the principal cause of urbanisation. People in the countryside in LEDCs move to towns and cities because they think that living standards are better there.
Counter-urbanisation is an important trend in MEDCs.	• In many MEDCs, the number of people living in conurbations and large cities is falling. Better-off people are moving out to the commuter belt and retiring to environmentally attractive areas. A few are moving to remoter rural areas.
Land-use patterns in cities are known as urban structure.	• Most cities have a central business district surrounded by distinctive zones, sectors and areas. There are clear differences in urban structure between cities in MEDCs and those in LEDCs.

Revision section: chapter summaries from Book 1

In MEDCs, population density generally declines with distance from the centre to the edge of the city.	• Few people live in the CBD in MEDCs. Densities usually peak in the inner city, and then fall steadily towards the edge of the city. • Urban renewal and gentrification have led to an increase in the population of some central areas of cities.
Different social, economic and ethnic groups locate in different parts of the city.	• Different groups become segregated according to income and ethnicity. High-income groups are able to choose areas with most advantages. Low-income groups have little choice and often suffer many disadvantages. Ethnic minorities may cluster together out of choice.

Book 1, Chapter 7 summary — Urban problems and planning

Cities in both the developed and developing worlds face a number of urgent problems.	The main urban problems are: • housing (shortages and sub-standard housing) • poverty • congestion and environmental pollution • urban sprawl The scale and seriousness of these problems is greater in LEDCs.
Planners have tackled the inner city problems of poverty, crime, unemployment, social exclusion and urban decay in MEDCs with a variety of measures.	Planning responses include: • urban renewal involving demolition of slums and the building of high-rise flats e.g. Hulme in Manchester and Sarcelles in Paris • urban improvement • enterprise zones, urban development corporations, partnership schemes.
Urbanisation and the increasing number of households in MEDCs has led to urban sprawl.	• In the UK, green belts have been used to curb urban sprawl. • Green wedges and corridors of growth have been preferred in some other European countries, such as Denmark, France and the Netherlands.
New towns have been developed to accommodate rising urban populations and increasing numbers of households in MEDCs.	• In the UK, new towns have been developed around major conurbations and beyond the green belt. In south-east England, a total of eleven towns have been built since 1945. • Paris has five new towns located in the city's twin axes of development.
Congestion in the central areas of cities in the MEDCs has led to the decentralisation of economic activities.	• New towns, e.g. Tama in Tokyo, have assisted decentralisation. Business centres, industrial estates and office parks (e.g. Makuhari in Chiba, Tokyo) have been developed outside the CBD or at edge-of-town locations in the last 30 years.
Air pollution from traffic is a serious hazard in the urban environment in all large cities.	• 50 years ago, air pollution was caused by burning coal, which caused winter smog. This is still a problem in poorer countries such as China. • In the MEDCs, air pollution is caused mainly by car exhausts. The result is photochemical smog and high concentrations of ozone. Restrictions on the use of private cars in cities, and new public transport schemes (e.g. rapid-transit, trams, etc.) aim to reduce both pollution and congestion.
Urban problems in the LEDCs concern the very survival of city dwellers.	• Rural-urban migration caused massive expansion of cities in LEDCs after 1950. City authorities do not have the resources to solve housing, employment and environmental problems. Most people rely on self-help. This is evident in the growth of (a) huge shanty towns and (b) informal employment based on small-scale services and businesses. Environmental concern has little priority at the moment in most cities.
There is an increasing awareness that cities and urban living should be sustainable.	• Many city authorities in MEDCs are planning to ensure the sustainable use of resources such as water supplies, clean air and green spaces. In the UK, the recycling of urban land – brownfield sites – will be given priority in an effort to protect the countryside from further urban sprawl.

Revision section: chapter summaries from Book 1

Book 1, Chapter 8 summary — Agricultural systems

Agriculture is an important economic activity.	• About 40 per cent of the world's economically active population works in agriculture. • 95 per cent of the agricultural workforce is in LEDCs. • Agriculture is the biggest user of land: 37 per cent of the world's land area is used for agriculture.
Agriculture is different from other economic activities.	• Agriculture relies heavily on the physical environment (especially climate). Farmers have limited control over climate. Agriculture also depends on the biological cycles of crops and animals.
Farms are both ecological and economic systems.	• Farms are systems, with inputs, outputs (yields) and food chains. • Outputs comprise crops and livestock products. • The main inputs are sunlight, precipitation, soil nutrients, agro-chemicals, labour and capital (machinery etc.). • The amount of inputs and outputs per hectare determine the intensity of farming.
Agriculture is just one link in a much larger integrated food system/chain.	• Food systems are chains of activities, from agricultural suppliers to supermarkets. • Farms are the production side of the food system, which includes industries supplying the farm, such as fertiliser and agricultural machinery makers, and purchasers of farm products, such as food processors and retailers.
Types of agricultural enterprise are defined according to several different criteria.	• The dominant enterprise – crops or livestock – defines agriculture. • Intensity of farming is determined by the level of inputs and outputs per hectare. • Commercial farming is a profit-making form of agriculture and non-commercial is a subsistence form.
On a global and national scale, climate is the major influence on agriculture.	• Climate determines the broad limits of what a farmer can grow. On a global scale, insufficient warmth and moisture explains the absence of farming in many high-latitude, hot desert and mountainous areas. • In the UK, the contrast between the arable east and pastoral west is largely related to climate.
Many traditional agricultural systems are sustainable but under threat.	• Shifting agriculture and nomadic herding are well adapted to harsh environments and are sustainable systems, i.e. they can be practised without long-term damage to the environment. However, both systems are under threat from population growth, habitat destruction, government policies and natural hazards, such as drought.
Modern farming in much of the economically developed world is agribusiness.	• Large-scale farming, based on scientific and business principles, is known as agribusiness. Agribusiness is capital intensive, employs expert farm managers and is geared to contract farming. In addition to local climatic and soil conditions, crops grown and animals reared are influenced by local market opportunities and government policies.
Farming in upland areas is strongly limited by the physical environment.	• Climate, soils and relief place strict limits on farmers' choice of enterprise in the uplands. • Farming in highland Britain is extensive and livestock based. Without assistance from the Common Agricultural Policy, most upland farms would not be profitable.

Book 1, Chapter 9 summary — Agriculture: problems and change

There are both high- and low-tech approaches to increasing agricultural production in LEDCs.	• In Adami Tulu, a low-tech aid project designed to increase food production and improve the environment has been successful. • The green revolution is a high-tech approach to increasing food production. It has increased food output significantly in some LEDCs. However, many small farmers, and many areas in the economically developing world, have not benefited.

Revision section: chapter summaries from Book 1

Low-tech improvements in agriculture in the economically developing world are likely to be more sustainable than high-tech ones.	• Low-tech improvements are affordable to poor peasant farmers. The use of simple technology allows the skills of local people to be used. • High-tech improvements are expensive and tend to benefit the rich more than the poor. • Low-tech improvements are more likely to be environmentally and economically sustainable.
There are chronic shortages of food in many of the poorest LEDCs.	• Food shortages stem from drought, rapid population growth, civil wars, etc. In times of famine many poor countries have to rely on food aid from MEDCs.
Population growth and agriculture have contributed to environmental degradation in the economically developing world.	• Rapid population growth has meant extending and intensifying agriculture to increase food production. This has put the environment under pressure in many parts of the economically developing world. The result is over-cultivation, overgrazing, deforestation, soil erosion, etc.
The Common Agriculture Policy has caused economic and environmental problems in the EU.	• The CAP encouraged the intensification and extension of farming. The result was huge food surpluses and environmental degradation (destruction of habitats, pollution, soil erosion, etc.). • In the 1980s and 1990s, the CAP introduced measures to reduce surpluses (e.g. set-aside) and improve the environment (ESAs, woodland and hedgerow planting, nitrate reduction, landscape/habitat protection and improvement).
Soil erosion invariably accompanies cultivation.	• Soil erosion is widespread in the UK and especially in the arable areas of eastern England. Wind and rain are responsible for most soil erosion. • Soil erosion not only results in the loss of topsoil, it increases the costs of cultivation and reduces crop yields.
There is a range of conservation measures that can significantly reduce rates of soil erosion.	• Wind erosion can be lessened by planting shelter belts. • Strip cropping, the growing of cover crops and leaving crop residues in the fields reduces both wind and water erosion. Contour ploughing helps to conserve soils on hill slopes. • Keeping soils fertile with the use of manure makes soil erosion less likely.

Book 1, Chapter 10 summary — Industrial activity and location

Economic activity can be divided into four sectors: primary, secondary, tertiary and quaternary.	• Primary activities such as agriculture and mining produce food and raw materials. Secondary activities manufacture goods e.g. iron-and-steel, motor vehicles. Tertiary and quaternary activities are service activities. The tertiary sector includes utilities and transport. The quaternary sector comprises services for other economic activities and individual consumers.
The importance of manufacturing industry varies in time.	• Manufacturing industry in the economically developed world has passed through a cycle in the last 200 years. The percentage of people working in manufacturing was low in the pre-industrial period. It reached its height in the industrial revolution. Today, in the post-industrial period, manufacturing employs between one-fifth and one-quarter of the working population.
The importance of manufacturing varies in different countries.	• Manufacturing is least important in the poorest countries of the economically developing world. In some newly industrialising countries (e.g. Taiwan, South Korea) manufacturing employs between one-quarter and one-third of the working population but in some of the world's richest countries (e.g. USA), it employs less than one-fifth of the working population.
Energy and raw materials influence industrial location.	• Heavy industries, using large amounts of energy and raw materials, (e.g. iron-and-steel), often locate close to sources of energy or raw materials to reduce transport costs.
The location of some manufacturing industries is explained by industrial inertia.	• Although the initial advantages of a location may have disappeared (e.g. local raw materials may no longer be used), some industries remain where they first started e.g. pottery. Inertia often keeps an industry in its original location because of the cost and difficulty of moving.
New reasons for retaining the location of an industry often replace the original ones.	• Acquired advantages such as skilled labour and linkages with nearby firms may also explain the survival of an industry in a region (e.g. pottery at Stoke-on-Trent), even though the initial advantages of the location have long since disappeared.

Revision section: chapter summaries from Book 1

Changes in technology and the source of materials may change an industry's location.	• In the UK, over the last 200 years, the iron-and-steel industry has shifted from locations on coalfields and iron ore fields to coastal sites. These moves reflect increasing reliance on overseas materials.
Some industries are successful only if they operate on a large scale.	• In the iron-and-steel and motor vehicle industries, large firms can lower their costs because of economies of scale. To achieve these economies, an industry may need a very large site and/or a huge, global market for its products.
Estuaries are often attractive industrial locations.	• Estuaries provide large areas of flat, reclaimed land for heavy, space-using industries like steel and oil refining. They also give access to imported raw materials and energy.
Some industries are dominated by trans-national corporations.	• Trans-national corporations (TNCs) are large international firms with factories and markets in many different countries. Car making and high-tech industries are dominated by TNCs.
Some manufacturing industries are footloose.	• Footloose industries, such as high-tech industries, are those that are not strongly influenced by traditional locational factors i.e. transport, materials, energy. However, footloose industries cannot locate anywhere. In the UK, they are highly concentrated in regions such as the South-east and central Scotland.
Governments influence the location of manufacturing.	• Governments give grants to foreign firms to encourage them to invest. In less-prosperous regions, grants are available to attract industry, which will provide jobs.
In LEDCs, cottage industries are important.	• Cottage industries are based in the countryside. They are usually small-scale, labour-intensive industries, and rely on simple technology.
Appropriate technology is the basis for industrial development in many LEDCs.	• Appropriate technology aims to benefit the people of LEDCs by using their skills and encouraging them to find their own solutions e.g. stove making in western Kenya. Large-scale, capital-intensive industrial development has often failed to benefit the people.

Book 1, Chapter 11 summary — Industrial change

Industrial activity is increasingly organised on a global scale.	• Rapid industrial development is occurring in parts of the economically developing world, such as the Asian Pacific Rim. TNCs are investing in production worldwide. A global market offers TNCs higher output and lower costs. Global operations also allow TNCs to overcome trade restrictions.
South Korea is a newly industrialising country.	• South Korea has few natural resources for industry. However, it has undergone rapid industrialisation since 1970. Its development has been based on a few very large companies (*chaebols*) e.g. Samsung. In the 1990s, these companies started to invest heavily overseas.
Old industrial regions have been hit hard by de-industrialisation.	• During the 1970s and early 1980s, basic industries such as steel, shipbuilding and coal mining declined steeply in regions like Central Scotland, South Wales and North-east England. The result was high rates of unemployment and widespread dereliction.
Re-industrialisation has transformed many old industrial regions.	• Since the mid-1980s, there has been massive investment by foreign companies in several old industrial regions in the UK. This investment, encouraged by government grants, has made Central Scotland, South Wales and the North-east leading centres of the electronics industry in the EU.
Regions that have attracted large-scale foreign investment are most vulnerable to global economic change.	• The financial crisis in East Asia (1997–98) and overcapacity in industries such as semi-conductors and motor vehicles, have led to the closure of factories operated by foreign TNCs e.g. Siemens on Tyneside, and Fujitsu in Aycliffe.
Areas in the UK worst affected by de-industrialisation have been given special status.	• The government has created Urban Development Corporations (UDCs) and Enterprise Zones (EZs) in inner-city and riverside locations. By improving the environment and offering tax breaks, they aim to attract manufacturing and services to run-down industrial areas.

Revision section: chapter summaries from Book 1

There has been an urban-rural shift of manufacturing industry in MEDCs.	• Since 1970, manufacturing has declined in conurbations and large cities. New factories have preferred to locate in small towns and rural areas. Lack of space and obsolete factory buildings in large urban areas largely explain this change. Some remote rural areas such as Mid-Wales and the Highlands and islands of Scotland, have benefited from this urban-rural shift. The growth of industry in these areas has been helped by special government agencies and grants.
Port functions and industries have undergone major locational change.	• Revolutions in cargo handling – bulk carriers for oil, coal, ore, grain etc., and containers for manufactured goods – have been responsible for a rapid downstream shift in port activities since 1960. Bulk cargoes and containers are transported in very large ships, which need deep water. These cargoes also require large areas of land for storage. Most of the available deep water and space is near river mouths e.g. Europoort near Rotterdam.
In MEDCs, retailing has grown in the suburbs at the expense of the CBD.	• Food superstores, retail parks and regional shopping centres have appeared in the suburbs of many British cities since 1980. Retailers have located in the suburbs to be nearer the better-off consumers, to obtain the space needed for large stores and parking, and to benefit from less congestion and lower land prices.
The growth of suburban retailing threatens town/city centres.	• Shopping centres in the suburbs compete with retailing in the town/city centre. Smaller town centres, such as Dudley, have suffered many shop closures. Well-known high street retailers have been replaced by second-hand shops and discount stores in the town centre.
Many office activities have moved out-of-centre in British cities.	• Offices have moved out-of-centre to purpose-built office parks in the suburbs. High rents and lack of space for new building have forced many financial services in the City of London to relocate in London's Docklands (e.g, in Canary Wharf).

Glossary

This glossary contains definitions of technical words, which appear in the book in bold, in the context in which they are used. They are given in alphabetical order.

abrasion Erosion caused by the scouring action of rock fragments carried by rivers, glaciers, waves and the wind.
accumulation zone That upper part of a glacier where the annual supply of snow and ice exceeds annual melting.
alluvium Rock particles (clay, silt, sand and gravel) deposited by a river.
aquifer An underground layer of water-bearing rock
arête A knife-edged ridge formed by glacial erosion and freeze-thaw weathering in glaciated uplands.
backwash The return of water from the beach to the sea, following the wave swash.
balance of trade The difference in value between a country's exports and imports.
barrier beach A long narrow beach that extends across a bay. Often the beach creates a lagoon on its landward side.
bayhead beach A crescent-shaped beach at the head of a cove or bay.
bilateral aid Aid given directly by a rich country to a poor country.
biofuels Fuels such as timber and biogas (i.e. methane) derived from living organisms. Biofuels are renewable forms of energy.
blowhole A vertical shaft connecting a sea cave to the cliff top.
bluff Slopes that rise steeply from a river's valley in its middle stage.
catchment (see drainage basin).
cliffs Steep, almost vertical slopes of solid rock. They are common along coastlines and in glaciated uplands.
constructive waves Low, gentle waves, which cause a net transport of sand and shingle onshore. These waves build up steep beaches.
core Prosperous, centrally located regions, e.g. South-east England and the Paris Basin in the EU.
counter-urbanisation The net movement of population from urban to rural areas. It has been the dominant migration movement within MEDCs since the 1960s.
crevasse A deep crack or fissure on the surface of an ice sheet or valley glacier.
crude birth rate The number of live births per 1000 people per year.
crude death rate The number of deaths per 1000 people per year.
delta A river mouth choked with sediment causing the main channel to split into hundreds of smaller branching channels or distributaries.
demographic transition The demographic change over a period of one to two centuries from high birth and death rates, to low birth and death rates.
dependency The economic and technological reliance of poor countries on rich countries.
depopulation An absolute decrease of population in an area. Its main cause is heavy out-migration.
deposit/deposition The process by which transported sediments are laid down by rivers, glaciers, waves and the wind.
destructive wave Waves that remove sand and shingle from beaches and transport them offshore. They create flat beaches.
discharge The volume of water flowing down a river channel in a given time (usually measured in cubic metres per second).
distributaries Finger-like river channels that branch away from the main channel in a delta.
drainage basin The area of land drained by a river and its tributaries. Also known as a river basin or catchment.
eco-tourism An alternative to mass tourism. It aims to be sustainable i.e. it causes no long-term damage to the environment.
emigration The migration of people out of a country.
erode/erosion The wearing away of the land surface by rivers, glaciers, waves and the wind.
estuary A broad, shallow, funnel-shaped river mouth.
evaporation The change in state of water from liquid to gas.
exports Goods and services produced by a country and sold to other countries.
famine Acute food shortages within a country/ region or among an economic/social group.
fission The splitting of uranium atoms inside a nuclear reactor. It results in the release of huge amounts of energy.
fjord A U-shaped valley (glacial trough) that has been flooded by the sea.
flood plain The wide, flat floor of a river valley. It consists of sediments (alluvium) deposited by the river.
food security When people have certain access to enough food to live a healthy life.
fossil fuels Non-renewable forms of energy such as oil, coal, natural gas and peat.
free trade The movement of goods and services between countries without restrictions like tariffs and quotas.
glacial trough A valley with a U-shaped cross-section carved by a valley glacier or ice sheet.
globalisation The worldwide location of production by large companies in order to serve global markets and reduce costs.
green revolution The successful introduction of high-yielding varieties of rice and wheat into LEDCs.
gross national product The total value of goods and services produced by a country both at home and overseas.
groundwater Water stored in porous rocks underground.
hanging valley A tributary valley to a larger, over-deepened glacial trough. Where the two valleys meet, the tributary valley may 'hang' several hundred metres above the main valley.
headland A promontory between two bays. It often consists of relatively resistant rock.
honeypot A popular attraction in an area of tourism (e.g. Windermere in the Lake District).
hydrograph A graph that shows changes in river flow over time (e.g. per hour, per month, per year).
immigration The movement of international migrants into a country.
imports Goods and services bought by a country from other countries.
incised meander A gorge-like meander formed by river rejuvenation.
interception When precipitation is trapped by the leaves, branches and stems of plants, preventing it from reaching the ground.
interdependence The interrelationships of trade, aid, investment, migration, etc., which exist between the economically developed and economically developing world.
interglacial A warmer spell between ice ages, lasting about 10 000 years.
interlocking spurs Areas of higher ground that project into V-shaped river valleys. They occur on alternate sides of a valley and result from rapid downcutting by upland streams and rivers.
isostatic change Vertical movements of the continents relative to sea-level. Often caused by the formation and melting of ice sheets during ice ages and interglacials.
landforms Natural features like mountains, valleys, cliffs, etc. formed by weathering, erosion, deposition, landslides, tectonic forces, etc.
lateral erosion Erosion by a river on the outside of a meandering channel. It eventually leads to the widening of the valley and the formation of the flood plain.
lateral moraine A narrow linear band of rock debris that runs along the margins of a valley glacier.
leisure Free time that we can spend on a non-work activity.
long profile The cross-sectional shape of a river's course from source to mouth.
long-term aid Aid given to support development projects like improvements in education, farming, water supply, etc.
longshore drift The movement of sediment by wave action and currents along a coastline. It produces landforms such as spits and bars.
Malnutrition Ill health caused by an unbalanced diet – most often a lack of protein.
mass tourism Tourism that caters for very large numbers of visitors and often has a damaging effect on the natural environment.
medial moraine A narrow, linear band of rock debris that runs down the centre of a valley glacier. It forms from the merging of two lateral moraines.
meltwater deposits Sand and gravel deposits (forming landforms like eskers and kames) left

Glossary

by streams and rivers that flow within and from ice sheets and glaciers.

migration The movement of people from one place to another to settle permanently.

mouth The place where a river reaches the sea.

mud flats Areas of mud deposited by tidal currents and exposed at low tide, common in estuaries.

multilateral aid Aid given to poor countries through the World Bank, IMF, UN etc.

natural arch A rock bridge formed by wave erosion on the coast. It is a stage in cliff destruction by erosion and weathering.

natural population change A growth of population caused by an excess of births over deaths.

net migration gain Where more people move into an area than move out.

net migration loss Where more people move out of an area than move in.

newly industrialising country A country that has undergone rapid and successful industrialisation in the last 30 years. e.g. South Korea, Taiwan.

non-governmental organisation Independent organisations, such as charities (e.g. Oxfam), that give aid to LEDCs.

non-renewable resource Resource such as fossil fuels and soils. Once used, these resources cannot be replaced.

notch A groove at the vase of a coastal cliff eroded by wave action. It creates an overhang and eventually leads to cliff collapse.

ox-bow lake A meander that has been cut off from the main river channel and abandoned.

periphery Areas that are geographically remote from a core region. The periphery is noticeably less prosperous than the core.

point bar Sand and gravel deposited by a river on the inside of a meandering channel.

polders Areas of wetland close to sea level that have been drained and reclaimed for agriculture and other land uses.

population pyramid A type of bar chart that shows the distribution of a population by age group and sex.

potholes Holes eroded in the solid rock of a river channel. They are drilled by pebbles caught in eddies in the river.

primary fuels Fuels burned to generate energy directly e.g. gas burned by a cooker, wood burned on an open fire.

primary product A raw material or foodstuff (usually for export) that has not been processed in any way e.g. iron ore, cocoa, etc.

pull factors The factors that create migrants to a particular place e.g. better job prospects, better educational opportunities, etc.

push factors The disadvantages within an area that force people to move out e.g. low wages, unemployment, wars.

pyramidal peak A sharp peak formed by the intersection of three corries.

quarrying A type of erosion where a glacier freezes on to rocks, and as it moves forwards removes (or plucks) them along rock joints.

raised beach An ancient beach that today lies several metres above sea level. Raised beaches usually form by uplift of the coast following the melting of ice sheets (see isostatic change).

refugee A person forced to leave an area and made homeless following a disaster such as war, famine, earthquake etc.

rejuvenation The renewed erosional activity of a river. It results either from an uplift of land or a fall in sea level (see incised meander).

rejuvenation terraces Matching benches on opposite sides of flood plains. They are the remains of an earlier flood plain removed through renewed erosion (rejuvenation) by a river.

renewable resource A resource that is either inexhaustible (e.g. solar energy) or follows a biological cycle (e.g. timber) or physical cycle (HEP) of continuous renewal.

ria An incised river valley and its tributaries flooded by the sea.

ribbon lakes Long, narrow lakes that occupy rock basins in U-shaped valleys in glaciated uplands.

river's regime The pattern of seasonal change in a river's flow over a year. The regime is influenced by precipitation, evaporation, transpiration, geology, etc.

roches moutonnées Isolated outcrops of rock eroded by glaciers and ice sheets. In cross-section, they have a characteristic gentle, smooth slope and a steep, rugged one.

rural-urban migration The movement of people from the countryside to live permanently in towns and cities.

salt marshes Vegetated mud flats on lowland coastlines that are only flooded at the highest tides.

secondary energy Energy manufactured by burning primary fuels such as coal, oil, gas and uranium (i.e. electricity).

sediment load Rock particles, ranging in size from boulders to clay, transported by a river.

shore (wave-cut) platform A rocky coastal platform covered at high tide and extending for some distance in front of cliffs.

short-term aid Aid given to help poor countries experiencing a crisis or emergency, e.g. famine.

snowline The altitude where permanent snow begins in mountainous regions.

source The place where a river starts e.g. a spring, a boggy area or moorland, etc.

spit A long, narrow beach that is joined to the land at only one end.

stack A steep-sided rocky islet located offshore from coastal cliffs. Stacks are the remains of old cliffs and natural arches.

sustainable When any economic activity (e.g. farming, tourism, etc.) can continue indefinitely without causing permanent harm to the environment (e.g. soils, water, landscapes, vegetation etc.).

swash The movement of water up a beach following the breaking of a wave.

terminal moraine A prominent ridge of debris that is dumped at the end of a glacier or ice sheet, and is made up of boulders, sand, gravel and clay.

tied aid Aid given to poor countries but with certain conditions attached (e.g. money must be spent on exports from the donor country).

till An unsorted mixture of boulders, gravel, sand and clay deposited by glaciers and ice sheets. Also known as moraine.

till plain An extensive lowland plastered with a thick layer of till e.g. Holderness in East Yorkshire.

tombolo A type of beach joining an island with the mainland e.g. Chesil Beach.

trans-national corporations Very large firms, such as IBM, Samsung and Shell, that have production facilities in many different countries.

transpiration Water produced by plants and evaporated into the atmosphere.

transport Rock particles (whether solid or in solution) carried by the agents of erosion – rivers, glaciers, waves and wind.

truncated spur A spur in a U-shaped valley that has been sliced off by glacial erosion.

U-shaped valley A glacial valley, with steep sides and a flat floor. U-shaped valleys are usually straight in planform, deep, and have hanging tributary valleys.

undernutrition Insufficient food to maintain body weight and good health. Prolonged undernutrition leads to death by starvation.

unsustainable Economic activity that damages the environment, degrades natural resources and that cannot continue in the long term.

urbanisation An increase in the proportion of a population living in towns and cities.

V-shaped valleys Narrow steep-sided valleys typically found in upland areas. They owe their shape to rapid vertical erosion by streams and rivers.

value added Increase in the value of goods or raw materials through the process of manufacturing.

water cycle The continuous movement of water (as liquid water, ice and water vapour) between the oceans, continents and atmosphere.

watershed The boundary separating two drainage basins.

weathering The chemical and physical breakdown of rocks by the action of moisture, heat and cold.

Index

abrasion
    coastal landforms 47, 61
    glacial landforms 28, 30
accumulation zones, glacial 26
acid rain 110–11
age, and population growth 72, 75–6, 77
    age-sex structure 73
    ageing (greying) populations 78–9
agribusiness, Dominican Republic 152–3
aid to poor countries 168–77
AIDS, Nigeria 73–4
alluvium 11
arches, coastal landforms 48
arêtes 30
Aswan High Dam 97–8
asylum seekers 91
attrition, coastal landforms 47, 61

backwash, wave formation 45
balance of trade 161
Bangladesh, population 76–7
bayhead beaches 50
beaches 49–50, 51, 52
    coastal erosion 57
    raised beaches 53, 54
bilateral aid 169
bilharzia 154
biodiversity, Galapagos Islands 134
birth rates, and population growth 69, 71, 72, 76–7
    immigrants to USA 89
Blackpool 119–21
blowholes, coastal landforms 48
borrowing, developing countries 147, 151

Cairngorms
    glaciation 24–5
    skiing 36–40
carbon emissions, United Kingdom 105
caves, coastal landforms 48
CBR and CDR (crude birth and death rates) 69
charities, foreign aid 170
cities
    global population distribution 67
    new towns in the Netherlands 99
cliffs 46, 47–8
    erosion in Holderness 56–9
climate
    and development in Italy 158
    and global warming 104
    and population distribution 63–4, 66
coasts 43–61
    coastal system 44, 61
    depositional landforms 49–52
    erosion in Holderness 56–9
    features 46–52
    sand dune erosion 60–1
    waves 44–5, 61
coffee 166–8
conservation
    glaciated uplands 39–41
    national parks 124
    recycling resources 115–16
    sand dunes 60
    tourism in the Galapagos 133, 135
    see also environmental management
contraception, and population growth 72
corries (cirques or cwms) 28–30
corrosion, coastal landforms 47, 61
counter-urbanisation (urban-rural migration) 81, 85–6
crevasses, glacial 26

Darwin, Charles 135
death rates, and population growth 69, 70, 71, 72
debt, developing countries 147, 151

deforestation
    and population growth 94–5
    problems caused by 153
deltas 13–14
demographic transition 71, 75
depositional landforms 49–52, 59
Derwent, River and Valley 17–20
development
    foreign development aid 169
    global 137–59
    in developing countries 147–56
    inequalities in 137–9
    Italy 156–8
    measured 139–41
    regional contrasts 156, 157
Devon, coastal features 46–52
DFID (Department for International Development) 171, 173, 174
discharge
    and population distribution 66–7
    river regimes 15–16
diseases, developing countries 153–5
Dominican Republic 151–3
Dorset, coastal features 46–52
drainage basins 5, 6
drumlins 34

eco-tourists 131
economy
    and environmental damage 96
    and global development 137–40, 141, 146–8, 158
    and natural resources 102–3
    and population distribution 66–7
    and tourism in Zimbabwe 131
    economic migration 88–9
    foreign aid 168–74
    international trade 161–2, 165–7
    population growth 72, 75
education, and development 150, 174
Egypt, population and resource base 96–8
electric power 22, 35, 114–15
emigration 81
employment
    and education in developing countries 150
    and global population distribution 66–7
energy
    alternative resources 113–15
    electric 22, 35, 114–15
    wave energy 44–5
environmental damage
    acid rain 110–11
    and agribusiness 152–3
    and population growth 93–6, 97
    global warming 106–7
    oil production 176
    ozone layer 109–10
environmental management
    coastal protection 58–60, 61
    Galapagos Islands 135–6
    glaciated uplands 40–1
    national parks 126–7
    natural resources 107–17
    oil spillage 55
    opencast mining 112
    sand dunes 60
erosion
    and deforestation 153
    coastal 47–8, 49, 56–61
    footpaths in national parks 126, 127
    glacial 28–31, 32
    rivers 4, 7–10
erratics 34
eskers 35
EU (European Union) 162
exports 161, 165–8, 169

family planning, population growth 72, 76–7, 78
famine 144
farming
    and population distribution 64–5
    Dominican Republic 152–3
    glaciated uplands 35
    Indian population growth 92–3
    Italy 158
    Nepal, food production 145–6
filariasis 154
fishing industry, and oil spillage 56
fjords 53
flood plains 11
floods
    coastal erosion 59
    rivers 16–20, 21
food
    and global development 143–6
    and population growth 92–5
    exports from poor countries 165–7
foreign aid 168–77
    interdependence 175–6
    local involvement 173–4
    long-term projects 171–4
    misuse of 170–1, 172
foreign investment 165
fossil fuels 101, 104–9
free trade 162
freeze-thaw weathering 28, 30, 32

Galapagos Islands, tourism 133–6
GATT (General Agreement on Trade and Tariffs) 162
Ghana 173–4
glacial landscapes 24–42
    deposition 32–5
    glaciers as systems 26–7
    human use of uplands 35–41
    ice age 24–5
    land-shaping agents 27–8
global population distribution 61–8
global warming 104–9
globalisation of trade 164
GNP (Gross National Product) 137, 139–40
green revolution 92–3, 144
green tourism 129–33
greenhouse effect 105
groundwater 15

hanging valleys 31
HDI (Human Development Index) 140–1
headlands, coastal landforms 46
health
    and population growth 72, 77
    and poverty 153–5
Holderness 56–9
housing, population growth 72
hydraulic action, coastal landforms 47
hydroelectric power
    glaciated uplands 35
    Kielder Water 22

ice age 24–5
    sea-levels 53
IEPs (international environmental problems) 107–11
Ilisu dam project 172–3
immigration 81
imports 161, 167
India, agriculture and population 92–3
industry
    coastal areas 43, 54–6
    developing countries 158, 165
    national parks 124
interlocking spurs 7
international trade 160–8
investment, developing countries 146–7

Index

isostatic change 53
Italy, development 156–8

kames and kame terraces 35
Kielder Water 21–3
Koshi Hills, East Nepal 144–6

Lake District, tourism 124–7
land degradation, and population growth 94–5
land reclamation, Netherlands 98–9
landscape formation
    glaciers 27–35
    rivers 4–23
    sea-level changes 53
LEDCs
    food security 144
    foreign aid 168–75
    foreign investment and trade 165–8, 175–6
    migration 81
    population 67, 69, 70, 71, 73, 77, 78–9
    rapid growth
    tourism 129–35
leisure *see* tourism
literacy 149–50, 174
living conditions, population growth 72
long profiles, rivers 6
longshore drift 51, 57

malaria 153, 154
malnutrition 143–4, 153
Malthusian theory 92
meanders 10, 12
MEDCs, and population 69, 70, 71, 77
    ageing populations 78–9
    and global warming 105, 107
    environmental damage 96
    foreign investment 170–5
    migration 81
    trade and investment 160–6
meltwater deposits 35
Mercosaur (Common Market of the South) 164
Mexico, migration to USA 88–90
migration
    and resources, population 81–100
    international 88–91
    refugees 90, 91
    rural-urban 81–4, 158
Milford Haven 54–6
mining, opencast 112
mobility and counter-urbanisation 86
moraines 32–3
mud flats 52

NAFTA (North America Free Trade Association) 164
national parks 122–7
natural resources
    and global population distribution 66–7
    management of 101–17
net migration gains and losses 85
Netherlands, land reclamation 98–9
NICs (newly industrialising countries) 149
Nigeria 73–4, 176
Nile, River 14, 96–7
nomads, population distribution 64
nuclear power 113–14

oil production 176
oil refineries, and the environment 54–6
oil reserves 102
outwash plains 35
over-cultivation, and population growth 94–5
ox-bow lakes 13
ozone layer 109–10

Pembrokeshire, sand dunes 60
Peru, rural-urban migration 82–4
pests, developing countries 154, 155, 156
physical factors, global population distribution 63–8

political factors
    developing countries 150
    population growth 72
    tourism in Zimbabwe 132
population 62–100
    change and development 70–2
    growth 68–9, 92
    growth and developing countries 148–9
    growth and environmental management 93–6
    human factors and global distribution 66–8
    migration and resources 81–100
    physical factors and global distribution 63–5
    problems of change 75–6
population pyramids 73–4, 84
potholes 7
poverty
    and global development 137–40, 143, 145–8, 151
    and health 153–5
primary products 165–7
push and pull factors, migration 81, 82, 85–6
pyramidal peaks 30

quarrying (or plucking) 28, 30

radioactive fuel disposal 114
rainfall
    acid rain 110–11
    and river regimes 15–16, 17
recreation *see* tourism
recycling 115–17
refugees 90, 91
regional contrasts, wealth and development 156, 157
religious factors
    and development 150
    population growth 72, 77
renewable and non-renewable resources 101–3, 114–15
resources
    for tourism 118–19, 131, 133–4
    population and migration 81–100
retirement, and counter-urbanisation 86
rias 53
ribbon lakes 31
river blindness 153, 155
river regimes 15–16
rivers 4–23
    drainage basins 5, 6
    landshaping 6–14
    sea-level changes 53
    the water cycle 15–17
roches moutonnées 31, 32
rural areas, migration and resources 81–8
rural-urban migration 81–4

salinisation of land, and population growth 95
salt marshes 52
sand dunes, erosion 60–1
sea
    coastal systems 44–5
    defences and coastal erosion 58–60
    level changes 53, 109
Sea Empress oil spill 54–5
sea walls 58
seaside resorts 119–21
sediment
    glacial 35
    rivers 4
shore (wave-cut) platforms, sea erosion 47, 48
Sierra Leone, water supplies 142–3
skiing, Cairngorms 36–40
skills shortages, developing countries 148
sleeping sickness 154
Snowdonia, glacial erosion 28–31
snowline, and glaciers 26
social factors
    and development 149–50
    population growth 72
soils, and population distribution 65

spits 51, 52
stacks
    coastal landforms 48
    sea erosion 47
swash, wave formation 45
Sweden, population 70, 74–5, 79

technological development
    and environmental damage 96
    and global population distribution 66–7
Teesdale, population change 86–8
terraces 11, 12
till (boulder clay) 32, 57
TNCs (trans-national corporations) 146–7, 152, 164, 165
tombolos 52
tourism 118–36
    advantages and disadvantages in national parks 126–7
    economically developing world 129–35, 152, 158
    Galapagos Islands 133–6
    glaciated uplands 35–41
    Pembrokeshire sand dunes 60–1
    problems of in Dominican Republic 152, 153
    sustainable tourism in the Cairngorms 41
    United Kingdom 119–28
    Zimbabwe 129–33
trade
    developing countries 148
    international 160–8
trade barriers 162
traffic management, national parks 127, 128
truncated valleys 31
TB (tuberculosis) 154

undernutrition 143
United Kingdom
    carbon emissions 105
    migration within the UK 84–8
    opencast mining 112
    tourism 119–22, 122–7
    water grids 21–2
urban areas
    migration and resources 81–8
    urban-rural migration (counter-urbanisation) 81, 85–6
urban populations 67
USA, Mexican (Latino) immigrants 88–90

valleys
    glacial 27, 28, 31
    river landscapes 8–9
    sea-level changes 53

water catchment, glaciated uplands 35
water cycle, rivers 15–17
water grids, United Kingdom 21–2
water resources, dams 97–8, 172–3
water supplies, global contrasts 142–3
waterfalls 8
watersheds 5
waves, coastal systems 44–5, 47, 57
weathering 4
    glaciated uplands 28, 29–31
wildlife
    Cairngorms 38
    Galapagos Islands 133–4
    Pembrokeshire 54–5, 60
    Zimbabwe and tourism 130, 131
wind, and wave energy 44–5
wind power 114–15
women, in developing countries 150
world commodity prices 167
WTO (World Trade Organisation) 162
Wyre, River 5–13

Zimbabwe, tourism 129–33
Zuider Zee, land reclamation 98

Published by Collins Educational
An imprint of HarperCollins*Publishers* Ltd
77–85 Fulham Palace Road
Hammersmith
London W6 8JB

www.CollinsEducational.com
On-line Support for Schools and Colleges

© HarperCollins*Publishers* Ltd 2001
First edition published 1996
Second edition published 2001

ISBN 0 00 711647 0

Michael Raw and Sue Shaw assert the moral right to be identified as the authors of the first edition upon which this edition is based. Michael Raw asserts the moral right to be identified as the author of the new material in this second edition.

All rights reserved. No part of this publication may be reproduced, stored in a retrieval system, or transmitted in any form or by any means, electronic, mechanical, photocopying, recording or otherwise, without either the prior permission of the Publisher or a licence permitting restricted copying in the United Kingdom issued by the Copyright Licensing Agency Ltd, 90 Tottenham Court Road, London W1P 9HE.
British Library Cataloguing in Publication Data
A catalogue record for this book is available from the British Library.

Edited by Louise Pritchard at Bookwork
Design by Kim Bale at Visual Image
Cover design by Jerry Fowler
Map artwork by Jerry Fowler
Illustrations by Barking Dog Art, Joan Corlass, Jerry Fowler, Jeremy Gower, Hardlines
Cartoons by Oliver Raw, Richardson Studio
Picture research by Diana Morris and Caroline Thompson
Production by Kathryn Botterill
Printed and bound by Printing Express, Hong Kong

Acknowledgements

Every effort has been made to contact the holders of copyright material, but if any have been inadvertently overlooked, the publishers will be pleased to make the necessary arrangements at the first opportunity.

Photographs
The publishers would like to thank the following for permission to reproduce photographs:

Aerofilms Ltd, Figs 3.23, 3.24, 6.14, 7.13; Airfotos Ltd, Fig. 1.34; Aviemore Photographic, Figs 2.31, 2.38, 2.39; Cambridge Committee for Aerial Photography: © Crown, Fig. 2.10; Christies Colour Library, Fig. 4.14; John Cleare Mountain Camera, Fig. 2.8; Bruce Coleman Ltd, Figs 2.3CR, 2.3BR; Eye Ubiquitous/R Kreutzman, Figs 1.1C, 2.3BL, P Thompson, Figs 5.28, 6.1, D Cummings, Figs 8.4, 8.18, James Davies Travel Photography, Figs 7.5, 7.31, 7.32, 7.33, 7.39, 7.41; Geoscience Features, Figs 2.23, 3.4, 4.6; GettyOne Stone, Fig. 4.20; Sean Smith/Guardian, Fig. 9.20; Robert Harding Picture Library/W Maughan, Figs 9.1TL, 9.1TR, 9.1BL, R Frerck, Figs 9.1BR, 9.7; Highlands Photo Library, Fig. 3.28; Hulton Getty Picture Collection, Figs 7.8, 7.9; Hutchison Photo Library, Figs 5.5, 5.27, 5.32, 5.33, 8.29; Images Colour Library, Fig. 1.1CL; Mansell Collection, Figs 2.7, 7.7; Alberto Nardi/NHPA, Fig. 3.39; Panos Pictures, Figs 4.30, 8.14, 8.16, Betty Press, Figs 4.18, 5.24, Paul Smith, Fig. 5.20, Giacomo Pirozzi, Fig. 8.9, Irene Slegt, Fig. 8.20; Michael Raw, Figs 1.7, 1.9, 1.13, 1.15, 1.17, 2.4, 2.3TCR, 2.12, 2.14, 2.15, 2.18, 2.21, 2.25, 2.33, 2.34, 2.35, 3.1, 3.7, 3.10, 3.11, 3.12, 3.13, 3.14, 3.17, 3.18, 3.19, 3.37, 5.10, 5.12, 7.1, 7.2, 7.19, 7.20, 7.21, 7.22, 7.23, 7.25, 7.27, 7.28; Renewable Energy Ltd, Fig. 6.28; Rex Features Ltd, Fig. 9.29; Science Photo Library, Figs 6.20, 8.31; Sue Shaw, Figs 2.41, 5.15, 5.16; Peter Smith Photography, Fig. 1.32; FSP/Gamma/R Gaillande, Fig. 9.16; Still Pictures/P Frismuth, Figs 1.1CR, 1.1B, M Edwards, Figs 3.3, 3.33, 4.4, 4.5, 4.7, 4.8, 4.11, 4.25, 5.2, 5.38, 6.2, 8.3, J Kaplan, Fig. 6.4, D Hoffman, Fig. 6.7, N Dickinson, Figs 6.11, 6.13, A Arbib, Fig. 7.4, R Seitre, Figs 7.35, 7.36, 7.40, J Schytte, Figs 8.5, 9.25, G & M Moss, Fig. 8.30; The Stock Market Photo Agency, Figs 1.1T, 1.21, 2.3CL, 2.19, 2.36, 2.37, 4.10, 5.3, 5.30, 5.42, 6.15, 7.3, 8.8, 8.33, 8.36, 9.11; TRIP/ B Gabsby, Fig. 5.6, H Rogers, Fig. 8.26, Bartos, Fig. 8.37; Woodfall Wild Images/ P Wilson, Figs 3.2, 3.29, 5.31, 6.12, D Woodfall, Fig. 6.26.

Cover photographs: G R Roberts, (top), C Palmer at www.buyimage.co.uk, (top centre), Michael Raw, (centre), Edison Mission Energy, (bottom centre), GettyOne Stone, (bottom).

Maps
Maps reproduced from Ordnance Survey mapping with the permission of The Controller of Her Majesty's Stationery Office, © Crown copyright, Licence Number 100018599.
Extracts showing: River Wyre (upper course) (Fig. 1.11); River Wyre (middle course) (Fig. 1.12); River Wyre (lower course) (Fig. 1.18); Snowdonia (Fig. 2.11); South Devon coastline (Fig. 3.20).